Advances in Organic Synthesis

(Volume 12)

Edited by

Atta-ur-Rahman, *FRS*
Honorary Life Fellow, Kings College, University of Cambridge, Cambridge, UK

Advances in Organic Synthesis

Volume # 12

Editor: Atta-ur-Rahman

ISSN (Online): 2212-408X

ISSN (Print): 1574-0870

ISBN (Online): 978-1-68108-680-4

ISBN (Print): 978-1-68108-683-5

©2018, Bentham eBooks imprint.

Published by Bentham Science Publishers – Sharjah, UAE. All Rights Reserved.

First published in 2018.

BENTHAM SCIENCE PUBLISHERS LTD.
End User License Agreement (for non-institutional, personal use)

This is an agreement between you and Bentham Science Publishers Ltd. Please read this License Agreement carefully before using the ebook/echapter/ejournal (**"Work"**). Your use of the Work constitutes your agreement to the terms and conditions set forth in this License Agreement. If you do not agree to these terms and conditions then you should not use the Work.

Bentham Science Publishers agrees to grant you a non-exclusive, non-transferable limited license to use the Work subject to and in accordance with the following terms and conditions. This License Agreement is for non-library, personal use only. For a library / institutional / multi user license in respect of the Work, please contact: permission@benthamscience.org.

Usage Rules:

1. All rights reserved: The Work is the subject of copyright and Bentham Science Publishers either owns the Work (and the copyright in it) or is licensed to distribute the Work. You shall not copy, reproduce, modify, remove, delete, augment, add to, publish, transmit, sell, resell, create derivative works from, or in any way exploit the Work or make the Work available for others to do any of the same, in any form or by any means, in whole or in part, in each case without the prior written permission of Bentham Science Publishers, unless stated otherwise in this License Agreement.
2. You may download a copy of the Work on one occasion to one personal computer (including tablet, laptop, desktop, or other such devices). You may make one back-up copy of the Work to avoid losing it. The following DRM (Digital Rights Management) policy may also be applicable to the Work at Bentham Science Publishers' election, acting in its sole discretion:

- 25 'copy' commands can be executed every 7 days in respect of the Work. The text selected for copying cannot extend to more than a single page. Each time a text 'copy' command is executed, irrespective of whether the text selection is made from within one page or from separate pages, it will be considered as a separate / individual 'copy' command.
- 25 pages only from the Work can be printed every 7 days.

3. The unauthorised use or distribution of copyrighted or other proprietary content is illegal and could subject you to liability for substantial money damages. You will be liable for any damage resulting from your misuse of the Work or any violation of this License Agreement, including any infringement by you of copyrights or proprietary rights.

Disclaimer:

Bentham Science Publishers does not guarantee that the information in the Work is error-free, or warrant that it will meet your requirements or that access to the Work will be uninterrupted or error-free. The Work is provided "as is" without warranty of any kind, either express or implied or statutory, including, without limitation, implied warranties of merchantability and fitness for a particular purpose. The entire risk as to the results and performance of the Work is assumed by you. No responsibility is assumed by Bentham Science Publishers, its staff, editors and/or authors for any injury and/or damage to persons or property as a matter of products liability, negligence or otherwise, or from any use or operation of any methods, products instruction, advertisements or ideas contained in the Work.

Limitation of Liability:

In no event will Bentham Science Publishers, its staff, editors and/or authors, be liable for any damages, including, without limitation, special, incidental and/or consequential damages and/or damages for lost data and/or profits arising out of (whether directly or indirectly) the use or inability to use the Work. The entire liability of Bentham Science Publishers shall be limited to the amount actually paid by you for the Work.

General:

1. Any dispute or claim arising out of or in connection with this License Agreement or the Work (including non-contractual disputes or claims) will be governed by and construed in accordance with the laws of the U.A.E. as applied in the Emirate of Dubai. Each party agrees that the courts of the Emirate of Dubai shall have exclusive jurisdiction to settle any dispute or claim arising out of or in connection with this License Agreement or the Work (including non-contractual disputes or claims).
2. Your rights under this License Agreement will automatically terminate without notice and without the need for a court order if at any point you breach any terms of this License Agreement. In no event will any delay or failure by Bentham Science Publishers in enforcing your compliance with this License Agreement constitute a waiver of any of its rights.
3. You acknowledge that you have read this License Agreement, and agree to be bound by its terms and conditions. To the extent that any other terms and conditions presented on any website of Bentham Science Publishers conflict with, or are inconsistent with, the terms and conditions set out in this License Agreement, you acknowledge that the terms and conditions set out in this License Agreement shall prevail.

Bentham Science Publishers Ltd.
Executive Suite Y - 2
PO Box 7917, Saif Zone
Sharjah, U.A.E.
Email: subscriptions@benthamscience.org

CONTENTS

PREFACE	i
LIST OF CONTRIBUTORS	ii

CHAPTER 1 STEREOSELECTIVE METHODOLOGIES FOR THE SYNTHESIS OF ACYCLIC POLYISOPRENOIDS 1
Didier Desmaële

INTRODUCTION	1
Natural Head-to-Tail Polyisoprenoids	2
Natural Tail-to-Tail Polyisoprenoids	5
SYNTHESES OF HEAD-TO-TAIL POLYISOPRENOIDS	6
General	6
ω-Functionalization of Short Natural Polyprenols	7
Synthesis of Bifunctional Mono-Isoprenyl Building Blocks	7
Synthesis of Bifunctional Polyisoprenyl Building Blocks	9
Synthesis of Oligoprenoids by sp^3-sp^3 Carbon Bond Formation	14
Construction of 1,5-Dienes via S_N2 Reactions	14
Construction of 1,5-Dienes via S_N2' Reactions	22
Enzymatic and Biomimetic Approaches	27
Synthesis of Polyisoprenoids by sp^2-sp^2 Carbon-Carbon Bond Formation	30
Synthesis of Polyisoprenoids by Wittig Reaction	30
Synthesis of Polyisoprenoids by Metathesis Reaction	31
Synthesis of Polyisoprenoids by sp^2-sp^3 Carbon Bond Formation	32
Miscellaneous Reactions	37
[3,3] and [2, 3]-sigmatropic rearrangements	37
[1.3]-Sigmatropic Rearrangements	40
Synthesis of Z-Polyisoprenoids	40
SYNTHESES OF TAIL-TO-TAIL POLYISOPRENOIDS: SQUALENE	41
Functionalization of Squalene	41
Synthetic Approaches to Squalene	43
FUNCTIONALIZATION OF CARBOCYCLES WITH ISOPRENOID SIDE CHAINS: ISOPRENOID QUINONES	47
Synthesis of Isoprenoid Quinones	47
Arylation of Polyprenyl Chains via Electrophilic Aromatic Substitution Reactions	47
Arylation of Polyprenyl Chains by Alkylation Reactions with Organolithium or Organomagnesium Derivatives	49
Arylation of Polyprenyl Chains via Transition Metal Catalyzed Coupling Reactions	51
Synthesis of Other Isoprenoid Substituted Carbocycles and Heterocycles	54
Direct Alkylation Reactions	54
Wittig Type Reactions	56
Sigmatropic Rearrangements	56
Organometallic-Catalyzed Cross-Coupling Reactions and New Organometallic Processes	57
POLYMERIZATION REACTIONS	60
CONCLUDING REMARKS	62
CONSENT FOR PUBLICATION	62
ACKNOWLEGMENTS	62
CONFLICT OF INTEREST	62
REFERENCES	62

CHAPTER 2 MONOSUBSTITUTED FERROCENE-CONTAINING THERMO- TROPIC LIQUID CRYSTALS 81
Irina Carlescu, Daniela Apreutesei Wilson, Gabriela Lisa, Nicolae Hurduc and *Dan Scutaru*
INTRODUCTION 82
LIQUID CRYSTALLINE FERROCENES WITH FERROCENE RIGIDLY CONNECTED TO MESOGENIC UNIT AND FLEXIBLE ALKYL CHAIN AS TERMINAL GROUP 87
MONOSUBSTITUTED LIQUID CRYSTALLINE FERROCENES CONTAINING A CHOLESTERYL GROUP 100
    Ferrocene Derivatives with Ferrocene Rigidly Connected to the Cholesteryl Unit 100
    Ferrocene Derivatives with Ferrocene Flexible Connected to Cholesteryl Unit 103
    Ferrocene Derivatives with Cholesterol Flexible Connected to Mesogenic Unit 106
    Symmetrical Derivatives with Two Monosubstituted Ferrocenyl Units 110
CONCLUSIONS 112
CONSENT FOR PUBLICATION 113
CONFLICT OF INTEREST 113
ACKNOWLEDGEMENTS 113
REFERENCES 113

CHAPTER 3 PROGRESS IN THE CHEMISTRY OF PHOSPHOROTHIOATES 117
Mihaela Gulea
INTRODUCTION 117
SYNTHESES OF PHOSPHOROTHIOIC ESTERS 119
    P-S Bond Formation from P-nucleophile and S-electrophile 119
    P-S Bond Formation from P-electrophile and S-nucleophile 122
    Metal-free Oxidative P-S Bond Formation 123
    Metal-catalyzed Processes Involving Phosphorus and Sulfur Sources 126
        Metal-catalyzed P-S Formation 126
        Metal-catalyzed Three-component Reactions 129
    From Phosphorothioic Acids or their Salts 131
        By S-alkylation 131
        By Metal-mediated C-S Bond Formation 133
REACTIONS OF PHOSPHOROTHIOIC ESTERS 133
    Rearrangement Reactions 134
        Anionic 1,2-Rearrangement 135
        Anionic 1,3-Rearrangement 136
        Anionic 1,4-Rearrangement 138
        Thiophosphate to Phosphoroamidate Rearrangement 138
    Anionic P-S Bond Cleavage with C-S Bond Formation 139
    Homolytic P-S Cleavage 141
    Metal-catalyzed P-S Cleavage; Thiophosphorylation of Triple Bonds 142
    Allylic Substitution of Phosphorothioate Group 142
CONCLUSION 144
CONSENT FOR PUBLICATION 144
CONFLICT OF INTEREST 144
ACKNOWLEDGEMENTS 144
REFERENCES 144

CHAPTER 4 KINETIC RESOLUTION USING DIASTEREOSELECTIVE ACYLATING AGENTS AS A SYNTHETIC APPROACH TO ENANTIOPURE AMINES 151
   Galina L. Levit, Dmitry A. Gruzdev and *Victor P. Krasnov*
   INTRODUCTION 152
      Basic Principles of Kinetic Resolution 153
      Acylative Kinetic Resolution of Racemic Amines 157
   DIASTEREOSELECTIVE N-ACYLATION OF RACEMIC AMINES 157
      Early Examples 157
      Derivatives of 2-Arylpropionic Acids as Diastereoselective Acylating Agents 161
      Derivatives of 2-Hydroxy Carboxylic Acids as Diastereoselective Acylating Agents 171
      Amino Acid Derivatives as Diastereoselective Acylating Agents 177
   CONCLUDING REMARKS 189
   CONSENT FOR PUBLICATION 190
   CONFLICT OF INTEREST 190
   ACKNOWLEDGEMENTS 190
   REFERENCES 190

CHAPTER 5 ADVANCES IN THE SYNTHESIS OF FUNCTIONAL α-ORGANYL *gem*-BISPHOSPHONATES FOR BIOMEDICAL APPLICATIONS 200
   Vadim D. Romanenko
   INTRODUCTION 200
   SYNTHESIS FROM PHOSPHITES AND MONOPHOSPHONATES *VIA* C-P BOND FORMATION 202
      New Variations in the Michaelis-Arbuzov and Michaelis-Becker Reactions 202
      Phosphorylation of α-Anionic Monophosphonates 204
      Synthesis *Via* Enolate Chemistry 207
   FUNCTIONALIZATION OF COMPOUNDS ALREADY CONTAINING THE P-C-P BACKBONE 209
      Alkylation of Methylenebisphosphonate Carbanions 209
      Synthesis Based on "Click" Methodology 213
      Radical Based Approach 215
      Miscellaneous Reactions 218
   SYNTHESIS *VIA* ALKENYLIDENE-1,1-BISPHOSPHONATES 221
      Addition Reactions 222
         Reactions with Organometallic Reagents 222
         Base-Promoted Addition of Carbon Nucleophiles 224
         Reactions with Alcohols, Thiols and Amines 232
         Reactions with Phosphorus Nucleophiles 238
      Cycloaddition reactions 238
   CONCLUDING REMARKS 242
   CONSENT FOR PUBLICATION 244
   CONFLICT OF INTEREST 244
   ACKNOWLEDGEMENTS 245
   REFERENCES 245

SUBJECT INDEX 253

PREFACE

This volume of **Advances in Organic Synthesis** presents some recent exciting developments in synthetic organic chemistry. It covers a range of topics including important researches on novel approaches to the construction of complex organic compounds.

The chapters are written by authorities in the field and are mainly focused on synthesis of acyclic polyisoprenoids, ferrocene-containing thermotropic liquid crystals, phosphorothioates, kinetic resolution using diastereoselective acylating agents as a synthetic approach to enantiopure amines and synthesis of functional α-organyl *gem*-bisphosphonates for biomedical applications.

The book should prove to be a valuable resource for pharmaceutical scientists and postgraduate students seeking updated and critically important information about synthetic organic chemistry.

I hope that the readers will find these reviews valuable and thought-provoking so that they may trigger further research in the quest for new developments in the field.

I am thankful to the efficient team of Bentham Science Publishers, especially Dr. Faryal Sami (Manager Publications), Mr. Shehzad Iqbal Naqvi (Editorial Manager Publications) and Mr. Mahmood Alam (Director Publications).

Atta-ur-Rahman, *FRS*
Honorary Life Fellow
Kings College
University of Cambridge
Cambridge
UK

List of Contributors

Dan Scutaru	Faculty of Chemical Engineering and Environmental Protection, "Gheorghe Asachi" Technical University of Iasi, 73 Prof. dr.docent Dimitrie Mangeron street 700050, Iași, România
Daniela Apreutesei Wilson	Systems Chemistry, Radboud University, Heyendaalseweg 135, 6525 AJ Nijmegen, The Netherlands
Didier Desmaële	Faculté de Pharmacie, Institut Galien Paris-Sud, UMR 8612, CNRS, Université Paris-Saclay, Châtenay-Malabry, France
Dmitry A. Gruzdev	Postovsky Institute of Organic Synthesis, Russian Academy of Sciences (Ural Branch), Ekaterinburg, Russia
Gabriela Lisa	Faculty of Chemical Engineering and Environmental Protection, "Gheorghe Asachi" Technical University of Iasi, 73 Prof. dr.docent Dimitrie Mangeron street 700050, Iași, România
Galina L. Levit	Postovsky Institute of Organic Synthesis, Russian Academy of Sciences (Ural Branch), Ekaterinburg, Russia
Irina Carlescu	Faculty of Chemical Engineering and Environmental Protection, "Gheorghe Asachi" Technical University of Iasi, 73 Prof. dr.docent Dimitrie Mangeron street 700050, Iași, România
Mihaela Gulea	Laboratoire d'Innovation Thérapeutique (LIT, UMR 7200), Université de Strasbourg, CNRS, 74 Route du Rhin, 67401 Illkirch, France
Nicolae Hurduc	Faculty of Chemical Engineering and Environmental Protection, "Gheorghe Asachi" Technical University of Iasi, 73 Prof. dr.docent Dimitrie Mangeron street 700050, Iași, România
Vadim D. Romanenko	Institute of Bioorganic Chemistry and Petrochemistry, National Academy of Sciences of Ukraine, Kiev-94, 02660, Ukraine
Victor P. Krasnov	Postovsky Institute of Organic Synthesis, Russian Academy of Sciences (Ural Branch), Ekaterinburg, Russia

CHAPTER 1

Stereoselective Methodologies for the Synthesis of Acyclic Polyisoprenoids

Didier Desmaële[*]

Institut Galien Paris-Sud, UMR 8612, CNRS, Université Paris-Saclay, Faculté de Pharmacie, Châtenay-Malabry, France

Abstract: Acyclic polyisoprenoids are ubiquitous in nature from bacteria to human cells. Beside, their leading role as precursor of thousands of cyclic terpenoids, they have also a tremendous importance as membrane constituents, protein modulators and nanoparticle carrier material. Their synthesis is a main topic since the dawn of organic chemistry, nevertheless today there is still no universal method to access these compounds and it remains space for finding original and efficient solutions. In this review we provided an overview of the synthetic methods available for the synthesis of head-to-tail and tail-to-tail 1,5-diene-containing polyprenyl derivatives, including alkylation reactions of organometallics and heteroatom stabilized carbanions, sigmatropic rearrangements, transition metal catalyzed methods and also biocatalytic syntheses. The synthesis of small difunctionnal building blocks from cheap naturally occurring polyprenols such as geraniol or farnesol are described. A special emphasis will be given on the coupling of polyisoprenoid chains to carbocycles including synthesis of isoprenoid quinones.

Keywords: Allylic Reductive Coupling, Bielmann-Ducep Coupling, Coenzyme Q_{10}, 1,5-Dienes, Farnesol, Geraniol, Geranylgeraniol, Isoprene, Menaquinone, Olefin Formation, Polyprenols, Polyisoprenoids, Polyprenyl Quinones, Shapiro Reaction, Solanesol, Stereoselectivity, Squalene, Suzuki-Miyaura Cross Coupling, Terpenes, Trisubstituted Double Bond, Vitamin K, Wittig Reaction.

INTRODUCTION

In the recent years the recognition of the crucial role of acyclic polyisoprenoid compounds as membrane constituent, protein modulators or nanoparticle carrier material has induced a renewal of interest for these derivatives beside their old known outstanding biologically activities. Among them, simple all-*E* oligoprenols from geraniol to decaprenol feature the archetypal carbon skeleton of head-to-tail polyprenyl compounds found also in many meroterpenes such as menaquinones

---

[*] **Corresponding author Didier Desmaële**: Institut Galien Paris-Sud, UMR 8612, CNRS, Université Paris-Saclay, Faculté de Pharmacie, Châtenay-Malabry, France; Tel: 33(0)1 46 83 57 53 ; E-mail: didier.desmaele@u-psud.fr

or coenzymes Q_3 to Q_{10}. Beside these regular polyprenyl compounds, a smaller group of naturally occurring terpenes displays two farnesyl moieties joined in a tail-to-tail fashion rather than in the head-to-tail fashion. It is the case of squalene and its derivatives such as achilleol or turbinaric acid. Squalene is the corner stone in the biosynthesis of most triterpenes including lanosterol and cycloartenol which in turn are the precursors of all steroids.

Chemical synthesis of polyprenoid compounds started with the dawn of organic synthesis. The first synthesis of farnesol was published by Ruzicka in 1923 and since that time an endless quest for efficacy and selectivity was started and countless clever solutions were proposed. However, almost a hundred years later and despite an extraordinary research effort, there is no universal method to apply for the synthesis of any polyisoprenoids and there is still space for the finding of new original solutions. Nowadays, the development of biomimetic cyclizations of open-chain polyisoprenoids into complex polycyclic structures using either organometallic catalysts or recombinant enzymes is a powerful incentive for the design of new synthetic access to polyisoprenoids. This review will attempt to provide the reader with the current status of synthetic methodologies to stereoselectively synthesize acyclic polyprenyl compounds focusing on the most recent developments.

Natural Head-to-Tail Polyisoprenoids

Isoprenoid compounds constitute one of the largest and most diverse groups of natural products, with greater than 35 000 identified members. Despite their amazing structural diversity, they are derived from two simple five-carbon precursors: isopentenyl diphosphate (IPP) and dimethylallyl diphosphate (DMAPP) (Scheme **1**). Until recently IPP and DMAPP were believed to originate from acetate by the mevalonate pathway. However, pioneering studies by Rohmer and Arigoni revealed an alternative pathway that operates in plant chloroplasts, algae, and bacteria [1, 2]. These enzymatic reactions catalyzed by prenyl-transferases proceed with high stereoselectivity and terminate precisely until the prenyl chains reach the requisite length depending on the enzymes. Combination of IPP and DMAPP produces geranylpyrophosphate that is either the substrate for various cyclase enzymes which catalyze the biosynthesis of all monoterpenes or it can be substrate of prenyl transferases that catalyzed the chain elongation into farnesyl pyrophosphate which is the progenitor of sesquiterpenes. Iteration of the process leads to geranylgeranyl pyrophosphate, which then cyclizes into diterpenes. Polyprenyl phosphates and pyrophosphates are the biosynthetic intermediates of all terpenoids, but they are also involved in the formation of the most primitive membranes [3]. The highest accumulation of polyprenols is observed in plant photosynthetic tissues, but they are also detected in wood, seeds and flowers and in bacterial cells. The high hydrophobicity of polyisoprenoids causes their localization in cellular membranes, e.g. mitochondria, chloroplast envelopes, Golgi membranes where they play a major role as cofactors in the biosynthesis of bacterial peptidoglycan and eukaryotic glycoproteins and as substrates for protein prenylation.

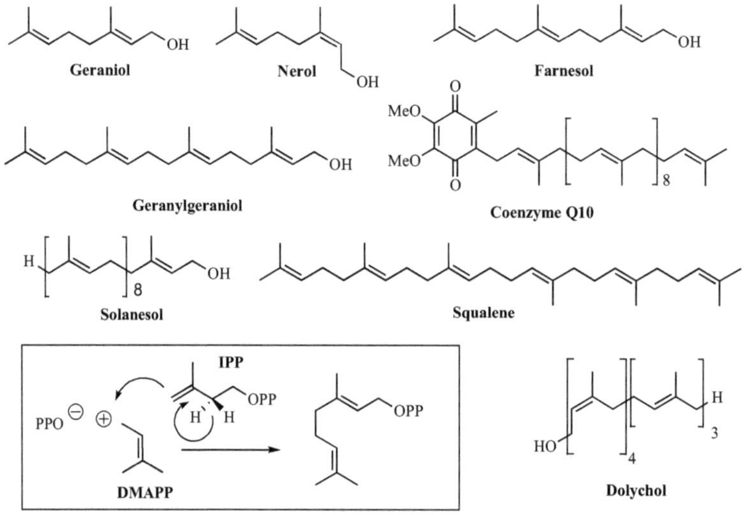

Scheme 1. Structure of typical polyisoprenyl compounds and general head-to-tail biosynthetic mechanism of polyisoprenoids.

It has been proposed that one central role of acyclic isoprenoids, in archaebacteria and prokaryote organisms, is to participate in the formation and reinforcement of biomembranes as surrogates of sterols. Archaebacterial membrane lipids are mainly constituted of phosphorylated polyisoprenyl glyceryl ethers whose chains possess the suitable length to achieve the amphipathic ratio required for spontaneous vesicle formation. In line with these findings, polyprenols and their phosphorylated derivatives are known to alter the structure of the phospholipid bilayer by promoting the formation of a nonlamellar structure, thus increasing the fluidity and permeability of membranes [4].

Polyisoprenoid alcohol chains are built of 2 to 40 isoprenoid units and more in natural rubber, creating oligomers that differ in the chain-length and/or geometrical configuration. Prominent members among polyisoprenoid alcohols are geraniol, farnesol and geranylgeranol that display respectively 2, 3 and 4 isoprenyl units. The geraniol market is primarily driven by cosmetics companies for flavoring soaps, detergents, personal care products *etc.* In food industry it is used as flavoring agent, taste and odor enhancer, furthermore it is an effective insect repellent. Geraniol is mainly obtained from rose oil, palmarosa oil, and citronella oil, and at a lesser extent, from geranium, lemon, and other essential oils. Natural sources are essential in many applications, but synthetic geraniol obtained by orthovanadate catalyzed isomerization of the cheaper linalool has been developed to meet the growing industrial demand [5]. Industrial annual production of geraniol exceeds 1000 metric tons and the demand is increasing [6]. A large array of biological and pharmacological activities are reported for geraniol but given its wide presence in cosmetic and household products one of the main concern is its potential allergenic activity [7]. Farnesol is found as minor component in many essential oils such as neroli, cyclamen, citronella, rose, and others. It is used in perfumery to accentuate the flagrances of sweet floral perfumes. Farnesol is a pheromone that female spider mites use to attract males for mating, used in combination with standard pesticides farnesol enhances their mite killing effect. Among the various pharmacological effects reported for farnesol, antimicrobial properties are probably the most promising, for example it was found to be a potential therapeutic for clinical *Staphylococcus epidermidis* biofilm infections [8]. Today, industrial production of farnesol relies mainly on isomerization of nerolidol which is abundantly found in neroli oil [9]. Likewise geranylgeraniol, which is an important starting material for producing vitamin K2 can be obtained from farnesyl acetone as a mixture of isomers. Separation methods to obtain pure *trans*-material have been developed [10].

Beside all-*trans* prenols such as geraniol, farnesol or solanesol (Scheme **1**), naturally occurring acyclic polyisoprenoid compounds belong to three-other distinct categories: (i) the dolichol type prenols with two *cis* double bonds and a saturated isoprene unit at the tail, which are common constituents of animal and yeast cells, (ii) the betulaprenols and (iii) the ficaprenol whose double bonds are *cis* except the two or three last ones respectively at the head extremity. Their structures, biosynthesis and functions have been reviewed [11]. The corresponding pyrophosphates bound to polysaccharides or glycoproteins, play an essential role for reconnaissance by enzymatic systems. For example, Lipid II a peptidoglycan bearing a pyrophosphate betulaprenyl chain is involved in the synthesis of the cell walls of bacteria, likewise complex glycans harboring a dolichyl chain are donor substrates for bacterial protein *N*-glycosidation [12]. Branched isoprenoids, first postulated by Ourisson to derive from polyprenyl present in biomembranes in primitive organisms are abundantly found in sediments and have been isolated from diatomaceous algae [13].

Beside these linear polyprenols and their phosphate esters, many cyclic compounds display a pending polyisoprenyl chain bound to a carbocycle or a heterocycle through a C-C bond. Among these compounds, phenolic meroterpenes such as ubiquinones, plastoquinones and menaquinones have a particular biological importance. These substances possess a quinone ring dedicated to the transport of electrons, playing a fundamental role in oxidation-reduction in living organisms and a long polyprenyl chain likely to be attached in cell membranes (Scheme **1**).

Finally, nitrogen, sulfur and phosphate bound polyisoprenyl chains are also widely found in nature. Alkylation of the thiol function of cysteine in peptides and proteins with a farnesyl or geranylgeranyl chain is a post-translational event that leads to increase lipophilicity [14]. Many proteins and peptides in both eukaryote and mammalian cells are geranylated or farnesylated by polyprenyl-transferase enzymes, causing localization of the resulting conjugates to the membranes and inducing biological changes in signal transduction pathways controlling cell growth and differentiation, cytoskeletal or membrane rearrangement.

Nitrogen bound polyisoprenyl derivatives are relatively rare in nature, the most prominent are probably the phytohormones 6-(γ,γ-dimethylallylamino) purine, zeatin or agelasin whose the essential role as regulator of various processes in plant growth and development has been recently reviewed [15].

Natural Tail-to-Tail Polyisoprenoids

The higher polyprenols are formed according to the same general biosynthetic pathway as the smaller terpenoids, *i.e.* geranylgeranyl pyrophosphate condenses with IPP to give pentaprenyl pyrophosphate etc, but the triterpenoids and steroids are synthetized by a completely different mechanism. The bio-synthesis of squalene from farnesyl pyrophosphate can be considered as a reductive dimerization tail-to-tail. Bacteria usually lack sterols but in yeasts, plants and animals, one of the terminal double bond of squalene is epoxidized to give 2,3-oxidosqualene which then cyclizes to yield lanosterol, which is further transformed into all kinds of steroids. Squalene is a valuable compound widely used in the food industry, cosmetic and vaccine. 2500 tons were produced each year with a commercial value of $100 million. For long time, shark liver oil was the largest source of squalene but nowadays botanical sources, including rice bran, wheat germ, and olives tend to replace it due to shark fishing regulation. Specific physico-chemical methods such as extraction with supercritical fluids have been developed to isolate it in pure-form from plant seed oils [16]. Owing to its clinical, cosmetic, and pharmaceutical importance, much effort has been directed towards enhancing biosynthesis in genetically modified organisms. Considerable progresses were recently made in this field and it is likely that polyisoprenoid compounds including squalene obtained by biotechnologies will be marketed in the near future [17, 18].

SYNTHESES OF HEAD-TO-TAIL POLYISOPRENOIDS

General

While polyisoprenoids do not possess any chiral center, the iterative synthesis of stereodefined olefins with uniformly high specificity remains challenging with the existing modern synthetic arsenal. The stereoselective synthesis of olefin is central to the construction of polyisoprenoids, but despite its importance, synthesis of trisubstituted alkenes, relies mainly, with the exception of metathesis, on methods introduced in the mid-twentieth century: Wittig (1949), Claisen-Johnson rearrangement (1970), Negishi carboalumination reaction (1978), *etc*. Not surprisingly, any advances in olefin synthesis lead to rapid application in the field of polyisoprenoids. When planning the synthetic route to any polyisoprenyl chain, there are conceptually four main disconnections to be considered, as depicted in Scheme **2**.

Disconnection ***a***: Allylic-allylic coupling with a sp^3-sp^3 carbon-carbon bond formation

Disconnection ***b***: Double bond formation

Disconnection ***c*** and ***d***: Organometallic sp^2-sp^3 carbon-carbon bond formation

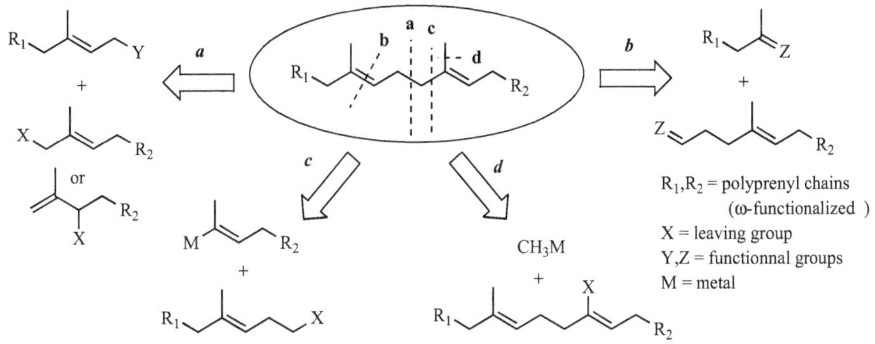

Scheme 2. Main disconnections for polyisoprenoid 1,5-diene construction.

In any case, it is necessary to design bifunctional building blocks with the terminal function protected or temporary masked in order to dispose of the requisite function for the next isoprene-unit elongation. These small bifunctional building blocks are usually obtained from isoprene or cheap commercially available natural terpenes such as geraniol or linalool. This topic will be first discussed.

ω-Functionalization of Short Natural Polyprenols

Synthesis of Bifunctional Mono-Isoprenyl Building Blocks

Though isoprene (**1**) is a readily available starting material, its functionalization to get a bifunctional building block remains a challenging task. Isoprene mono-epoxide (**2**) is a corner stone to access most derivatives. It is usually obtained by cyclization of the corresponding bromohydrin formed by treatment of isoprene (**1**) with NBS in water [19]. When using peracids, including m-CPBA as epoxidation reagent, a mixture of the two possible epoxides with a 10:1 ratio is obtained together with polymeric materials [20, 21]. Alternatively, Jacobsen type epoxidation using Mn(*t*-Bu-salen)Cl/PhIO gives regioselectively the mono-epoxide **2** but with a modest 34% yield [22]. The ring opening of isoprene mono-epoxide (**2**) gives access to a large array of useful building blocks. For example, reaction of **2** with CO in the presence of [Pd(C_4H_7)Cl]$_2$, NaBr, and maleic anhydride affords ethyl (*E*)-5-hydroxy-4-methyl-3-pentenoate (**4**) in 86% yield *via* carbonylation of the intermediate η^3-allylpalladium intermediate [23]. Acetyl chloride treatment of **2** in the presence of LiCl gives a 3:2 mixture of chloroacetate **6**, together with the isomeric 2-chloro derivative **7**. Initially assigned as the *E*-form [21], the major product **6** is now definitively identified to be the (*Z*)-4-chloro-2-methylbut-2-en-1-yl acetate [24]. On the other hand, the corresponding (*E*)-alcohol **5** can be obtained by TiCl$_4$-mediated regio- and stereoselective ring-opening reaction of isoprene mono-epoxide (**2**) [25]. The regioisomeric (*E*)-4-chloro-3-methylbut-2-en-1-yl acetate (**8**) is available by reacting isoprene with *tert*-butyl hypochlorite in glacial acetic acid, followed by allylic rearrangement with cupric ions in sulfuric acid [26] (Scheme 3).

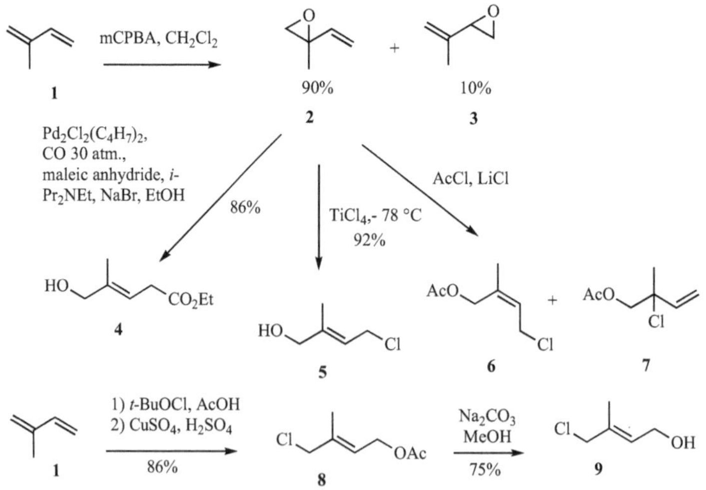

Scheme 3. Synthesis of monoisoprenyl bifunctional building blocks from isoprene.

Isoprene is the obvious starting material to synthesize these bifunctional building blocks, however alternative routes have been proposed. For example the stereoselective carbometalation reaction of propargylic alcohol **10** with methyl Grignard delivers after acetylation the pure *E*-amine **11**, which upon dealkylation with ethyl chloroformate provides chloro acetate **12** [27]. (*E*)-4-hydroxymethylallyl diphosphate (**15**) is a pivotal intermediate in the biosynthesis of IPP and DMAPP and thus in the biosynthesis of most isoprenoids. A straightforward synthetic route to **15** involves the addition of vinyllithium to pyruvaldehyde dimethyl acetal (**13**) to give tertiary alcohol **14**, which upon acidic treatment in the presence of cupric chloride and sodium borohydride reduction rearranges into pure (*E*)-4-chloro-2-methylbut-2-en-1-ol (**5**) [28]. The allylic chloride **5** is then readily converted to diphosphate **15** by nucleophilic substitution using tris(*n*-tetrabutylammonium) hydrogen pyrophosphate [29] (Scheme 4).

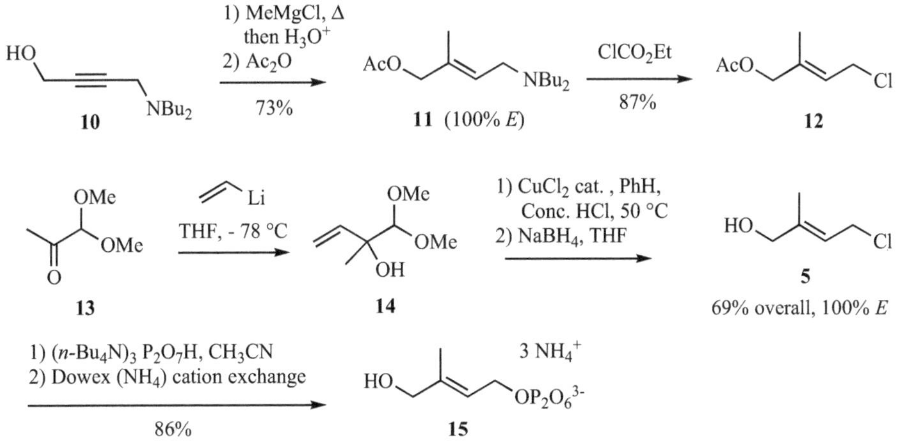

Scheme 4. Syntheses of monoisoprenyl bifunctional building blocks.

The control of the stereochemistry of the double bond is the main hurdle to access bifunctional monoisoprenoid building blocks. Stereoselective selenium dioxide allylic oxidation of the terminal *trans*-methyl group of polyisoprenyl compounds is one of the most convenient methods. For example, the alcohol **17** which is a direct precursor of the aforementioned diphosphate **15** can be prepared by SeO_2 oxidation of prenyl acetate (**16**) albeit in a modest yield [30]. Alternatively the Wittig reaction of chloroacetaldehyde (**19**) with *tert*-butyl 2-(triphenylphosphanylidene) propanoate provides after deprotection the carboxylic acid **20** with a 9:1 *E:Z* selectivity [31]. Likewise the orthogonally protected diol **22** can be obtained from silyl protected allylic alcohol **21** with a 15:1 *E/Z*-selectivity using Wittig reaction [32, 33] (Scheme 5).

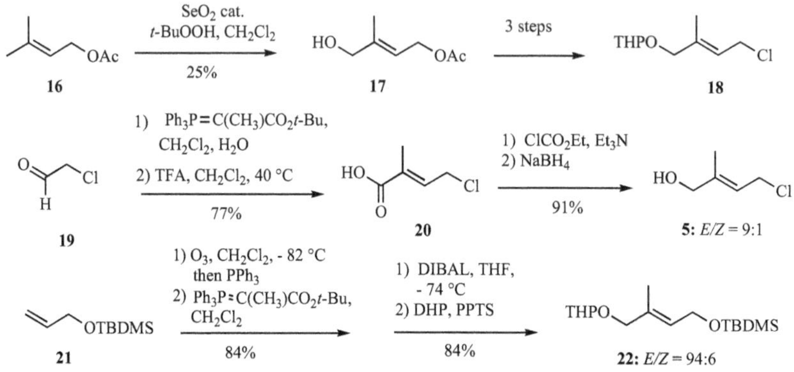

Scheme 5. Syntheses of monoisoprenyl bifunctional building blocks.

Owing to the importance of the sulfone chemistry for the synthesis of polyisoprenoid compounds using sulfur stabilized carbanion (Biellmann-Ducep coupling), the control of the regio- and stereochemistry during the synthesis of allylic sulfone such as **24** is an important topic. Generally, the direct nucleophilic substitution reaction of sodium sulfinate with allylic halides gives a mixture of S_N2 and S_N2' products. However, it has been shown that direct reaction of the (*E*)-allylic acetate **23** with sodium sulfinate in the presence of Pd(0) catalyst and Me$_4$NBr as phase transfer agent in methylene chloride/water biphasic mixture affords the desired *trans*-material with high regio- and steroselectivity, dppf ligand giving the best result [34] (Scheme **6**).

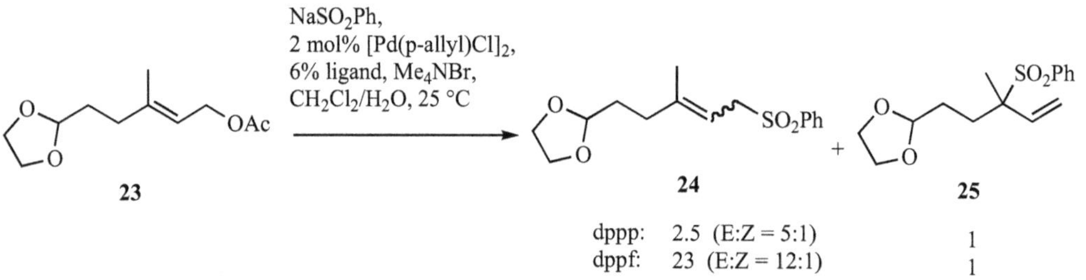

Scheme 6. Synthesis of sulfone-containing monoisoprenyl building block.

Synthesis of Bifunctional Polyisoprenyl Building Blocks

The functionalization of the remote double bond of small natural prenols such as geraniol and farnesol is a milestone step in numerous syntheses of larger polyprenols. These small bifunctional building blocks allow the rapid elongation of the growing chain in stepwise syntheses of polyprenols. Unfortunately, due to the weak nucleophilic difference between the terminal olefin and the $\Delta^{2,3}$-double bond, the direct oxidation of these compounds is highly challenging. Fringuelli *et al.* have extensively studied the epoxidation of geraniol with most available reagents [35]. Although epoxidation of the 2,3-double bond can be carry out easily with many reagents (mCPBA in emulsion, $MoO_5 \cdot HMPA \cdot py$, n-Bu_2SnO/t-BuO_2H, $VO(acac)_2$/t-BuO_2H, etc) the epoxidation of the 6,7-double bond is much more difficult to achieve chemoselectively. mCPBA in CH_2Cl_2 solution is poorly selective while the Grieco reagent ($PhSeO_3H$) provides mainly the distal epoxide **28** along with some 2,3-epoxide **27**. Monoperoxyphthalic acid (MPPA) in strongly basic media is not selective but a reasonable selectivity for **28** can be achieved at pH 8.3. The addition reaction of hydracids to geraniol acetate is equally unselective, however addition of $TiCl_4$ at low temperature gives a chlorotitanium species resulting of chemoselective addition on the remote double bond of geranyl acetate that cleanly affords the 6-(^2H), 7-Cl deuterated derivative on quenching with D_2O (Scheme 7) [36].

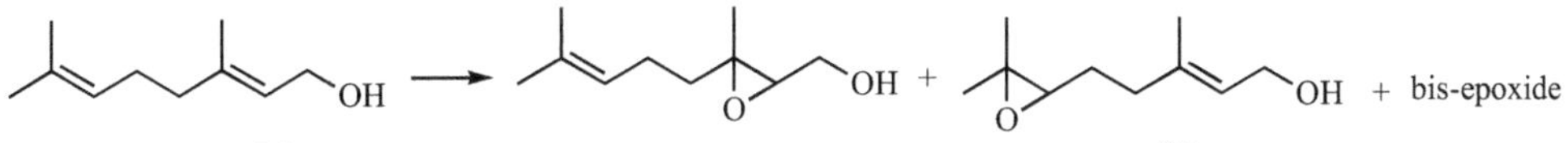

	27	28	bis-epoxide
$PhSeO_3H$, MeOH	13	87	0
MCPBA, CH_2Cl_2	30	57	13
MPPA, pH 8.3, H_2O	0	72	28
MPPA, pH 12.5, H_2O	92	0	8

Scheme 7. Epoxidation of geraniol.

SeO_2 allylic oxidation is the mostly used method to carry out the oxidation of the terminal *trans*-methyl group of geraniol or farnesol. Initially performed with a stoichiometric amount of SeO_2, the reaction is much more convenient using a catalytic amount of SeO_2 together with *tert*-butyl hydroperoxide as stoichiometric oxidizing agent. The use of selenium dioxide impregnated silica with microwave activation further improves the yield [37, 38] (Scheme **8**). However, if this method gives pure *trans*-compound (**30**), the yield is relatively modest, furthermore a mixture of aldehyde and allylic alcohol is generally obtained and reductive workup with $NaBH_4$ is usually needed. The influence of the alcohol protecting group on both the product distribution and the yield has been studied [39]. With longer isoprenoid compounds the SeO_2 oxidation procedure is much less effective. Farnesol gives a 3:1 mixture of all-*trans* 12-OH farnesol along with the secondary alcohol resulting of the allylic oxidation of the internal double bond, and geranylgeraniol silyl ether affords the terminal alcohol in only 10% yield [40, 41]. Regioselective ozonolysis of geranyl acetate has been proposed by Stork as an expeditious way to functionalize the distal double bond of geranyl acetate [42]. The addition of 1 equiv. of pyridine greatly improves the process. For example, epoxidation of geranyl benzyl ether in methylene chloride in the presence of 1 equiv. of pyridine delivers the corresponding aldehyde **32** in 43%. Reintroduction of the double bond using Wittig reaction followed by AlH_3 reduction gives the (*E*)-alcohol **33** together with 4% of *cis*-isomer [43, 44]. The palladium(II) catalyzed allylic oxidation of geranylacetone using *p*-benzoquinone as stoichiometric oxidizing agent and *o*-methoxyacetophenone as auxiliary ligand affords ω-acetyl-geranylacetone but with moderate selectivity [45]. The chlorosulfenylation/β-elimination sequence is a useful alternative. For example, the phenythioether **35** obtained by treatment of geranyl THP ether (**34**) with PhSCl and triethylamine gives the 8-hydroxygeranyl benzyl ether (**36**) upon sulfoxide/sulfenate rearrangement. High *E*-selectivity is claimed, but the exact value is not reported. Starting from farnesyl benzyl ether, the same process provides the corresponding ω-hydroxy farnesol benzyl ether in 58% overall yield [46]. Replacement of benzenesulfenyl chloride by methyl 2-(chlorosulfanyl) acetate gives the corresponding sulfide **37** which upon base-catalyzed [2, 3]-sigmatropic rearrangement and *in situ* desulfuration furnishes the homologated geranyl ester **38** [47] (Scheme **8**).

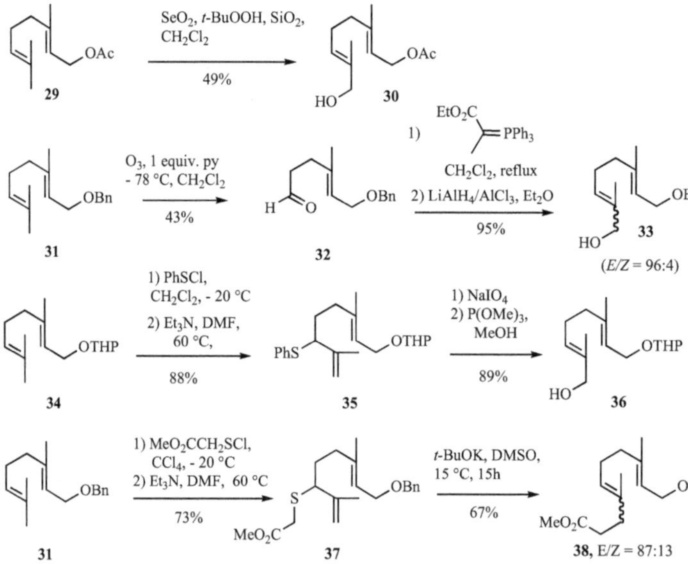

Scheme 8. Functionalization of the terminal double bond of geranyl derivatives.

Many reagents are available for the ene-type halogenation of the terminal isopropylidene unit of isoprenoid derivatives. Among which the use of SO_2Cl_2 [48], Cl_2O [49], *t*-BuOCl, SiO_2 [50], PhSeSePh/NCS [51] are the most useful. Electrogenerated Cl^+ ions can be used to induce the ene-type chlorination of geranyl and farnesyl acetate with 75-80% yield, while chlorination with chlorine gas in refluxing hexane appears suitable for industrial process [52, 53].

Chlorination of the terminal double bond of polyprenol derivatives using calcium hypochlorite and dry ice in CH_2Cl_2/H_2O two-phase mixture is very convenient for small scale reactions [54]. This remarkable method combined with a Cope elimination was applied by Sato *et al.* to prepare stereoselectively (*E,E*)-8 - hydroxy-geranyl acetate (**30**) from geranyl acetate (Scheme **9**) [55].

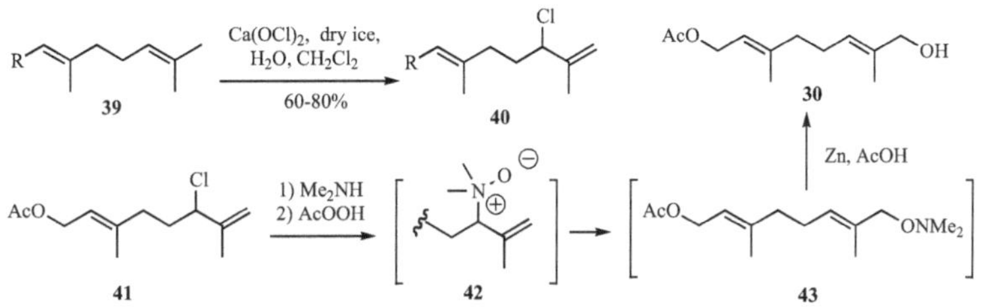

Scheme 9. Functionalization of the terminal double bond of geranyl derivatives by chemo- and regioselective chlorination.

Polymer-supported selenenyl bromide in combination with *N*-chlorosuccinimide (NCS) provides an extremely powerful way to achieve allylic chlorination of the terminal double bond of polyprenol derivatives. Remarkably, the reaction is also effective with farnesyl or geranylgeranyl acetate (**44**) and even solanesyl acetate that possesses nine double bonds, affords 45% of the terminal allylic chloride along with 15% of internal halogenated derivatives [56]. Corey has explored the intramolecular epoxidation reaction of polyprenol derivatives grafting a peroxy acid function on the alcohol head-group through a silyl ether spacer (**46**). This design allows orienting the steric course of the intramolecular epoxidation on the double bond situated at four or five units of the terminal active head as depicted in the approach **47**. High-dilution conditions are needed to avoid intermolecular side reactions (Scheme **10**) [57].

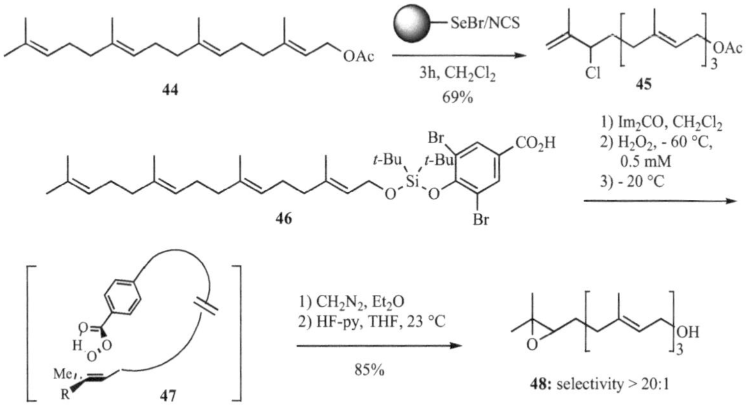

Scheme 10. Position selective epoxidation of geranylgeranyl derivatives.

Asymmetric dihydroxylation of geranyl acetate (**29**) using AD-mix-α and AD-mix-β reagents gives (*R*)- and (*S*)-6,7-dihydroxygeranyl acetate in 90% and 92% ee respectively. In both cases only a small amount of dihydroxylation of the $\Delta^{2,3}$ double bond was observed (7.5% and 4.5% respectively) [58 - 60]. Such selectivity is supposed to result of the lower nucleophilic character of the $\Delta^{2,3}$ double bond due to the electron withdrawing effect of the acetoxy group. As a consequence, it is much more difficult to discriminate the $\Delta^{6,7}$ and $\Delta^{10,11}$ double bonds of farnesyl acetate. Indeed, the dihydroxylation of farnesyl acetate (**49**) provides the expected 10,11-dihydroxyl farnesyl acetate (**50**) in only 34% yield (ee 92%) together with a mixture of starting material, internal diols and tetraols. A modified catalyst specially designed to achieve good regioselectivity of the dihydroxylation is described in which the catalytic pocked is closed by large ether groups that prevent the central part of the polyprenyl chain to access the active osmium atom. With this advanced catalyst dihydroxylation of farnesyl acetate (**49**) occurs with a 120:1 selectivity for the terminal isopropylidene group and with a 96% ee. The procedure can be extended with similar efficiency to more challenging geranylgeranyl acetate (**44**), squalene and even solanesol dimethylcarbamoyl with nine double bonds (Scheme **11**) [61].

Looking for a simpler catalyst, Corey and Zhang replaced the biscinchona alkaloids backbone of the standard Sharpless catalyst by the monocinchona ligand **52**. With this catalyst, farnesyl acetate (**49**) is dihydroxylated in 72% yield and 97% ee with formation of only 4% of the internal diol (Scheme **11**) [62].

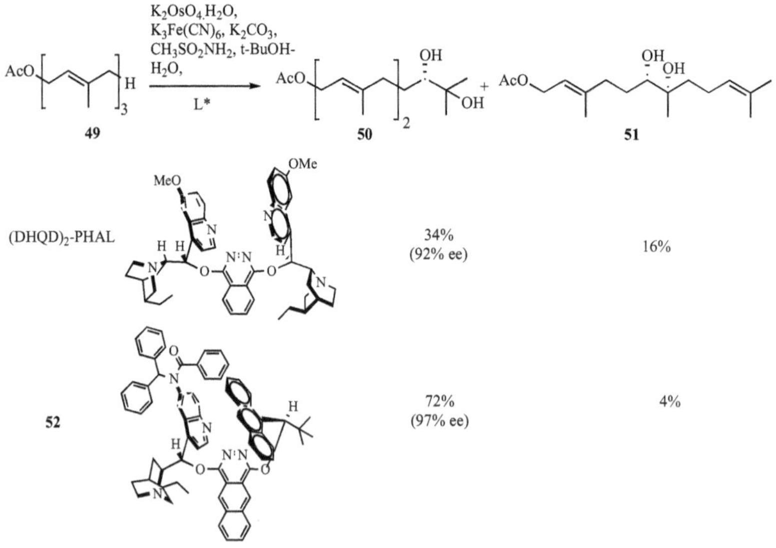

Scheme 11. Enantioselective dihydroxylation of farnesyl acetate.

Enzymatic processes to functionalize the remote double bond of geraniol and farnesol have been reported. For example, the Baker's yeast reduction of the ketone group of 8-hydroxy-7-oxo-geraniol prepared in four steps from geraniol, affords the corresponding 6,7-diol in ee > 95% [63]. A more direct method relies on the microbiological dihydroxylation of the distal double bond of geranyl phenycarbamoyl with *Aspergillus Niger* (49% yield, ee > 95%) [64].

Synthesis of Oligoprenoids by sp^3-sp^3 Carbon Bond Formation

Most of the reported syntheses of polyprenols involve a 5-carbon homologation of a shorter chain, mimicking the condensation of isopentenyl pyrophosphate with dimethylallyl pyrophosphate. The simplest way to achieve this task is the condensation of a prenyl-like nucleophilic synthon with an electrophilic poly-prenyl system bearing a leaving group, either through a S_N2 mechanism or alternatively through a S_N2' alkylation process To achieve this goal the nucleophilic mono-isoprene building block must carry a masked hydroxyl group at its non-nucleophilic terminus. The main hurdle to overcome in this strategy is to modulate the metal and the activating group to favor the α-alkylation over the γ–alkylation that leads to branched undesired derivatives (Scheme 12).

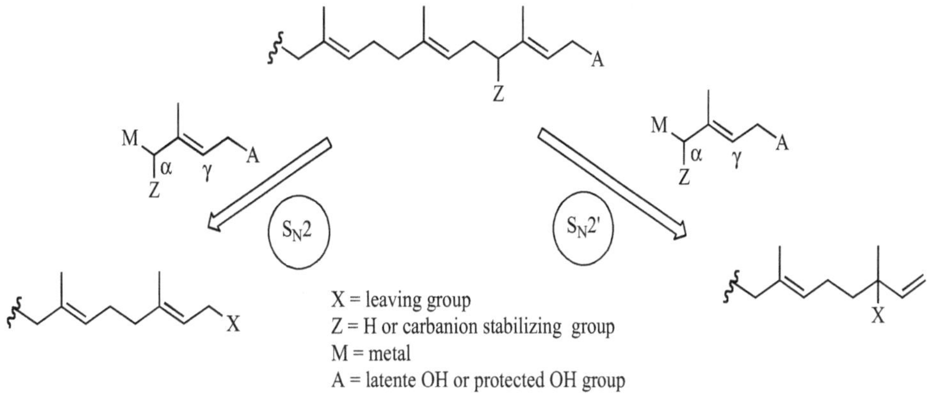

Scheme 12. α/γ selectivity in alkylation of isoprenyl halide with monoisoprenoid building blocks.

Construction of 1,5-Dienes via S_N2 Reactions

S_N2 Coupling Involving Sulfur Stabilized Carbanions

The most extensively used methods for building polyisoprene chain are based on the alkylation of sulfur stabilized carbanions (Scheme 12, Z = SPh, SO_2Ar). Nevertheless, a simple ester group can also be used. The lithiated dianion of 3-methyl-2-butenoic (**53**) gives a roughly equimolar mixture of α- and γ-alkylation products **54/55** upon reaction with prenyl bromide, however the corresponding copper derivative obtained by transmetalation with CuI undergoes higher γ-selectivity but without control of the geometry of the double bond [65, 66]. In a similar manner the copper dianion obtained by deprotonation of tiglic acid **56** with LDA followed with CuCl transmetalation, reacts with allylic chloride **57** to give acid **58** with a 3:1 *E/Z* ratio (Scheme **13**) [67].

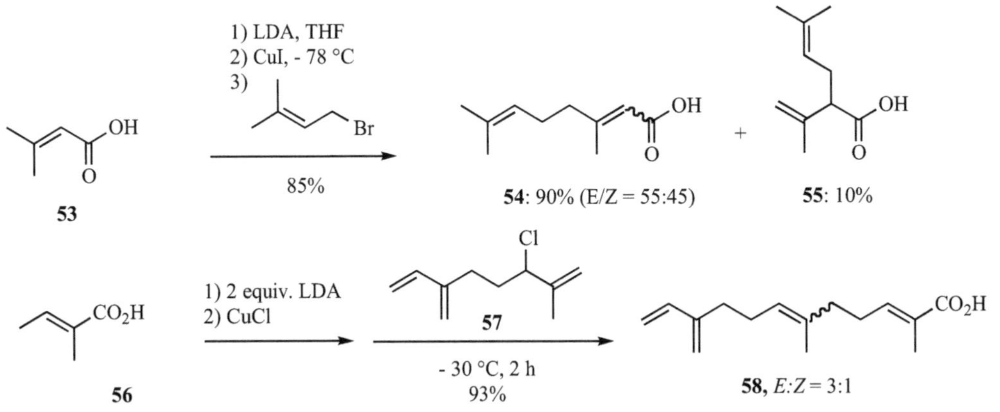

Scheme 13. Alkylation of senecioic and tiglic acids.

The discovery that the reaction of sulfur-stabilized carbanions with allyl halides proceeds *via* an exclusive S_N2 mechanism (Biellmann Ducep coupling) has opened the way to the synthesis of higher polyisoprenoids as exemplified by the synthesis of geranylgeraniol (**62**). Thus, alkylation of phenyl geranyl sulfide anion (**59**) with the allylic halides **60**, obtained by functionalization of the terminal double bond of geranyl benzyl ether upon selenium dioxide oxidation (see: Scheme **8**), provides sulfide **61** in 76% yield. Lithium ethylamine desulfuration of adduct **61**, then produces geranylgeraniol (**62**) with around 10% of isomer **63** arising from conjugated reduction [68, 43]. The reducing agent must be carefully chosen to minimize isomerization upon non-selective reprotonation of the allylic anion. For example Na/iPrOH or Li/NH$_3$, initially chosen as reducing agents gives a significant amount of double bond migration.

Replacement of the arylthio activating group for a *p*-toluenesulfonyl group allows an easier preparation and a more powerful activating effect. Furthermore, this group is easily removed using Li/EtNH$_2$, provided that no hydroxyl function stands in the close vicinity of the sulfone [69]. As a general rule, the sulfone alkylation of η^3-allylpalladium complex is more efficient that direct alkylation with allyl halide, furthermore the use of allylic carbonate as electrophilic partner avoids adding base as exemplified in the synthesis of farnesylacetone (**70**) (Scheme **14**) [70].

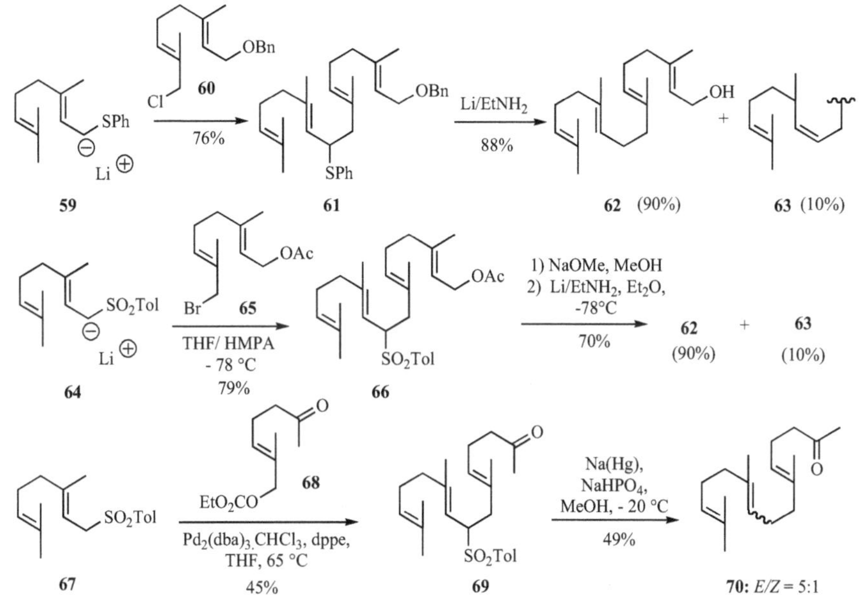

Scheme 14. Isoprenoid chain elongation using sulfur stabilized carbanions.

All-*trans* polyprenol are widely distributed in the plant kingdom. For example solanesol (**77**) which has 9 isoprene units is isolated from tobacco leaves and polyprenols with chain up to 10 units are found in mulberry leaves. Solanesol is also present in the leaves of other *Solanaceae* plants including tomato, potato, eggplant and pepper. It is known to possess useful medicinal properties such as anti-bacterial, anti-inflammation and anti-ulcer activities. Industrially, solanesol is extracted from *Solanaceae* leaves (about 450 tons in 2008) and is a key intermediate in the synthesis of coenzyme Q10 and vitamin K analogues [71].

Early syntheses of solanesol suffered from the low stereoselectivity in the formation of the *trans*-double bonds, leading to laborious purification. Sulfone alkylation with bifunctional farnesyl building block obtained by the chlorophenyl-sulfenylation method [46], permitted the first fully stereoselective synthesis of solanesol, the *trans*-stereochemistry present in the starting subunits being preserved during all the process. Full chain extension before final desulfonylation leads to a significant economy in the number of steps (Scheme **15**) [72]. The same strategy is applied for synthesis of solanesol, but using SeO$_2$ allylic oxidation to access the bis-functional geranyl building block. Careful analysis of the crude product of the Li/EtNH$_2$ desulfonation evidences that 10% of double bond isomerization occurs that can be removed by chromatography on silver nitrate-impregnated silica gel [69]. It is clear that many reported syntheses using the Li/EtNH$_2$ desulfonylation procedure suffer from the same drawback which is often neglected by the authors. This problem is now efficiently addressed by using the more efficient LiBHEt$_3$/PdCl$_2$(dppp) desulfonylation method. This procedure is much more selective than the standard sulfone reduction with Li/NH$_3$, Li/EtNH$_2$ or sodium amalgam leading invariably to undesirable isomerized products [73, 74]. The use of sodium borohydride in DMSO, together with Pd(OAc)$_2$/dppp appears as a still more efficient reductive desulfonylation procedure [75].

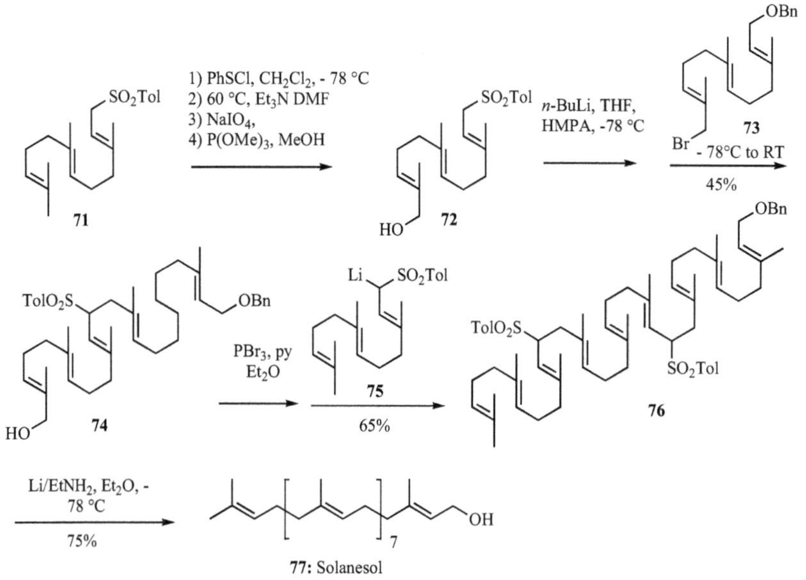

Scheme 15. Sulfone mediated synthesis of solanesol.

Stepwise chain extension of a polyprenyl compound using standard sulfone chemistry requires the use of a bifunctional isoprenyl synthon with a latent sulfone group at one extremity that can be unmasked after being bound on the extending chain. In most strategies this group is a protected alcohol that must be transformed into the sulfone group through several steps. A more straightforward method relies on simple phenyl thioether as latent sulfone group. Remarkably, the reaction of the sulfone anion **78** with thiophenyl allylic bromide **79** provides the alkylated compound **80** without any product arising from deprotonation in the α-position of the thioether group. Li/EtNH$_2$ desulfonylation and oxidation of the sulfur atom with LiNbMoO$_6$/H$_2$O$_2$ give the corresponding one-isoprene unit homologated sulfone **81,** ready for a new elongation sequence (Scheme **16**) [76, 77].

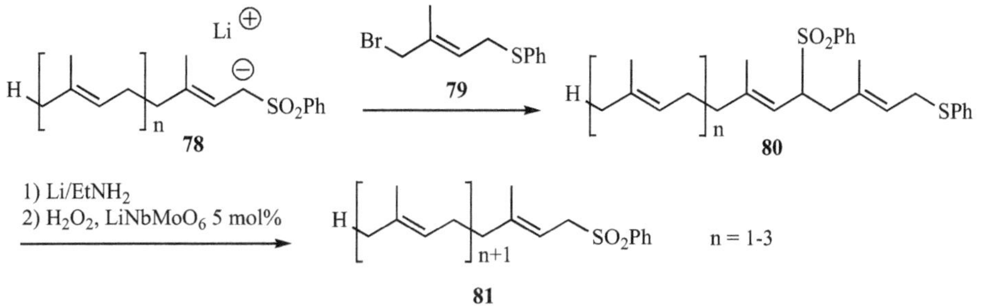

[Scheme 16 structures]

Scheme 16. Isoprenoid chain elongation using sulfur stabilized carbanions.

Ubiquinone-7 which is available in large quantities from the fermentation of *Candida utilis* has been converted into homologated coenzyme Q10 (CoQ10) the most abundant member in animals. The functionalization of the terminal double bond is a prerequisite before chain elongation. Beside SeO_2 allylic oxidation or chlorophenylsulfenylation, bromohydrin formation is a facile and selective method, first described by van Tamelen for squalene [78]. Thus, NBS bromination in wet DME of heptaprenyl derivative **82** gives the corresponding bromohydrin which is converted into epoxide **83** by K_2CO_3 treatment. β-Elimination with LDA affords the corresponding allylic alcohol which is next converted into the allylic chloride **84** through $S_N i'$ process. Biellmann-Ducep elongation with geranyl sulfone then provides CoQ10 (**86**) after reoxidation with Fe(III) (Scheme **17**) [79].

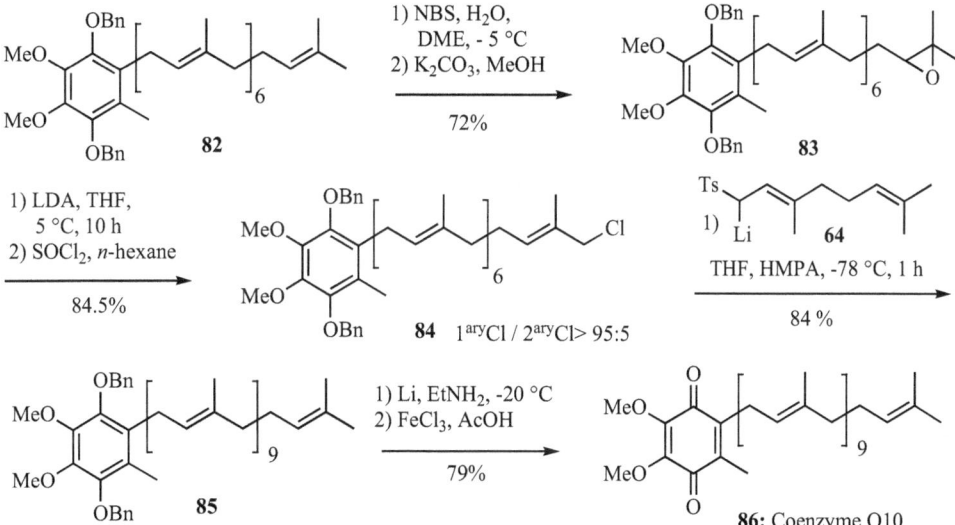

Scheme 17. Chain extension of ubiquinone-7 to ubiquinone-10.

The main drawback when using sulfone chemistry to build polyisoprenoid chains is the reductive desulfonylation step that induces generally a partial isomerization of the allylic double bond with concomitant loss of the stereochemistry. In order to circumvent this problem, umpolung carbonyl synthons can be used. Thus, dithiane **89** available by alkylation of the lithium anion of **87** with ω-bromo-farnesyl acetate (**88**) gives hexaprenol **91** after deprotection with Dess-Martin reagent and Wolff-Kishner reduction of the intermediate enone. An overall yield of 77% of the presumed all-*trans* hexaprenol (**91**) is reported but without details on the stereochemistry and position integrity of the double bond [80] (Scheme **18**).

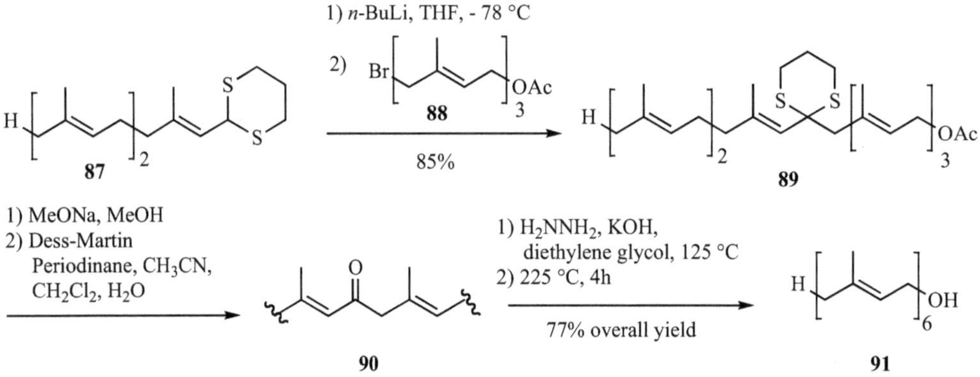

Scheme 18. Synthesis of hexaprenol using dithiane alkylation reaction.

The polymer-like structure of polyisoprenol makes them good candidates for solid-phase chemistry since it provides facile purification by washing away the excess of reagent. Several syntheses using sulfone chemistry take advantage of the solid-phase to access long chain polyisoprenoid compounds. For example, a synthesis of solanesol is reported in 50% overall yield starting from geraniol bound to Merrifield's resin through a tetrahydropyran linker [81]. Polystyrene-bound sodium sulfinate is a convenient primer to grow a polyisoprenyl chain. Thus, alkylation with farnesyl bromide affords the polymer-bound farnesyl sulfone **92** which is then engaged in a conventional sulfone chemistry using dimsyl sodium as base. Final desulfonylation with $LiBHEt_3$/Pd(0) releases the all-*trans* heptaprenol (**96**) from the resin. This method can also be applied to polyisoprenoids that possess double bonds with different geometries such as betulaprenols or ficaprenols [82] (Scheme **19**).

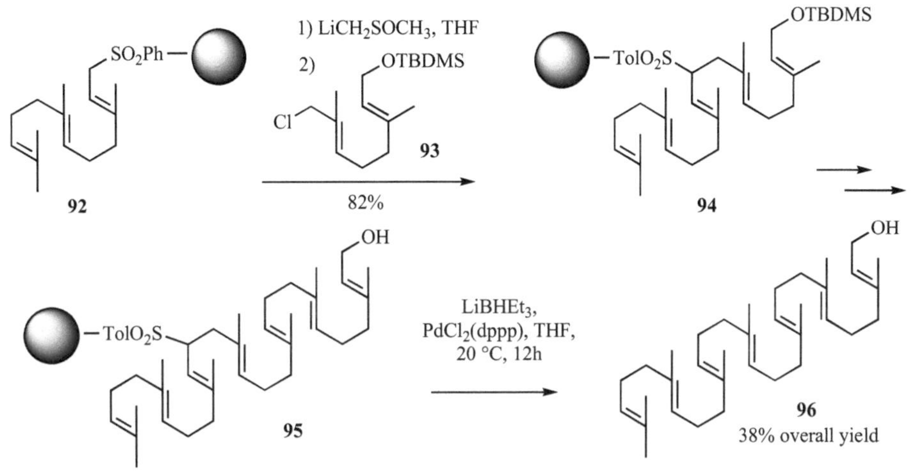

Scheme 19. Polymer supported synthesis of all-*trans* heptaprenol.

S_N2 Coupling with Unstabilized Organometallic Reagents

The main hurdle when using allylmetal reagents is their inherent stereochemical instability. For example *E*-crotylmagnesium bromide isomerizes rapidly into a *E/Z* mixture at temperature as low as -80 °C [83]. Furthermore, α-selectivity with prenyl halides is difficult to achieve, for example only 8% of geranyl acetate resulting of α-attack are obtained upon treatment of chloride **8** with the cyanocuprate derived from prenylmagnesium chloride [84]. As expected, the isomeric chloride **98** gives nearly exclusively geranyl acetate (**29**) resulting of S_N2' process, but with the *Z*-configuration. On the other hand, the displacement of sulfone **99** by prenylmagnesium bromide in the presence of a catalytic amount of copper(II) acetylacetonate gives rise to *E*-geraniol (**26**) in 60% yield [85]. The condensation of Grignard reagents with allylic diphenylphosphates in the presence of Fe(III) ions affords highly selectively α-alkylation products. Thus starting with terminal allylic phosphate such as **100,** the chain elongation can be achieved through S_N2 process with a total α-selectivity (Scheme **20**) [86]. By contrast, the cross-coupling reaction of prenylzinc bromide with allylic halide is poorly selective giving a mixture of all four regioisomers (α,α'/α,γ'/γ,α'/γ,γ'). Interestingly high γ,α'-selectivity is achieved in DMPU [87].

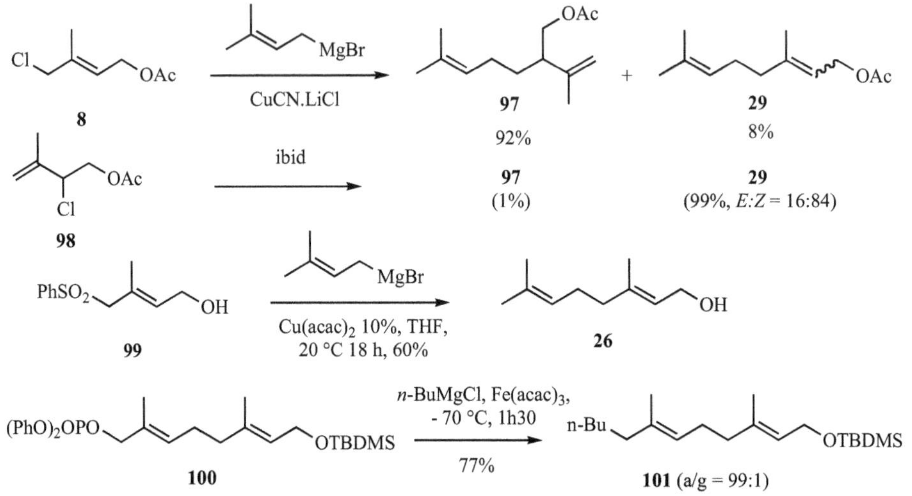

Scheme 20. Coupling of allylic derivatives with prenyl organocopper and organoiron reagents.

It is also possible to dispense with any carbanion stabilizing group, thus for example metalation of 3-methyl-3-butenol (**102**) with *n*-BuLi-TMEDA gives the di-lithiated species **103** which further reacts with prenyl bromide to afford the γ-methylene alcohol **104**. Oxidation into carboxylic acid followed by NaH isomerization give (*E*)-geranic acid **54** [88] (Scheme **21**).

In the quest for more selective methods, Sato in the early 70s showed that the reaction of η^3-allylnickel complexes with allylic halides, previously reported by Corey as a versatile tool for building cyclic terpenes, is also useful to synthetize polyisoprenoid chains with high *E*-stereoselectivity despite the lack of entropic factor. Nonetheless, this remarkable method suffers from the need to use highly toxic and volatile nickel tetracarbonyl [89, 90].

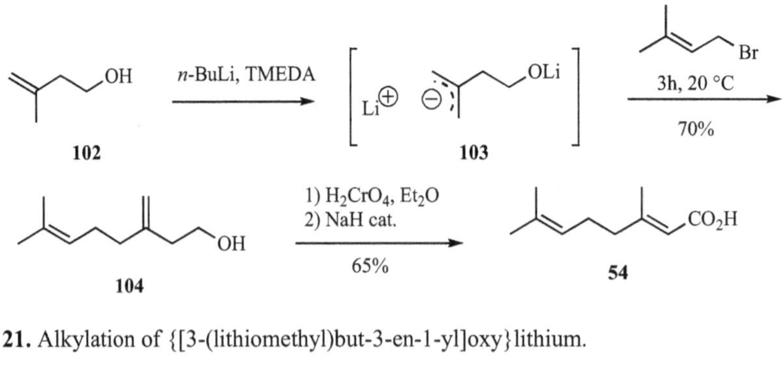

Scheme 21. Alkylation of {[3-(lithiomethyl)but-3-en-1-yl]oxy}lithium.

The ambident nature of allylic organometallic reagents has compelled to develop heteroatom stabilized carbanions (PhS, $PhSO_2$, PPh_3, $PhSO_2CHCO_2R$, etc) able to direct the alkylation reaction on the α-carbon center. In this regards, the discovery that allylbarium reagents retain the regiochemistry present in the starting halide is a significant breakthrough. Initially introduced for the homocoupling of two farnesyl units to obtain squalene, the reductive alkylation using Rieke barium can be extend to the synthesis of polyisoprenyl compounds. These stable allylbarium reagents prepared from the corresponding allylic chlorides and Rieke barium, react with allylic bromides according to a pure S_N2 process to give homologated chains without loss of the *E*-stereochemistry of both isoprenoid units [91, 92]. Thus condensation of geranylbarium reagent **108** with the ω-bromogeranyl-silyl ether **109** gives the silyl ether of all *trans*-geranylgeraniol (**110**) in 61% yield. Stabilization of the anionic species through a π-allylbarium complex such as **108** may explain the unusual stereospecificity of the process [93]. Reaction of the barium reagent derived from farnesyl chloride with carbon dioxide is a stereoselective method to perform the one-carbon homologation of the polysoprenoid chain without loss of the stereochemical integrity of the natural *E*-$\Delta^{2,3}$ double bond [94] (Scheme **22**).

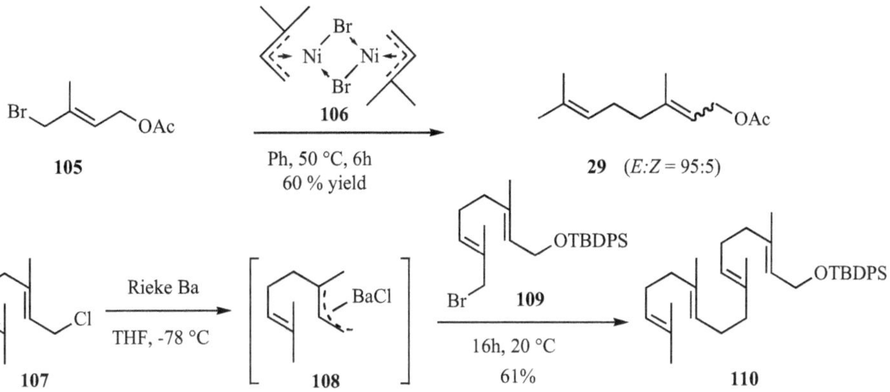

Scheme 22. Synthesis of geranyl acetate by using η³-allylnickel complexes and allylbarium reagents.

Construction of 1,5-Dienes via S_N2' Reactions

S_N2' Reaction Involving Organometallic Reagents

The S_N2' reaction of organocopper reagents with secondary allylic phosphates is a powerful method to build regio- and stereoselectively polyisoprenyl chains. For example, copper catalyzed addition of prenylmagnesium bromide to phosphate **111** gives geranyl benzyl ether (**31**) with less than 2% of the α-regioisomer [95], contrasting with the iron-catalyzed addition of Grignard reagents occurring mainly *via* a S_N2 process (see Scheme **20**). Likewise, complete γ-selectivity with high *E/Z* ratio can be achieved by reaction of allylic phosphate (**112**) with the cyanocuprate made from prenylmagnesium bromide and CuCN.2LiCl [84]. The similar high γ-selectivity (γ/α = 99.8:0.2) is observed by reaction of allylic chloride with organozinc reagent **114** derived from geranyl chloride (*E/Z* 70:30 mixture) in the presence of coper bromide. However, because allylic zinc reagents rapidly equilibrate into an *E/Z* mixture, the $\Delta^{10,11}$ double bond is not controlled in the process [96]. Following Julia's previous findings [85], Masaki studied sulfur containing leaving group in the same process. Both sulfoxides and sulfones (*i.e.***115**) are suitable leaving groups for the γ-substitution with dialkylcuprates, sulfones giving slightly higher regio- and stereoselectivity. Unfortunately the *E/Z*-ratio is less satisfactory than using phosphate leaving groups (Scheme **23**) [97]. 2-Benzothiazolyloxy is another potential leaving group in such processes, but a disappointing α/γ selectivity is observed [98]. Replacement of the phosphate leaving group with an acetate group does not improve the stereoselectivity since the key issue of the stability of the organomagnesium reagent is not addressed [99].

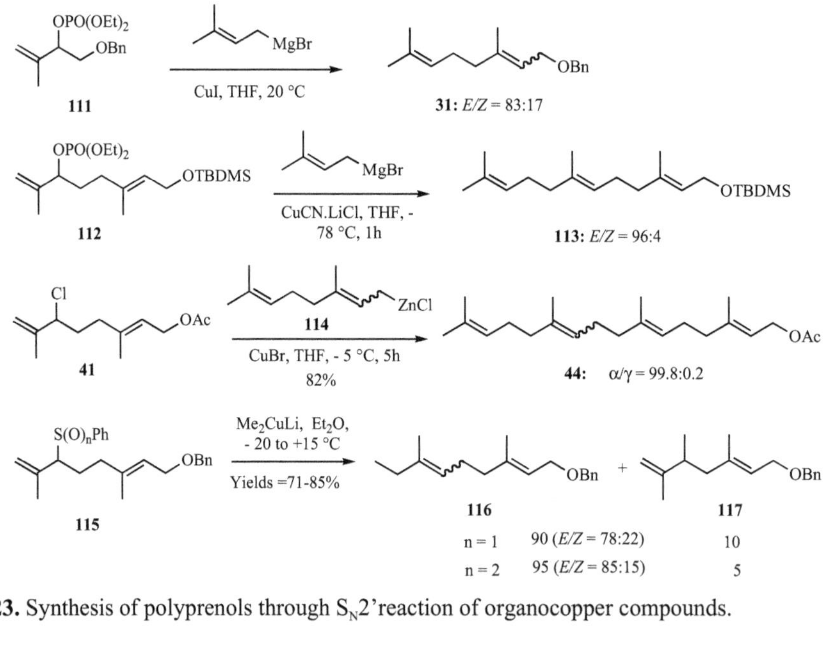

Scheme 23. Synthesis of polyprenols through S_N2' reaction of organocopper compounds.

Secondary allylic phosphates clearly emerge as the most useful leaving groups for polyisoprenoid synthesis. They are easily obtained from the terminal isopropylidene group by the sequence: epoxidation, base catalyzed ring opening and phosphorylation. Starting from the fully protected allylic phosphate **118**, a total synthesis of CoQ10 based on iterative S_N2' substitution of allylic phosphate is reported (Scheme **24**). Thus, condensation of allylic phosphate **118** with the Grignard reagent **119** in the presence of CuCN.2LiCl provides the vitamin Q6 precursor **120** in 82% yield. High stereoselectively (E/Z = 97:3) is observed at the newly formed double bond, unfortunately the poor geometrical stability of the organonomagnesium reagent leads to the 2:1 mixture a $\Delta^{9,10}$ double bond. Conversion of the terminal isopropylidene into the allylic phosphate **121** paves the way for a second coupling with the Grignard reagent **119** that delivers the decaisoprenoid **122**. Final, oxidative deprotection with CAN furnishes CoQ10 in 71% yield [100, 101].

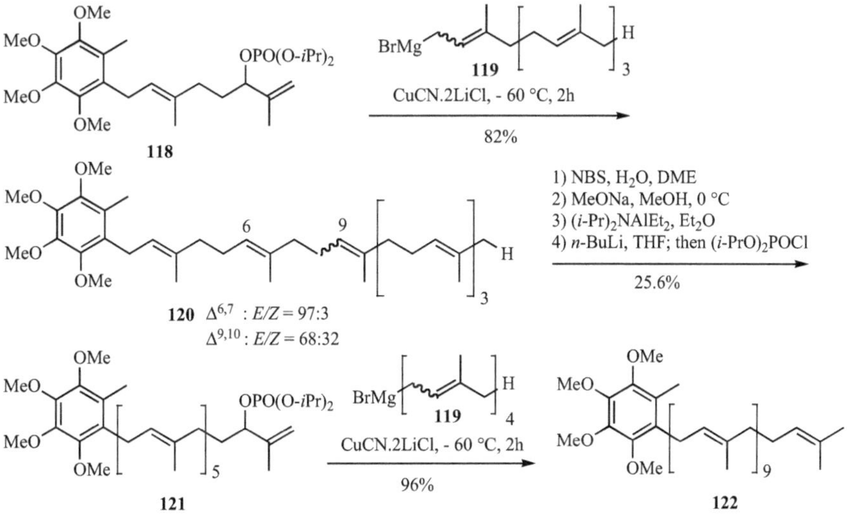

Scheme 24. Iterative chain elongation of isoprenoid using S_N2' organocopper reaction with allylic phosphates: Synthesis of coenzyme Q10.

4-(Prop-1-en-2-yl)oxetan-2-one (**124**) is a versatile 5–carbon building block for polyisoprenoid chain extension. Condensation of the cupromagnesium reagent derived from geranyl chloride with β-lactone **124** provides homofarnesyl acid (**125**) in 68% yield but with a moderate E/Z isomeric ratio at the newly created double bond (Scheme **25**) [102].

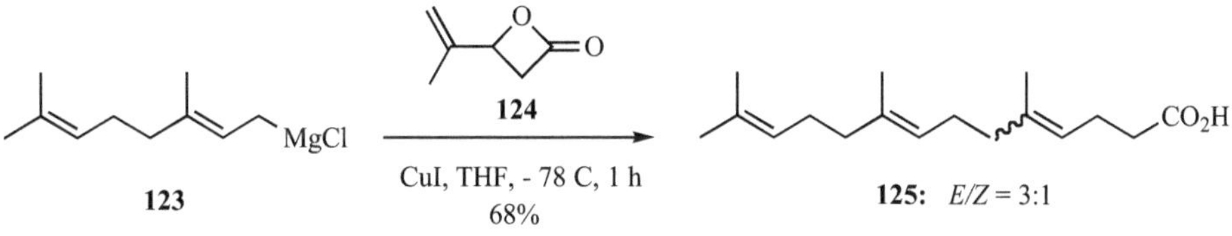

Scheme 25. Five-carbon elongation reaction using allylic β-lactones.

S_N2' Reaction Involving Sulfur Stabilized Carbanions

A fully stereoselective synthesis of ubiquinone 10 was achieved by Keinan et al. by iterative alkylation of the geranyl building blocks **127** [103]. A geranyl appendage is first introduced on the tetramethoxybenzene ring by alkylation of its cyanocuprate with geranyl bromide and is further elaborated into allylic carbonate **126**. The efficient access to this compound opens the way for chain elongation involving sequential Tsuji-Trost reactions. Thus, condensation of allylic carbonate **126** with the β-sulfone ester **127** in the presence of tetrakis (triphenylphosphine)palladium(0) provides the adduct **128** with a complete control of the newly formed double bond. The nature of the leaving group plays a major role to permit a fast reaction rate while controlling the geometry of the double bond. Cl, OAc or $OPO(OMe)_2$ leaving groups are unsatisfactory, when $MeOCO_2$ allows a good reactivity and a complete stereoselectivity without adding base. The E-stereochemistry of the formed double bond is supposed to be controlled by the nucleophilic attack of the stabilized carbanion on the more stable syn η^3-allylpalladium complex. Repeating the process three times, smoothly delivers the decaprenyl derivative **130** that displays the full chain of CoQ10. Completion of the synthesis requires the removal of both the methoxycarbonyl and the p-toluenesulfonyl groups without affecting the geometry of the vicinal double bonds. This non-trivial task is achieved by treatment with 4-aminothiophenol in the presence of cesium carbonate to remove the ester groups. Metal dissolving reduction is the usual process to get rid of the sulfonyl activating group in Biellmann-Ducep coupling. Unfortunately, some isomerization of the vicinal double bonds is usually observed. However, the treatment of the tetra-sulfone **131** with $LiBHEt_3$ in the presence of a catalytic amount of $PdCl_2(dppp)$ affords the full decaprenyl chain with 86% yield, based on previous findings of Mohri et al. establishing that allyl sulfones can be reduced with retention of stereochemistry by a hydride source in the presence of palladium(0) [74]. The synthesis is finally completed by cerium ammonium nitrate reoxidation of the aromatic ring to benzoquinone **86** in 76% yield (Scheme **26**).

This remarkable synthesis completed in 10 steps and 27.4% overall yield from allylic carbonate **126** remains the more efficient total synthesis of coenzyme Q10 to date.

The Keinan's strategy can been extended to the synthesis of simple polyprenols. Thus, the five-carbon elongation reaction of the farnesyl β-sulfonyl ester **132** using isoprene monoepoxide (**2**) as allylic partner in the Tsuji-Trost reaction furnishes exclusively the all trans-geranylgeraniol (**62**) after decarboxylation and desulfonylation with $LiBHEt_3/PdCl_2(dppp)$ (Scheme **27**) [104].

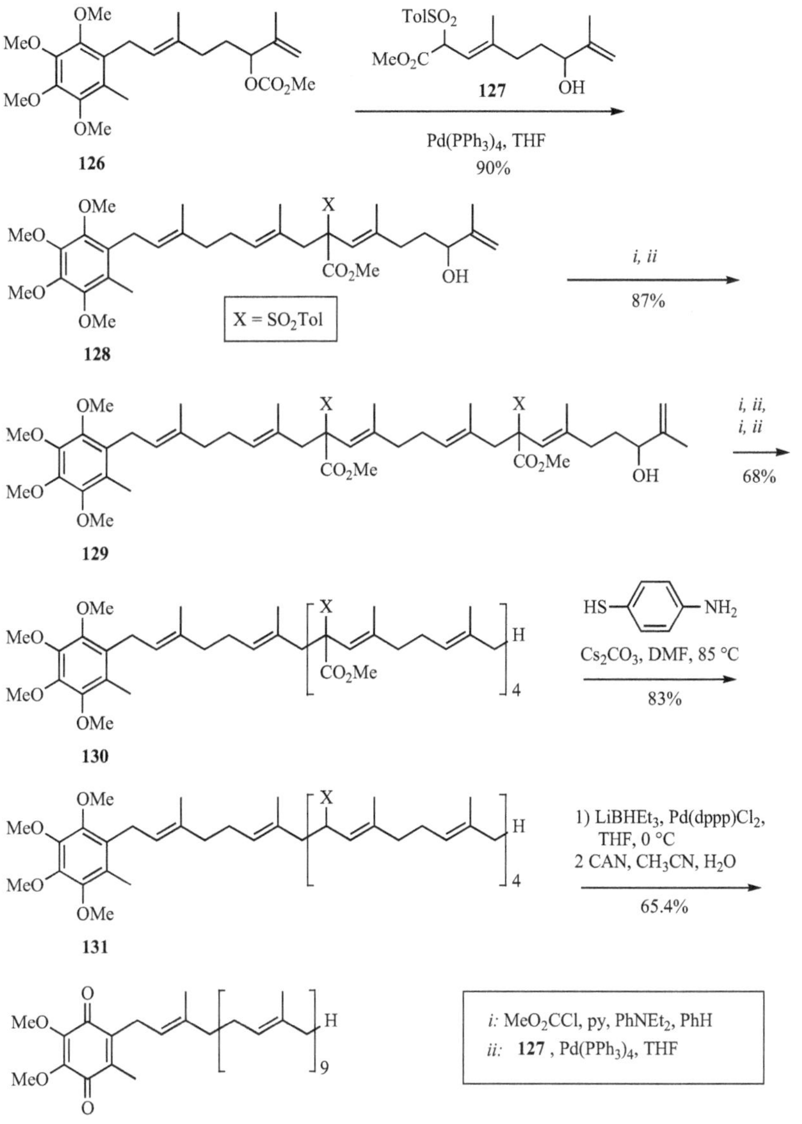

Scheme 26. Iterative chain elongation using Tsuji-Trost reaction of allylic carbonates: Keinan's synthesis of coenzyme Q10.

The *E*-stereochemistry of the double bond is attributed to the addition of the hydride on the less hindered side of the π-allyl complex in its more stable *syn*-configuration according to the approach **134**, as outlined in Scheme **27** [104].

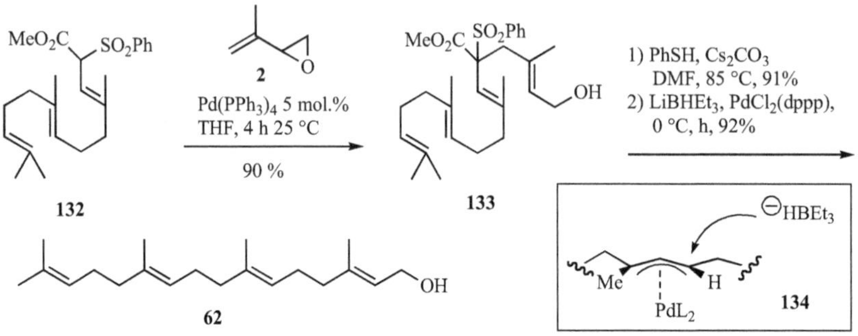

Scheme 27. Synthesis of geranylgeraniol using Tsuji-Trost reaction of isoprene oxide.

Polyprenol Synthesis via Free Radical Additions

Beside anionic and organometallic allyl-allyl cross-coupling reactions, the radical chemistry has been briefly explored to prepare polyprenol chains but with limited success. For example, irradiation of a mixture of geranyl bromide (**135**) and thioether **136** in the presence of hexamethylditin delivers farnesyl acetate (**49**) as a mixture of isomers suggesting a low preference for the attack on the less substituted terminus of the intermediate allylic free radical (Scheme **28**) [105].

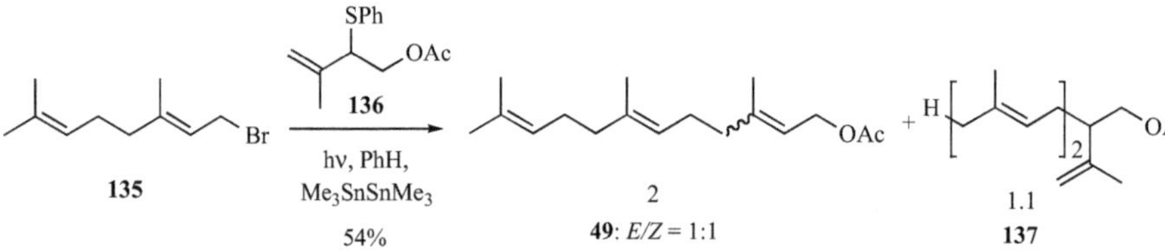

Scheme 28. Synthesis of farnesyl acetate using free radical process.

Enzymatic and Biomimetic Approaches

Polyisoprenoid compounds are biosynthesized by successive condensation of isopentenyl diphosphate (IPP) and dimethylallyl diphosphate (DMAPP) subunits catalyzed by *trans*-isoprenyl diphosphate synthases. The possible formation of polyprenoids from simple olefin containing alcohols mimicking IPP and DMAPP in prebiotic conditions has been addressed. Thus, incubation for several days of a mixture of dimethylallyl alcohol and isopentenol with clays delivers a mixture of geraniol (**26**) (3%), *trans*-isogeraniol (46%), 6-methyl-2-methylidenehept-5-en-1-ol (24%) and other isomers. Despite the low yields, this result supports the hypothesis that polyprenoids may be formed under prebiotic conditions [106].

The main problem associated with enzymatic production of prenol pyrophosphates is the lack of specificity of the prenyltransferase enzymes leading to a mixture of all possible compounds. The discovery of specific enzymes is a major step beyond. For example, crude cell-free extract from flowerheads of *Rosa dilecta* was used to prepare ^{14}C-labelled geraniol with an extremely low contamination by farnesol [107]. Several farnesyl diphosphate synthases (FPPase) including porcine and *Bacillus stearothermophilus* FPPases have been over-expressed and purified for synthetic purposes [108]. Even unnatural homologs of IPP are acceptable substrates for this thermostable FPPase as illustrated by the catalysis of the condensation of isopentenyl diphosphate and dimethylallyl diphosphate analogs **138** and **139** to give a mixture of polyprenyl pyrrophosphates **140** and **141** [109]. The synthesis of highly complex glycinoprenols **143** from phytyl diphosphate (**142**) at the sub-milligram scale using undecaprenyl diphosphate synthase expressed in *Escherichia coli* is a remarkable illustration of the potential interest of the enzymatic synthesis of the polyisoprenoids (Scheme **29**) [110].

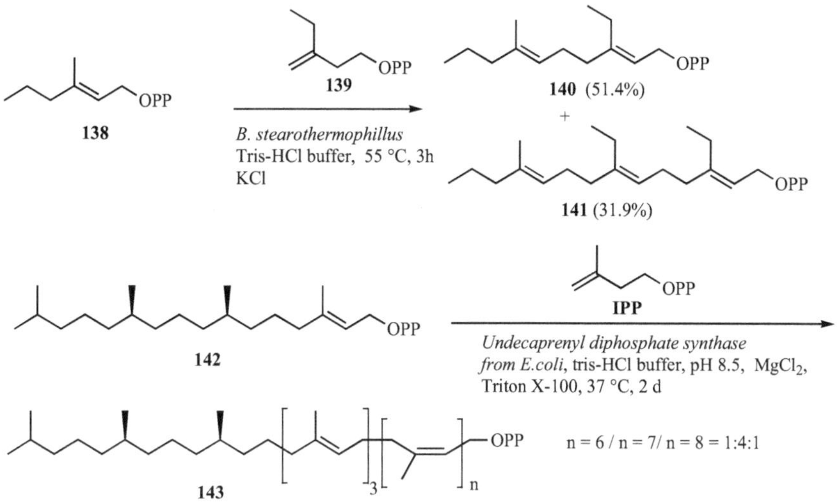

Scheme 29. Syntheses of oligoprenoids using purified enzymes.

In vitro biosynthesis of higher polyisoprenoids is more challenging because higher homologues of DMAPP become too lipophilic to be substrate of the enzymes and the use of non-aqueous solvents induces a drastic loss in activity. This problem can be overcome using thermostable enzymes from hyperthermophilic archaeon that keep their activity in *n*-BuOH/water mixture up to 65 °C. With this system, farnesyl- and geranylgeranyl-diphosphate are produced and accumulate in the organic phase [111]. The major limitation to the enzymatic synthesis of higher polyprenols is the consummation of farnesol by squalene synthase that blocks biosynthesis of regular polyprenols at the geranyl stage. Hence to maximize isoprenoid production, mutated squalene synthases were expressed in Saccharomyces *cerevisiae*. These mutants produce enough squalene to avoid sterol supplementation but allow accumulating farnesyl pyrophosphate that becomes available for the biosynthesis of higher polyisoprenoid compounds [112]. Because the industrial chemical synthesis of polyisoprenoid compounds remains challenging, their extraction from natural sources laborious, expensive and facing the rising of environmental concerns, there is an increasing interest for their direct production in so-called *microbial cell factories* [113]. It is now possible to reconstruct a complete synthetic pathway for a given terpene in *Escherichia coli* or *Saccharomyces cerevisiae* for large scale production. The various strategies used by metabolic engineering and synthetic biology to produce isoprenoids in bacteria and plants by over-expression of native or foreign genes have been recently reviewed [114].

Mimicking the attack of dimethylallyl diphosphate on the cation arisen from IPP, Corey *et al.* have described a biomimetic synthesis of regular isoprenoids implementing a Hosomi-Sakurai reaction that affords oligoprenol with high stereoselectivity. As a specific illustration, the condensation of all-*trans* farnesaldehyde dimethyl acetal (**144**) with the allylsilane reagent **145** in the presence of BF$_3$.OEt$_2$ gives after deprotection the methyl ether **146**. Li/EtNH$_2$ reduction of the lithium alkoxide provides the all-*trans* pentaprenol (**147**) in 96% yield [115]. Alternatively, the corresponding allylic tin reagent **149** reacts with geranial (**148**) to give alcohols **151** but accompanied by a substantial amount of primary-secondary coupling product **150** in 2.3:1 ratio. After protecting group exchange and Li/EtNH$_2$ reduction, geranylgeraniol (**62**) is obtained in 42% overall yield (Scheme **30**) [116].

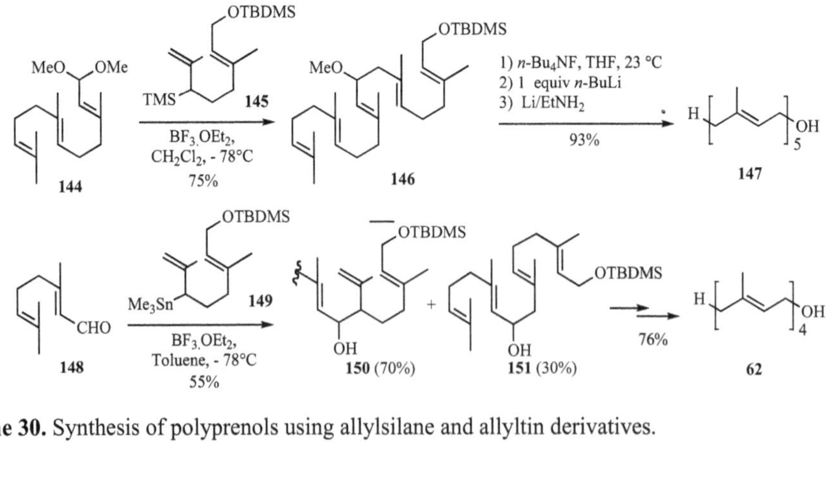

Scheme 30. Synthesis of polyprenols using allylsilane and allyltin derivatives.

Synthesis of Polyisoprenoids by sp^2-sp^2 Carbon-Carbon Bond Formation

Synthesis of Polyisoprenoids by Wittig Reaction

Iterative methods suitable to build polyisoprenyl ketones are quite rare. Among them, the use of the unstabilized ylide derived from phosphonium salt **152** must be emphasized. Thus, after Wittig reaction, the terminal alkyne of the obtained product is hydrated with mercuric sulfate in sulfuric acid to produce methyl ketone **153**, ready to be reengaged into a new Wittig reaction. The main drawback of the process is the complete absence of selectivity for the *E*-isomer [117]. Horner-Wadsworth-Emmons (HWE) olefination of geranyl- or farnesyl-acetone, readily available natural building blocks, gives moderate *E*/*Z*-stereoselectivity. Although the desired material could be obtained by chromatography on silver nitrate impregnated silica gel, the process is poor yielding and time consuming [118 - 120]. A more *E*-selective procedure (*E*:*Z* = 10:1) relies on the use of NaH as base in the presence of a sodium chelating crown ether as illustrated by the synthesis of ester **154** [121]. Synthetically useful *E*-selectivity can also be achieved using the more hindered diisopropyl phosphonoacetate [122]. Nevertheless, high *E*-selectivity is more easily achieved by building the chain in the opposite direction. For example HWE olefination reaction of diethyl phosphonopropionate (**156**) with aldehyde **155** using the Masamune-Roush conditions provides the triene **157** with a 96:4 selectivity in favor of the *E*-isomer (Scheme **31**) [123, 124].

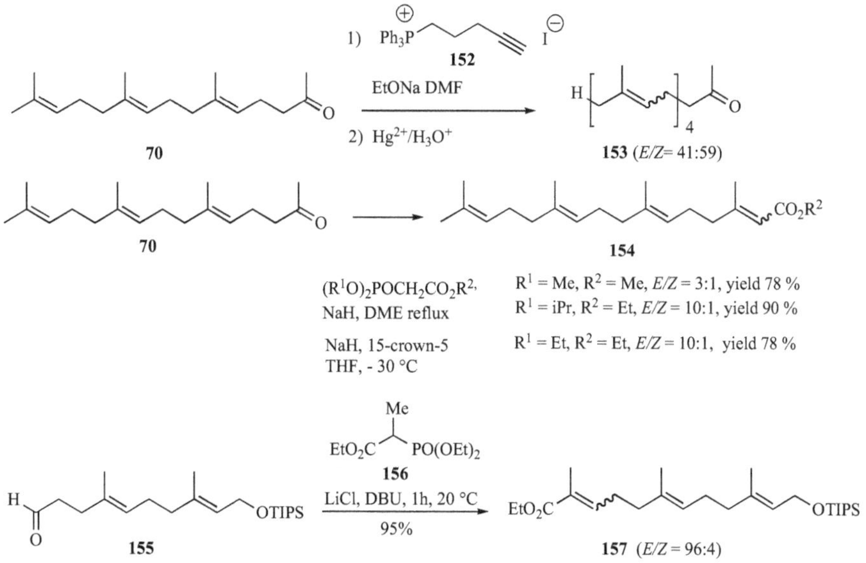

Scheme 31. Synthesis of all-*trans* polyisoprenoids using Wittig reaction.

Synthesis of Polyisoprenoids by Metathesis Reaction

The synthesis of trisubstituted olefins by cross-metathesis is usually difficult because a mixture of geometrical isomers is generally obtained. For these reasons, few syntheses of open chain polyisoprenoids involving cross-metathesis are reported. Among them, the homologation of the prenyl chain of the protected hydroquinone **158** into of α-tocopheryl acetate is noteworthy. Thus, reaction of the prenylated hydroquinone **158** with two equivalents of (*E*)-phythyl acetate (**160**) in the presence of 5 mol% of Grubbs II catalyst (**159**) gives the adduct **161** in 82% yield as a 70:30 *E/Z*-isomer mixture, this ratio reflecting roughly the thermodynamical equilibrium [125]. A synthesis of CoQ10 involving cross-metathesis reaction of allyl arene **162** with polyisoprenyl chain **163** in the presence of Hoveyda-Grubbs II catalyst (**159**) is patented. The dimethyl ether of dihydro CoQ10 **122** is thus obtained in 75% yield, but the stereochemistry of the newly form olefin is not reported (Scheme **32**) [126].

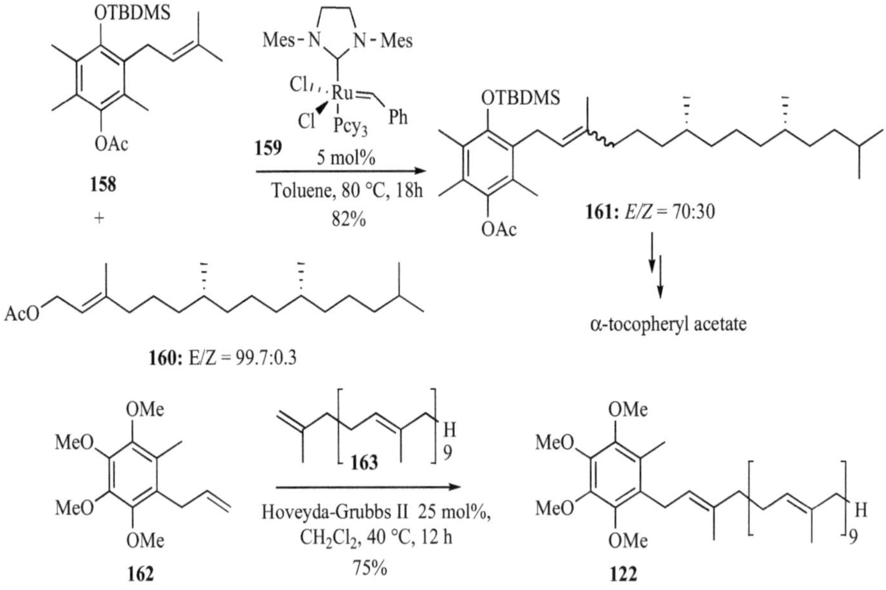

Scheme 32. Elaboration of polyisoprenyl chains using cross-metathesis.

Polyprenyl ansa chains are suitable target for macrocyclization using ring closure metathesis (RCM) reaction because the geometrical constraints allow efficiently controlling the stereochemistry of the newly formed double bond. For example, the RCM reaction of allyl cyclopentenone **164** using Grubbs II catalyst provides macrocycle **165** which displays the full carbon core of terpestacin in 44% yield. Only a small amount of *Z*-olefin is formed in the process. On the other hand,

RCM of polyene **166** using a terminal vinyl group as relay olefin is the key step for closing the highly strained chain of bicyclic compound **167**, a direct precursor of the anti-inflammatory diterpene hypoestoxide (Scheme **33**) [127 - 129].

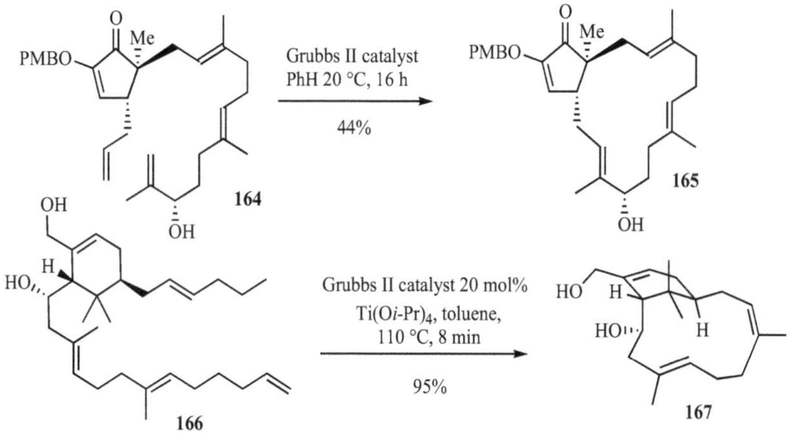

Scheme 33. Synthesis of polyisoprenoid ansa-chains using ring closure metathesis.

Synthesis of Polyisoprenoids by sp^2-sp^3 Carbon Bond Formation

The direct construction of the trisubstituted double bond is an attractive method to address the crucial problem of the control of the stereochemistry of polyisoprenoid compounds avoiding the usual difficulties encountered in the coupling of two-allylic units such as allylic transposition and homocoupling process. However, the Wittig reaction is unsatisfactory in most cases. On the other hand, configurationally stable vinyl organometallic reagents are now accessible by various means. The homologation of geraniol (**26**) into all-*trans* farnesol reported by Posner constitutes one of the first examples of this potential solution. Thus, (*E*)-3-iodo crotonic acid (**170**) prepared by the method of Le Noble [130] is reduced and protected as *tert*-butyldimethylsilyl ether **171**. Lithium iodide exchange provides the corresponding vinyllithium reagent which is then transformed into first-order cuprate **172** whose condensation with homogeranyl iodide (**173**) gives *E,E*-farnesol silyl ether **113** in 80% yield [131]. Similarly, the ketal **177**, a precursor of (*S*)-geranyllinalool can be stereoselectively obtained by reaction of the chiral tosylate **176** with the higher cuprate reagent derived from *Z*-vinyl iodide **175**. The latter is easily accessible upon zirconium-catalyzed carboalumination of alkyne **174** followed by trapping of the intermediate alane with iodine (Scheme **34**) [132].

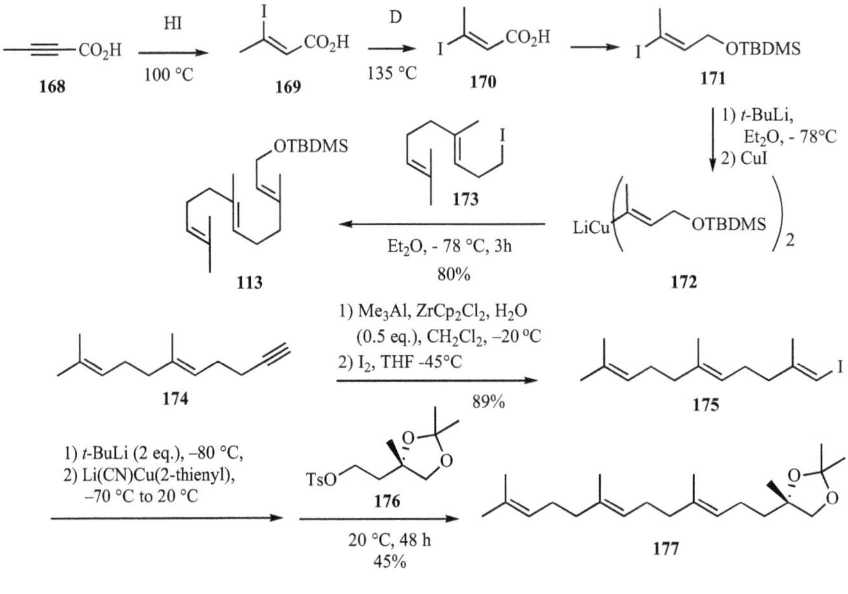

Scheme 34. Synthesis of polyisoprenyl chains by alkylation of vinyl copper reagents.

The reaction of enol phosphates and enol triflates with organometallic reagents is a versatile tool to build isoprenyl units. They are easily available from enolate of β-keto-esters. For example, NaH deprotonation of β-keto-ester **179**, prepared by alkylation of the dianion of methyl acetoacetate (**178**) with geranyl bromide (**135**), gives pure Z-enol phosphate **180** after trapping with diethylchlorophosphate, thanks to the intramolecular chelation of the metal cation. Dimethyl cuprate addition on **180** takes place with full retention of stereochemistry to provide *E,E*-methyl farnesoate **181** with 98% stereoselectivity [133, 134]. Alternatively, the use of enol triflate further improves the process by enlarging the C-C bond formation to copper reagents [135] and to palladium-cross-coupling reactions of organostannane [136], but also to Suzuki-Miyaura reaction, avoiding the use of highly toxic tetramethyltin, as illustrated by the conversion of enol triflate **182** into unsaturated ester *E*-**181** [137]. The steric course of the triflation reaction is highly dependent of the nature of the solvent. Deprotonation of **179** with KHMDS in THF gives a 95:5 *Z/E*-mixture of the two enol triflates, whereas the *E*-isomer is exclusively obtained in polar solvent such as DMF. Isomerization to the more stable *E*-enolate by disruption of the potassium chelate by competition with the carbonyl of DMF has been evoked to explain this result [138]. The selectivity for the Z-enol triflates can be further improved by quenching the potassium enolate of the β-ketoester with *N*-(5-chloro-2-pyridyl)bis (trifluoromethane-sulfonimide) [139].

Although the cross coupling reaction of enol triflate with organometallic reagents is a one of the most powerful methods, the stereospecific copper catalyzed displacement of the phenylthio group of β-thiophenyl-unsaturated esters, illustrated by the synthesis of the synthetic precursor of C-18-juvenile hormone **185** from methyl alkynoate **183**, is a useful alternative (Scheme **35**) [140, 141].

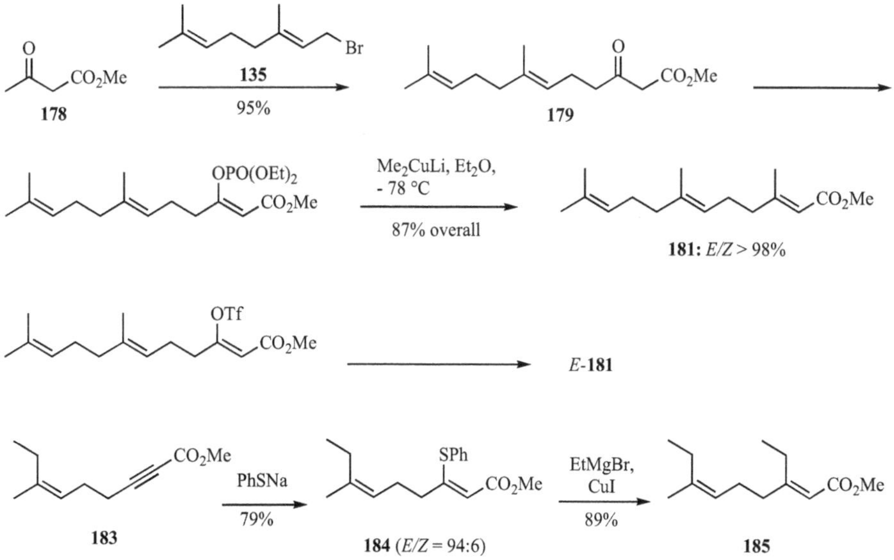

Scheme 35. Coupling of isoprenoid enol phosphates and enol triflates with organocopper reagents and boronic acids.

The main drawback of enol phosphates is their poor stability thwarting the removal of the small amount of undesired stereoisomer by chromatography. The more stable enol *p*-toluenesulfonates prepared from β-keto-ester significantly improve the method. The bisfunctional enol *p*-toluenesulfonate **187** is also much more accessible than its volatile dibromo counterpart **188** [142]. Alkylation of the dianion of methyl acetoacetate with bromide **187** followed by quenching of the enolate with TsCl affords the bis enol *p*-toluenesulfonate **189** with a Z,Z-selectivity usually higher than 95%. The double Negishi cross-coupling with diethylzinc and CyPF-*t*-Bu as palladium ligand occurs with complete retention of stereochemistry delivering the synthetic precursor of insect the juvenile hormone JH-0 **185** in 84% yield [143] (Scheme **36**). A straightforward synthesis of deuterium labelled geranylgeraniol implementing the same method was reported [144].

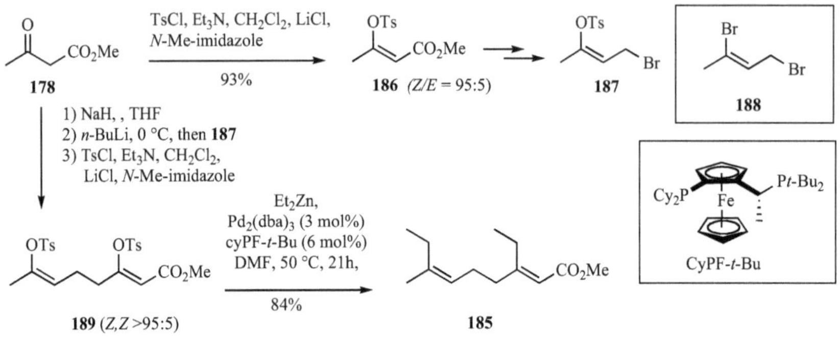

Scheme 36. Coupling of isoprenoid enol sulfonates with organozinc reagents.

Combination of olefin hydroboration and Suzuki-Miyaura cross-coupling reaction provides a versatile method for the stereospecific one-isoprene unit homologation of polyisoprenoid chains. For example, ethyl (*E*)-3-bromocrotonate (**192**) smoothly reacts with borane **191** in the presence of PdCl$_2$(dppf) to give ethyl farnesoate **193** in 69% yield (Scheme **37**) [145]. A similar strategy has been employed by Corey *et al.* to synthetize the acyclic precursor of the pentacyclic triterpene germanicol [146].

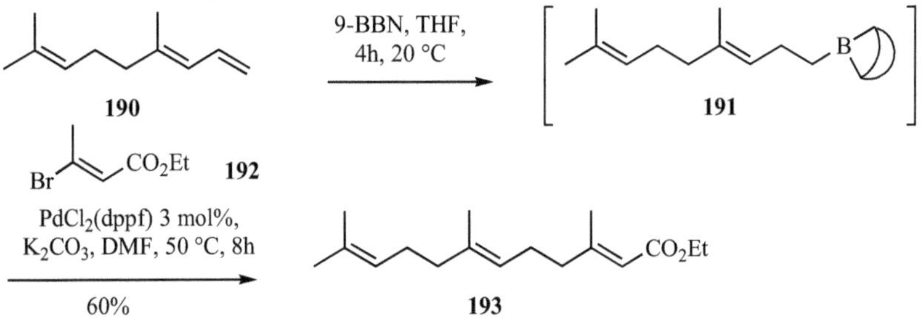

Scheme 37. Polyisoprenoid chain homologation using Suzuki-Miyaura cross-coupling.

On the occasion of the synthesis of menaquinones and coenzymes Q3 and Q10 Negishi disclosed a general methodology for the iterative construction of polyisoprenoid chains based on early findings showing that isoprenyl chains can be elaborated by using palladium catalyzed cross-coupling of organozinc compounds with vinyl iodides [147, 148]. Thus, zirconocene dichloride catalyzed carboalumination of 3-butynol (**194**) followed by trapping with iodine of the intermediate alkenylalane delivers the diiodide **196**. Remarkably, the palladium catalyzed cross-coupling of the organozinc reagent **197** with diiodide **196** occurs exclusively with the vinylic iodide group, preserving the primary iodide for the final capping reaction of the chain. Iteration of the process delivers organozinc reagent **201**, which is then coupled with the two-isoprene building block **202**.

Functional group conversion of the obtained alkyne **203** into vinyl iodide **204** and a new condensation with organozinc reagent **201** followed by a final carboalumination of alkyne **205** provides the vinylalane **206**, also prepared by Lipshutz for the synthesis of CoQ10 [149]. Throughout the overall synthesis no other stereoisomers are detected indicating that the stereoselectivity of each step is close of 100% (Scheme **38**). Interestingly, the same process can be modified to afford the corresponding all-*cis* polyprenols starting with the *Z*-isomer of diiodide **196** easily accessible through thermal equilibration of the primary carboalumination product of 3-butynol into the more stable 1,2-oxaluminine derivative [150].

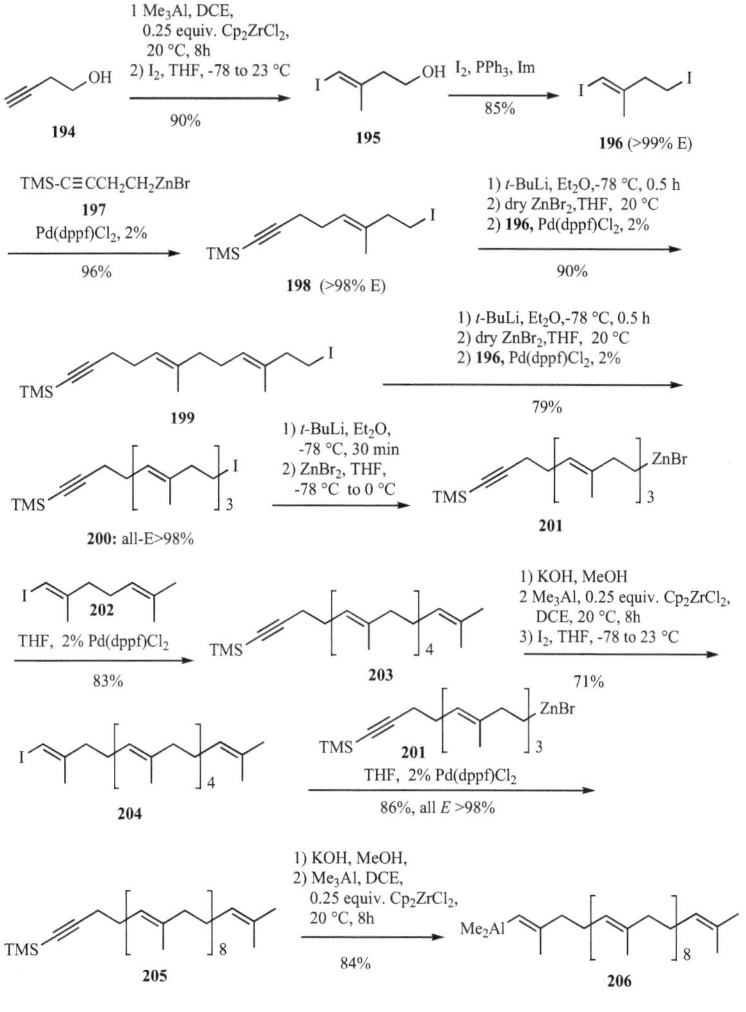

Scheme 38. Synthesis of long chain polyisoprenoids using carboalumination/Negishi cross-coupling sequence.

True iterative processes allowing polyprenol chain elongation in a stereocontrolled manner are quite rare. The nickel-catalyzed coupling of Grignard reagents with 2,3-dihydrofuran is one of the most effective method proposed so far, giving *E*-stereoselectivity greater than 97% at each iteration. The method is based on the reactivity of the 5-substituted 2,3-dihydrofuran **209** easily available by alkylation of 5-lithio-2,3-dihydrofuran (**211**). Alkylative ring opening of the dihydrofuran ring of **209** with methyl Grignard in the presence of a catalytic amount of a Ni(0) provides an efficient access to homogeraniol (**210**) with almost total control of the stereochemistry of the double bond. Conversion of this alcohol into the corresponding iodide **173** set the stage for the next iterations to homogeranylgeraniol (**214**) (Scheme **39**) [151, 152].

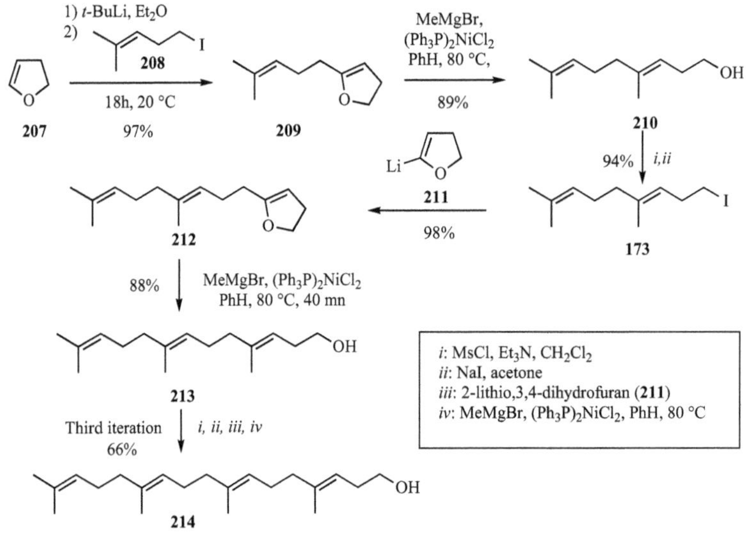

Scheme 39. Iterative approach to polyisoprenoid chains *via* nickel-catalyzed coupling of methyl Grignard with 2,3-dihydrofurans.

The same strategy was implemented for the synthesis non-natural isoprenoid diphosphates as chemical tools to explore the relevance of chain length and flexibility on FTase and GGTase activities [153].

Miscellaneous Reactions

[3,3] and [2, 3]-sigmatropic rearrangements

Sigmatropic rearrangements proceed generally through highly organized transition states allowing a good stereochemical control of the newly formed double bonds. An original sequence involving the Lewis acid addition of prenylstannane (**215**) to isoprene oxide (**2**), followed by Cope rearrangement gives geraniol (**26**) in 80% yield. Unfortunately, the high temperature required completely ruins the *E/Z*-stereoselectivity [154] (Scheme **40**). The same lack of selectivity is observed for the tandem [2, 3]-Wittig rearrangement/oxy-Cop--Claisen rearrangement of bis-allylic ethers and for the [2, 3]-sigmatropic rearrangement of *O,N*-carbenes produced upon heating allylic alcohol with dimethylformamide dimethylacetal [155, 156].

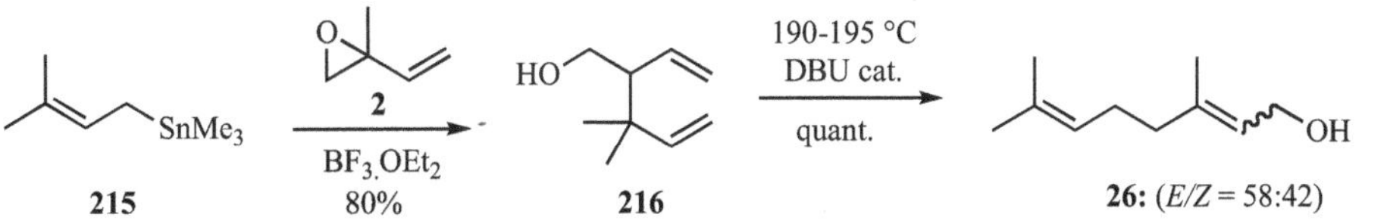

Scheme 40. Elaboration of the 1,5-diene system of polyisoprenoids through sigmatropic rearrangement.

The simple Claisen-Johnson rearrangement of allylic alcohols using triethyl orthoacetate provides an efficient means for the elongation of the polyprenyl side-chain of ubiquinones up to ubiquinone-9 [157]. Firstly introduced by Johnson for the total synthesis of squalene, the ketal Claisen rearrangement is a particularly appropriate method to achieve stereoselective, iterative chain elongation of polyprenoids [158]. For example, heating allylic alcohol **218** with ketal **219** gives the elongated ω-functionalized polyisoprenyl ketone **220**, with almost complete *E*-selectivity, which is further transformed into homologated allylic alcohol **221** ready for a new iteration (Scheme **41**) [159].

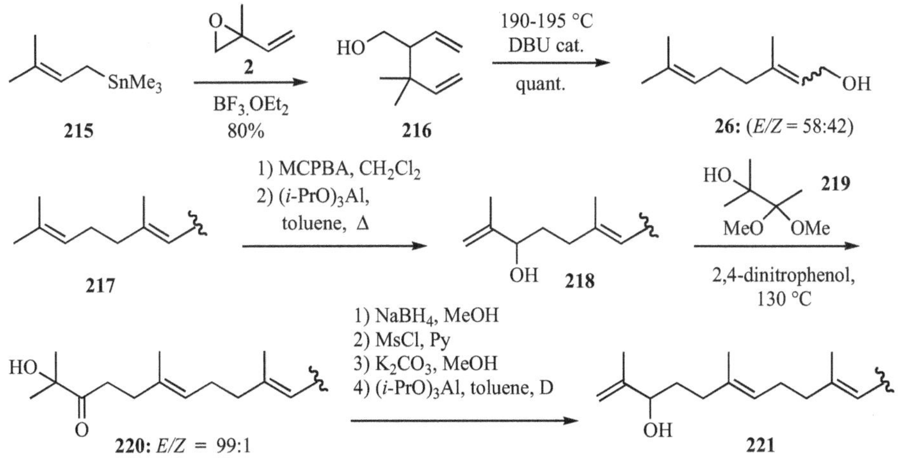

Scheme 41. Elaboration of the 1,5-diene system of polyisoprenoids through ketal Claisen rearrangement.

The reaction of allylic alcohols with vinyl ketals such as **223** is another variation of the Claisen rearrangement employed for the construction of trisubstituted double bonds of polyprenols as illustrated by the preparation of enone **224**. Reductive transposition of the corresponding tosylhydrazone then affords dendrolasin (**225**) in 79% yield (Scheme **42**) [160].

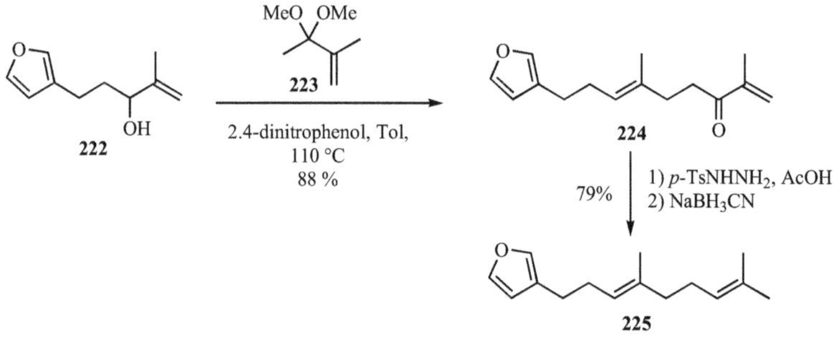

Scheme 42. Synthesis of dendrolasin through vinyl acetal Claisen transposition reaction.

The Sommelet-Hauser rearrangement of ylide **227** was used for the one-isoprene unit homologation of the allylic chloride **57** in order to synthetize β-sinensol (**229**). The stereochemistry of the newly formed isoprenyl units is sensitive to the nature of the base used. Moderate *E*-selectivity is achieved using *t*-BuOK in polar solvent. Quaternization of the nitrogen atom and sodium amalgam reduction complete the synthesis [161]. Likewise, the thioether obtained by displacement of allylic chloride **230** with thioglycolic methyl ester, gives the homologated ester **231** through a base-catalyzed [2, 3] thio-Wittig rearrangement and *in situ* desulfuration (Scheme **43**) [162].

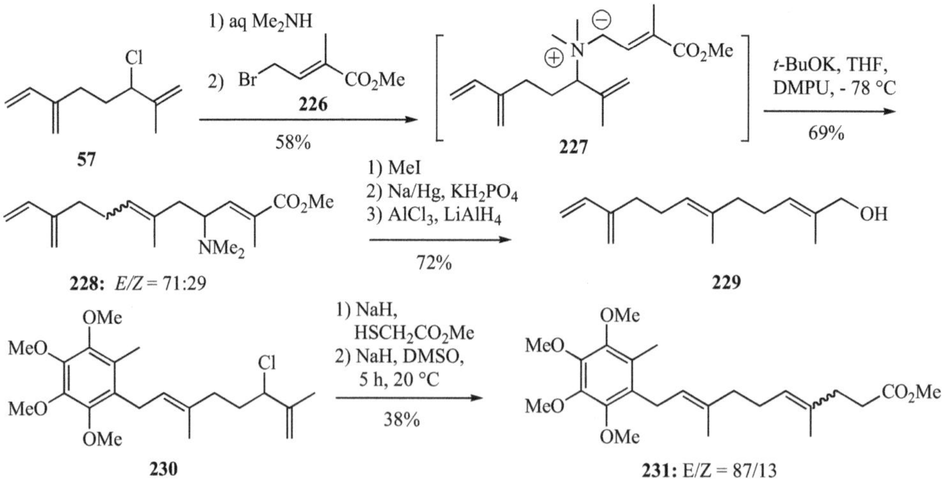

Scheme 43. Elaboration of isoprenoid chains through heteroatomic [2, 3]-sigmatropic rearrangements.

[1.3]-Sigmatropic Rearrangements

Allylic isomerization of vinyl carbinol was early envisioned as a quick way to control the stereochemistry of the isoprene-units double bond. Furthermore linalool, nerolidol or geranyllinalool are easily accessible by addition of a vinylmetal to the corresponding polyprenylacetones. However, the treatment of geranyllinalool **232** with PBr_3 provides the rearranged geranylgeranyl bromide **233** with a modest *E*-selectivity (*E/Z* = 3:1) contrary to early claims [163]. A slight improvement is observed using ZnI_2/TMSBr at -15 °C [164]. The direct rearrangement of (*E*)-nerolidol using $MoO_2(acac)_2$/Ac_2O is poorly selective whereas a 10.6:1 *E/Z* ratio is obtained using pyridine-2,6-dicarboxylic acid oxovanadium isopropoxide together with a supported lipase and vinyl acetate [165, 166]. The palladium catalyzed direct isomerization of geranyllinalyl acetate (**234**) provides the *E*-isomer of geranylgeranyl acetate (**44**) with an improved selectivity (*E/Z* = 5.6:1) [164, 167]. A more rewarding method involves Tsuji-Trost addition of diethylamine on the less hindered extremity of the π-allylpalladium complex derived from **234**. Deamination reaction with ethyl chloroformate of the obtained amine affords finally geranylgeranyl acetate (**44**) with a 10:1 *E:Z*-selectivity (Scheme 44) [168].

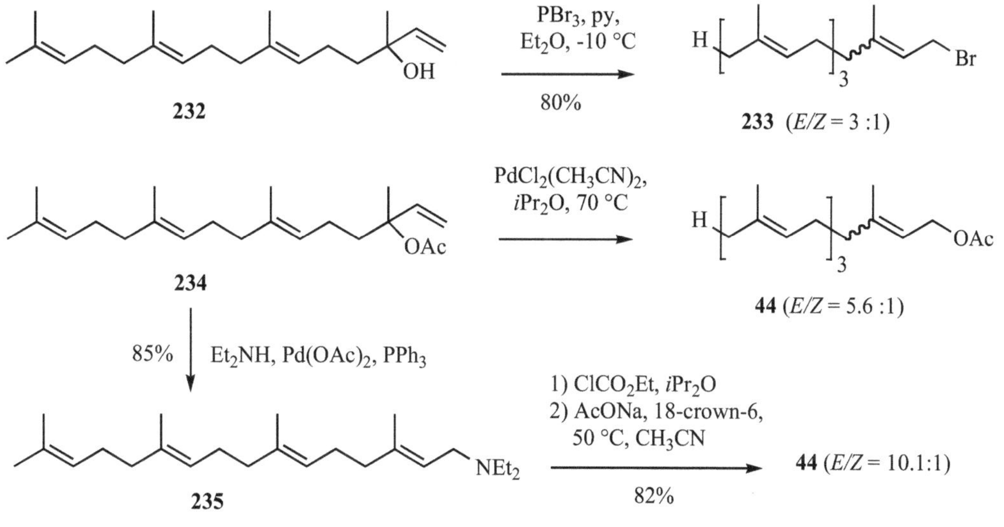

Scheme 44. Elaboration of isoprenoid chains through [1, 3]-sigmatropic rearrangements.

Synthesis of *Z*-Polyisoprenoids

Early attempts to build all-*cis* polyprenols involved alkylation of commercially available nerol derivatives. For example condensation of the dianion of butynoic acid (**236**) with neryl bromide (**238**) gives methyl ester **239** in 35% yield, together with the allene derivative **240**. Stereoselective dimethylcuprate addition to **239**, followed by AlH_3 reduction delivers 2Z,6Z-farnesol (**241**) [66]. Considering the intrinsic Z-stereoselectivity of the Wittig reaction, many syntheses of Z-polyprenols involve unstabilized ylides. Furthermore, higher selectivity is observed with α-alcoxy-acetone such as **243** (Scheme **45**) [169, 170]. In many cases, once (Z)-isoprenyl subunits are built, sulfone chemistry is used to access higher polyprenols. For example, sequential Wittig reaction with α-alcoxyacetone followed by Ducep-Biellmann coupling allow to alternate segments with *trans* and *cis* double bonds to access betulaprenols or ficaprenols found in natural glycoside conjugates involved in bacterial peptidoglycan biosynthesis [171 - 173].

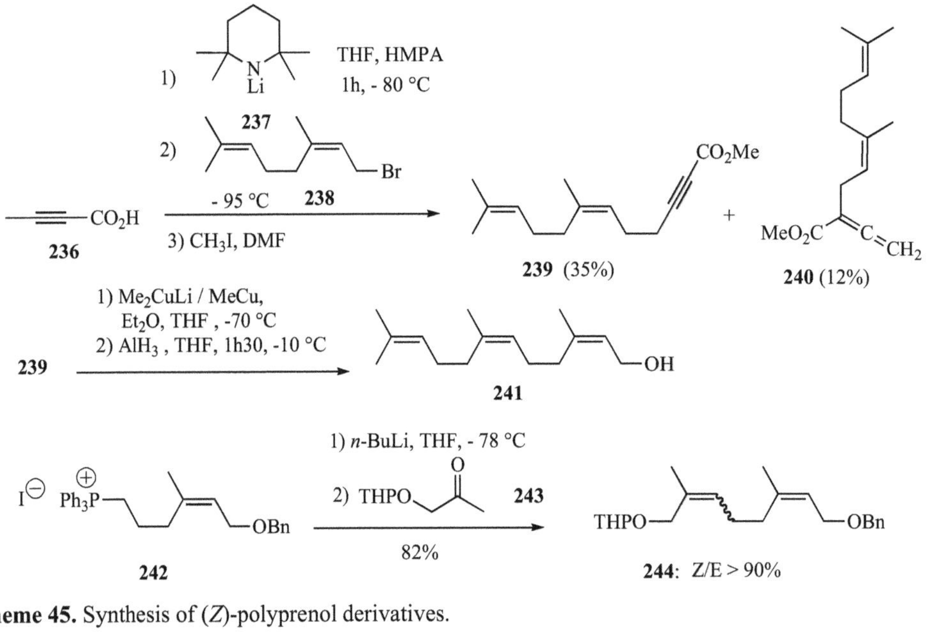

Scheme 45. Synthesis of (Z)-polyprenol derivatives.

SYNTHESES OF TAIL-TO-TAIL POLYISOPRENOIDS: SQUALENE

Functionalization of Squalene

Direct functionalization of squalene is a very important process, since (3S)-2,3-oxidosqualene is a milestone compound in the cholesterol biosynthesis. The research of potential inhibitors of oxidosqualene cyclase provided a strong impetus for the research of efficient squalene functionalization. From a chemical point of view, this transformation is highly challenging because the two terminal double bonds must be differentiated. Furthermore squalene is a highly lipophilic compound that reacts sluggishly with polar reagents. To date the most satisfactory procedure still relies on the finding of van Tamelen in the early 60's that NBS/H$_2$O treatment of squalene gives the bromohydrin **246** with a reasonable selectivity but a modest yield, the main byproduct being the ω,ω'-bis-bromohydrin. Potassium carbonate treatment of **246** then provides the racemic 2,3-oxidosqualene (**247**) (Scheme **46**) [78, 174, 175]. The compact conformation of squalene in the aqueous medium, exposing mainly its extremities to the aqueous phase may account for this selectivity. This hypothesis is strongly supported by the numerous observations that most epoxidation reactions performed in apolar organic solvents using peroxy acids or dimethyldioxirane give a mixture of internal epoxidation products [176, 177]. The asymmetric dihydroxylation of squalene with (DHQD)$_2$-PHAL gives low regioselectivity for the terminal olefins (2.4: 1.8: 1) but with high enantioselectivity (95% ee) [178]. The use of the ligand **52** (Scheme **11**, page 13) specially designed for the dihydroxylation of polyprenols, is more regioselective, providing the 2,3-dihydroxysqualene in 30% yield with 90% ee [62].

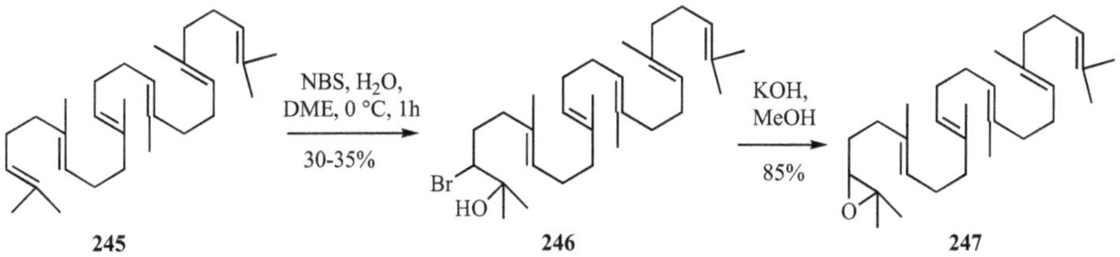

Scheme 46. Synthesis of dl, 2,3-oxidosqualene.

Recent advances in laser technology and mathematical image processing have transformed Raman spectroscopy into a powerful technology for the intracellular tracking of drugs or nanomedicines. Deuterium labelling improves the contrast for the simultaneous detection of both the drug and the cell components, as deuterium atom exhibits a significant Raman signal in a so called "silent region" of most biological molecules. In this context, ω-di-(trideuteromethyl)-squalene (**253**) was prepared as a potential Raman probe to follow the intracellular uptakes of squalenoyl nanomedicines [179]. The replacement of the terminal double bond of squalene by the isopropylidene-d_6 moiety was achieved taking into profit the Shapiro reaction of trisylhydrazone **249** derived from acetone-d_6. Thus, the treatment of **249** with two equivalents of *n*-BuLi and warming to 0 °C, afforded the perdeuterated 2-propenyllithium reagent **250** which upon condensation with trisnor-squalenaldehyde (**251**), easily obtained from 2,3-oxidosqualene (**247**) by oxidative cleavage with periodic acid [78], furnished the allylic alcohol **252** in 59% yield. Conjugated reduction of the hydroxyl group of **252** was straightforwardly achieved in 47% yield by sequential treatment with a large excess of thionyl chloride followed by LiAlD$_4$ reduction (Scheme **47**) [180].

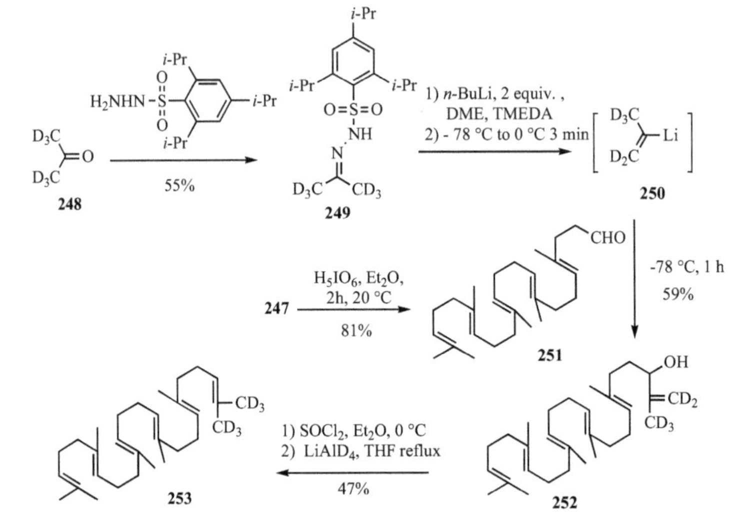

Scheme 47. Synthesis of ω-di-(trideuteromethyl)-squalene.

Synthetic Approaches to Squalene

Contrary to regular polyprenols, squalene displays two farnesyl residues linked in tail-to-tail fashion. Accordingly, many synthetic approaches take advantage of such a centro-symmetrical structure to implement a two-directional strategy. The early squalene synthesis by Cornforth already used addition of homogeranyl-lithium on both ends of the symmetrical 3,6-dichloro-octan-2,7-dione [181]. Following this pioneering work, Rapoport proposed a fully stereoselective synthesis of squalene starting from dimethyl 2,7-octa-2,6-diene (**254**). Selenium dioxide allylic oxidation of both ends of the diene **254** provides the corresponding bis aldehyde with complete *trans-trans* selectivity. A standard reduction-bromination sequence delivers the bis-bromide **255** which is then subjected to a double alkylation with the ylide **256** derived from tri-*n*-butylgeranyl phosphonium. This process gives the all-*trans* diphosphonium salt **257** whose stereochemistry is remarkably preserved during the lithium methylamine reduction to squalene [182].

In the early 70s, Biellmann and Ducep reported a very short hemi-synthesis of squalene involving the coupling of the lithium anion of phenyl farnesyl sulfide. The key step involves alkylation of carbanion **258** with farnesyl bromide (**259**) taking advantage of the stereochemical stability of the allylic anion. The α-alkylation product **260** is thus obtained with less of 2% of γ-alkylation adduct **261**. Desulfuration of **260** with lithium ethylamine provides all-*trans* squalene in 70% yield, together with 16% of stereoisomers and some dehydrosqualene [183, 184]. Following this seminal work, an improved yield was obtained using farnesyl sulfone as nucleophilic partner and LiBHEt$_3$/PdCl$_2$(dppp) as desulfonylation reagent (Scheme **48**) [185].

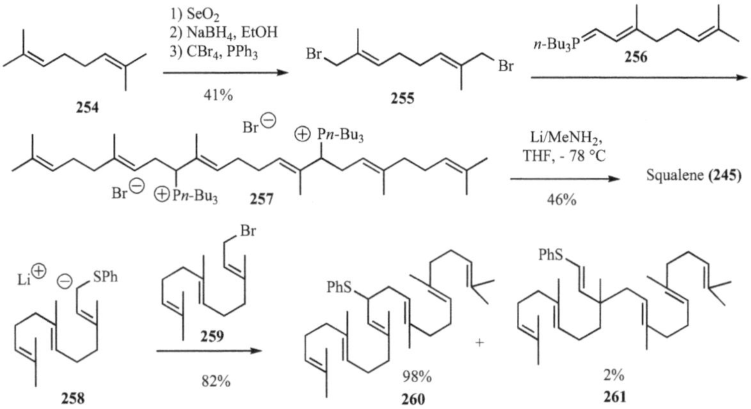

Scheme 48. Synthetic approaches to squalene using stabilized farnesyl carbanions.

The protected β-hydroxysulfone **262** resulting of the Julia condensation of farnesylsulfone (**71**) with farnesal can be reduced with Pd(dppe)Cl$_2$/LiBHEt$_3$ to give squalene in 77% yield. Interestingly, both the allylsulfone and the allylic ether functions are reduced in the process, the sulfone at room temperature whereas reduction of the allyl ether requires more drastic conditions (Scheme **49**) [186].

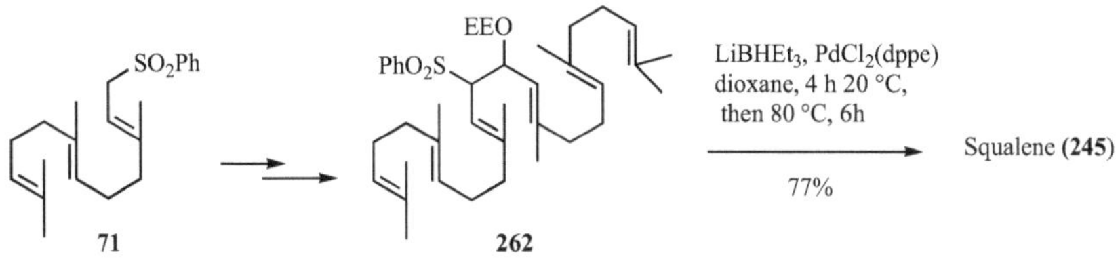

Scheme 49. Synthetic approaches to squalene using Julia's reaction.

A very efficient two-directional synthesis of squalene implementing sequential Claisen-Johnson rearrangements is reported. Thus the diol **263** obtained by addition of 2-propenyl lithium to succinaldehyde when subjected to a double Claisen rearrangement with triethyl orthoacetate gives the diester **264**. Adjustment of the oxidation degree, followed by a new addition of 2-propenyl lithium delivers the diol **265** which is then subjected to another two-directional orthoester Claisen reaction that gives diester **266** in 84% yield. Chain termination using a Wittig reaction provides all-*trans* squalene in 13 steps, each Claisen rearrangement occurring with *E*/*Z*-selectivity higher than 98:2 [158]. The use of 3-dimethoxy-2-methylbut-1-ene (**267**), or 2-chloro-3,3-dimethoxy-2-methylbutane, instead of triethyl orthoacetate affords still shorter synthetic routes, delivering squalene in 7 steps from succinaldehyde (Scheme **50**) [187, 188].

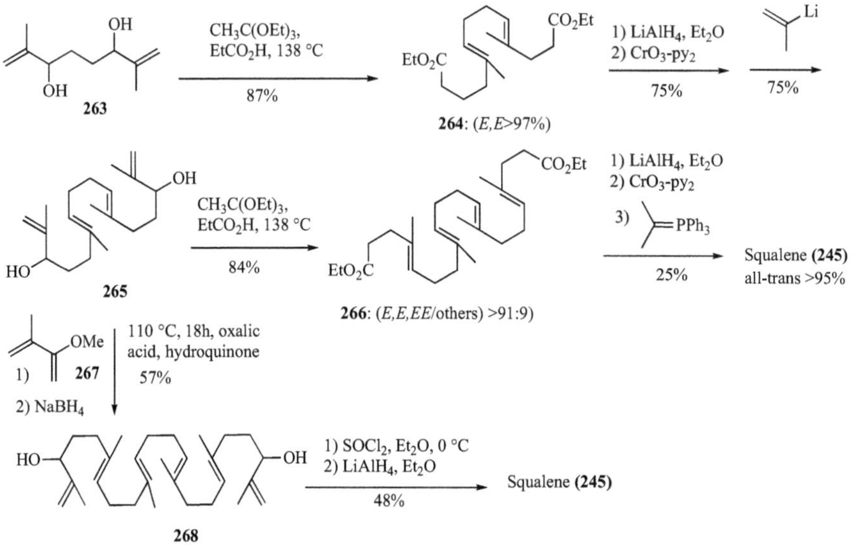

Scheme 50. Synthesis of squalene through two-directional Claisen-Johnson rearrangements.

Various synthetic approaches of the symmetrical central core of squalene are reported. For example, the diester **271**, featuring the C-9/C-16 central segment of squalene can be prepared by $TiCl_4$ catalyzed γ-dimerization of silylvinylketene **270** obtained from the dienolate of tiglic acid methyl ester (**269**) [189]. Alternatively, selective ozonolysis of 1,5-cyclooctadiene followed by two-directional HWE olefination of the *in situ* formed succinaldehyde, affords the same diester **271** which is then reduced to diol **273** (Scheme **51**) [33, 190].

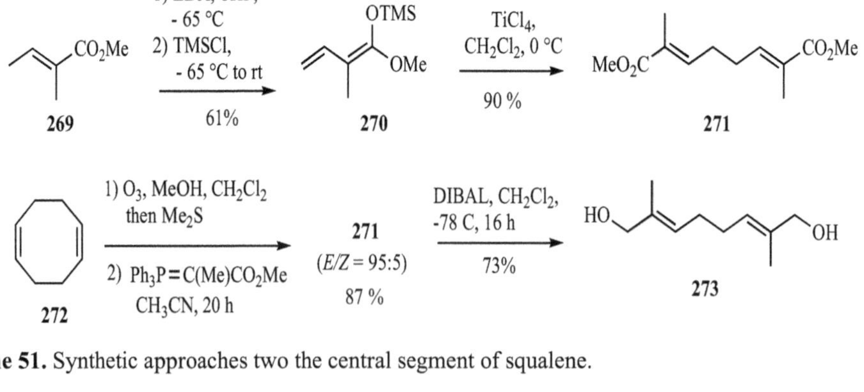

Scheme 51. Synthetic approaches two the central segment of squalene.

The peculiar symmetry of squalene (**245**) prompted many research groups to explore the reductive homocoupling of two farnesyl chains. This method is the most straightforward one, provided that it is possible to control the regio-and the stereoselectivity of the coupling. Initial attempts using low-valent titanium reagent (TiCl$_3$/LiAlH$_4$) gave a disappointing mixture of α,α' and α,γ' coupling products in low yield [191]. TiCl$_3$/MeLi affords a higher proportion of the requisite α,α' coupling product (7:1, for the homocoupling of geraniol (**26**)) [192]. The reaction of farnesyl bromide with CoCl(PPh$_3$)$_3$ provides the all-*trans* squalene in a better yield but with a large amount of α,γ' and γ,γ'coupling product [193]. Zn/Pd(PPh$_3$)$_4$ [194] and Mo(CO)$_6$/Zn-Cu [195] reagents are similarly, poorly selective. The copper derivative of 1-lithiopyrrolidine induces the coupling of farnesyl bromide to give squalene (**245**) in 50% yield. The α,α'/α,γ' selectivity is not reported, but a low *E/Z*-selectivity is usually observed with this reagent [196]. Allylbarium reagents once introduced by Yamamoto react with allylic halides according to a pure primary- with a complete retention of the stereochemistry of the starting materials. Barium is generally highly superior to most other alkali metals or low valent transition metals to achieve the reductive coupling of allyl halides [91]. Taking into profit this unique behavior, the condensation of the *E,E*-farnesylbarium reagent with farnesyl chloride gives in one step all-*E* squalene with 47% as a 97:3 mixture of α,α' and α,γ' products. Interestingly only 1% of *Z*-isomer is found in the mixture [91, 93, 197].

Table 1. Metal catalyzed tail-to-tail homocoupling of two farsenyl units.

Reagent	X	α,α' (%),	α,γ'(%)	γ,γ'(%)	Ref
TiCl$_3$, LiAlH$_4$	OH	33	15	_	191
CoCl(PPh$_3$)$_3$	Br	55	22	12	193
Zn/Pd(PPh$_3$)$_4$	OAc	73 (*E/Z* = 89:11)	27	_	194
Mo(CO)$_6$, Zn-Cu	OAc	80 (*E/Z* = 42:58)	20	_	195
Li/Np	Br	69 (*E/Z* = 97:3)	31	_	91
K/Np	Br	78 (*E/Z* = 97:3)	22	_	91
CrCl$_3$, LiAlH$_4$	Br	74 (*E/Z* = 89:11)	26	_	91
Mn	Br	74 (*E/Z* = 89:11)	26	_	91
Rieke Ba	Br	97 (*E/Z* = 99:1)	3	_	91
Cp$_2$TiCl$_2$/Mn(0)	Br	88 (*E/Z* = 75:25)	12	_	198
Cp$_2$ZrCl$_2$/Mn(0)	Br	82 (*E* only)	18	_	198

Np = Naphthalenide

Single electron processes induced by catalytic amount of titanocene and zironocene reagents have been thoroughly explored by Barrero *et al*. The zirconium reagent obtained by treatment of Cp$_2$ZrCl$_2$ with manganese dust is highly effective to prepare squalene from farnesyl bromide without E/Z-isomerization of the allylic double bond [198] (Table **1**).

Starting from farnesyl chloride and chiral geranyl bromo-epoxide **277** the reductive coupling using Rieke barium can be enlarged to hetero-coupling of allyl halides to target (3*S*)-2,3-oxidosqualene **247** which is a pivotal intermediate in the biosynthesis of sterol (Scheme **52**) [199].

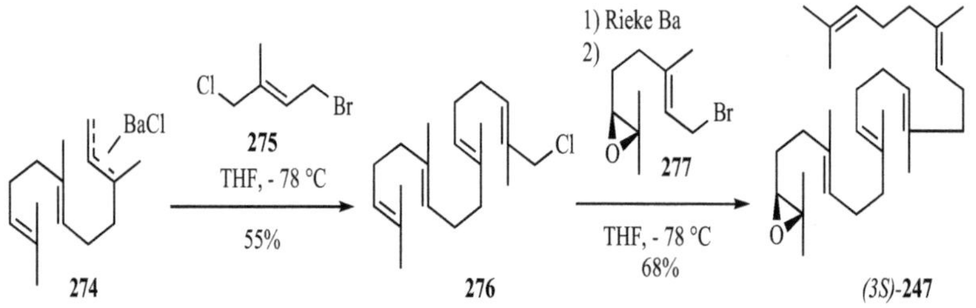

Scheme 52. Synthesis of (3*S*)-2,3-oxidosqualene using farnesylbarium reagent.

FUNCTIONALIZATION OF CARBOCYCLES WITH ISOPRENOID SIDE CHAINS: ISOPRENOID QUINONES

Synthesis of Isoprenoid Quinones

Arylation of Polyprenyl Chains via Electrophilic Aromatic Substitution Reactions

Isoprenoid quinones such as menaquinones (MK-n, vitamin K2), plastoquinones and ubiquinones (coenzymes Qn), constitute a group of benzo- or naphthoquinones substituted by long polyisoprenoid chains up to twelve units. Widely distributed in plant and animal kingdom, these fat-soluble compounds are mainly found in mitochondria, microsomes, chloroplasts and bacteria. They play an essential role in cellular metabolism of mammalian or bacteria as electron transfer species producing energy for living cells. CoQ10 appears suitable for use in the treatment of many diseases including heart failure, diabetic neuropathy, Huntington's disease, *etc*. Furthermore, CoQ10 supplementation confers health benefits by preventing chronic oxidative stress associated with cardiovascular and neurodegenerative diseases [200]. The global sales of CoQ10 as dietary supplement or pharmaceutical CoQ10 reached 1300 tons in 2017 with a market value of 370 million US$. The CoQ10 currently available on the market is produced by fermentation process, but efficient synthetic routes have been devised. There are roughly two strategies to access such compounds whether the whole chain is coupled directly to the activated aromatic ring or through a partial isoprenyl segment already bound on the aromatic ring. In most cases, transient reduction of the quinone moiety into protected hydroquinol is required (Fig. **1**).

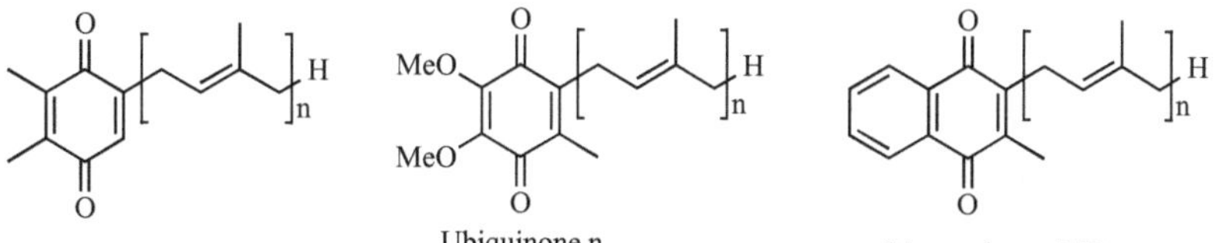

Fig. (1). Structure of main isoprenoid quinones.

Early attempts to directly attach polyisoprenoid side chain to an aromatic ring involved electrophilic aromatic substitution reactions with polyprenols and Lewis acids [201 - 203]. However these methods suffer from the formation of side products and require tedious purification steps. Furthermore the assignment of the stereochemistry is highly challenging with such polyenes and the stereoselectivity of the processes is rarely reported. Nevertheless, condensation of 2,3-dimethox--5-methyl-benzoquinone (**278**) with tertiary allylic isodecaprenol **279** in the presence of $ZnCl_2$ gives CoQ10 (**86**) in 60% yield [204]. Likewise, the borate complex of the 2-methyl-5,6-dimethoxy-1,4-hydroquinone supported on silica reacts with ^{14}C-labelled decaprenol to afford 3-^{14}C-CoQ10 in 26% yield after lead dioxide reoxydation [205]. Allylstannanes in the presence of $BF_3.OEt_2$ are highly efficient γ-allylating reagents. For example, Naruta prepared a series of CoQn (n = 1-10) by Lewis acid catalyzed addition of (polyisoprenyl)trimethylstannanes **281** to quinone **278** [206]. Remarkably, the stereochemical integrity of the $\Delta^{2,3}$ double bond is preserved in the process giving rise to coenzymes Q_{2-10}, albeit some isomerization is observed in the case of CoQ10 (E/Z = 86:14) (Scheme **53**).

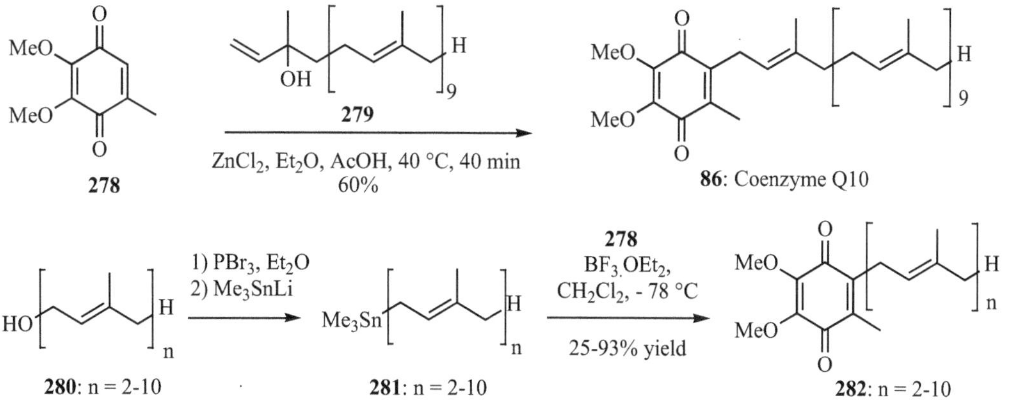

Scheme 53. Syntheses of coenzymes Q_n using electrophilic substitution reactions.

Friedel-Craft type reaction between dimethoxy-naphthalene **283** and allylic chloride **284** affords sulfone **285** in 30% yield, with a full retention of the *E*-stereochemistry of the double bond. Conventional sulfone homologation of the side chain as originally delineated by Terao [79] completes the synthesis of menaquinone MK-7 (**286**) [207]. The use of scandium triflate as catalyst drastically improves the yield for coupling the 4-(benzenesulfonyl)-3-methylbut-2-en-1-yl appendage with tetramethoxybenzene **287** providing a convenient access to CoQ10 [208]. A similar Friedel-Craft reaction was reported by Min *et al.* to prepare various ubiquinones and menaquinones [209] (Scheme **54**).

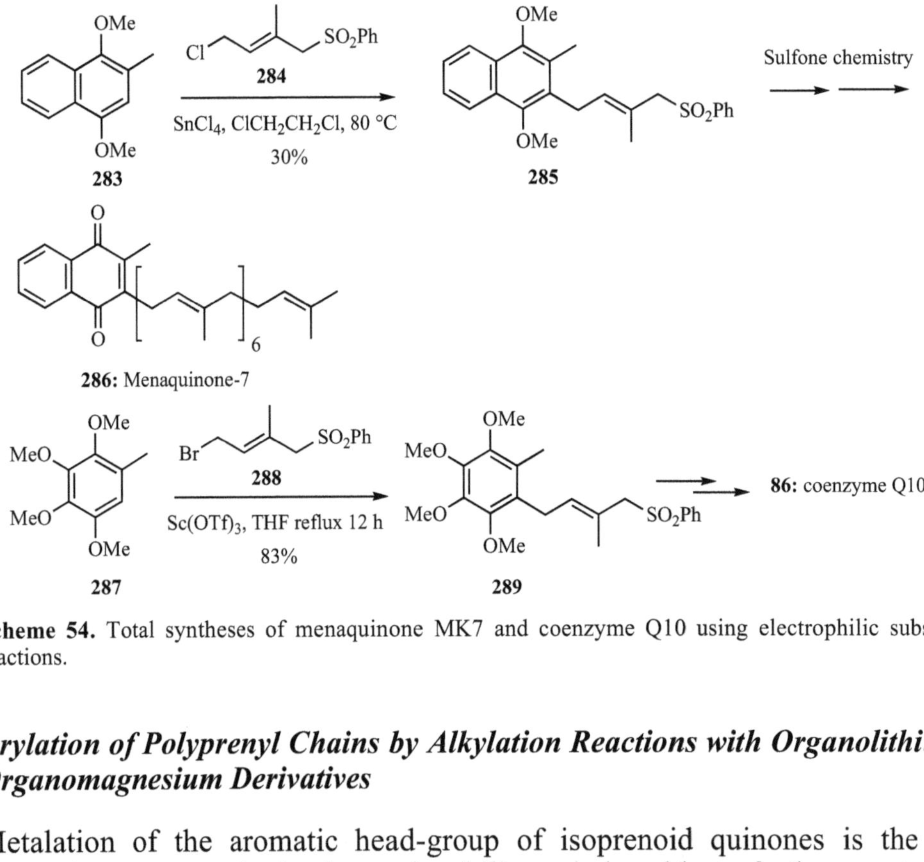

Scheme 54. Total syntheses of menaquinone MK7 and coenzyme Q10 using electrophilic substitution reactions.

Arylation of Polyprenyl Chains by Alkylation Reactions with Organolithium or Organomagnesium Derivatives

Metalation of the aromatic head-group of isoprenoid quinones is the more convenient way to obtain the nucleophilic moieties able to further react with a polyisoprenyl halide chain. Unfortunately, the use of magnesium or lithium reagents leads to a substantial amount of γ-alkylation, the copper derivatives are therefore preferred. Aryl cuprate reagents obtained by exchange with copper salts react smoothly with allylic halides to give S_N2 products in high yield. Syntheses of CoQ10 and vitamin K2 analogues implementing this method are described [210, 211].

This chemistry allows scaling-up as illustrated by the condensation of aryl magnesium bromide **290** with allylic chloride **284** in the presence of copper (I) salt to give the sulfone **291** on hundred gram scale. Subsequent alkylation of sulfone **291** with solanesyl bromide then brings the missing nonaprenyl chain to access CoQ10 (**86**) [212]. The same strategy but, reversing the partners of the chain elongation using solanesyl sulfone anion **295** was explored by Chen *et al.* In this case, a careful examination of the final product reveals that less of 4% of isomerization occurs during the lithium ethylamine reduction step [213] (Scheme **55**). THP ethers derived from polyprenols are good leaving groups for the regioselective alkylation of aryl-copper magnesium reagents [214]. A synthesis of jaspaquinol, a 15-lipoxoxygenase inhibitor, is reported using this method [215].

The protection of the phenol groups in these processes must be carefully planned since they must resist ortho-lithiation, and deprotection in the presence of double bonds. It has been shown that benzyl groups are the most convenient protecting groups since they can be removed using Na in 2-butanol without affecting the oligoprenyl side chains [216, 217].

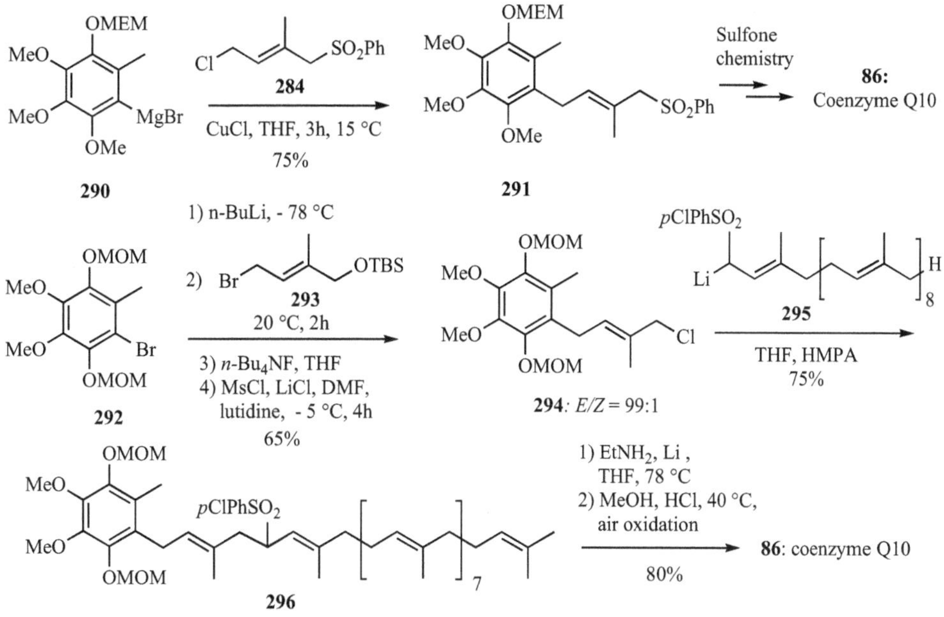

Scheme 55. Total synthesis of coenzyme Q10 using alkylation of lithium or copper reagents.

A straightforward three-step synthesis of menaquinone-4 (**300**) involves the hydroxyl alkylation reaction of lithio naphthalene **297** with *trans*-geranylgeranial (**298**). The reduction of the benzylic alcohol without scrambling of the double bond is a crucial issue. In the event, after silylation of the hydroxyl group, sonication of **299** with lithium metal followed by reoxydation to quinone with cerium ammonium nitrate provides menaquinone-4 (**300**) with a complete retention of the *E*-configuration of the $\Delta^{2,3}$ double bond (Scheme **56**) [218]. The temporary protection of the benzoquinone ring to facilitate the introduction of the polyprenyl side-chain was also explored for the synthesis of ubiquinone-5 and -9. Thus, transformation the dimethoxy-benzoquinone (**278**) into cyclopentadiene Diels-Alder adduct allows an easy alkylation of the enolate of the keto group with solanesyl bromide. Retro Diels-Alder reaction in refluxing toluene then unmasks the quinone ring [219].

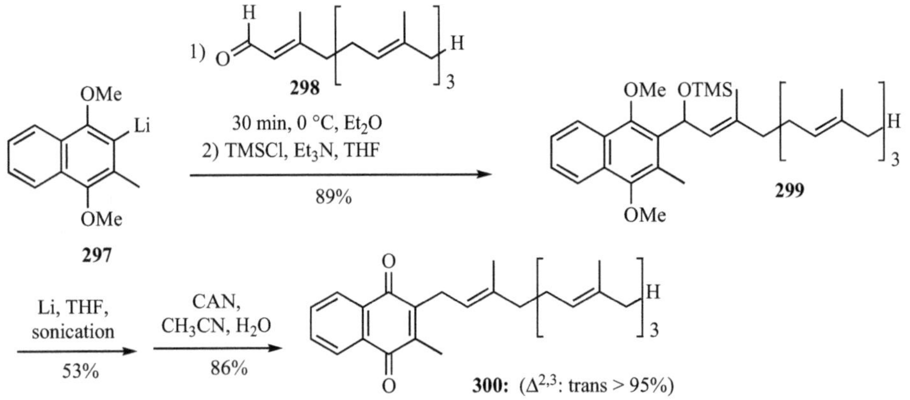

Scheme 56. Hydroxyalkylation reaction of lithio-naphthalene with prenylaldehyde.

Arylation of Polyprenyl Chains via Transition Metal Catalyzed Coupling Reactions

As an extension of the Ni(0) catalyzed synthesis of polyprenol introduced by Sato in the early 70s [90], a straightforward synthesis of Vitamin K2 MK-9 (**305**) is described involving the condensation of the η^3-allylnickel complex **302** derived from solanesyl bromide (**301**), with bromonaphthalene **303**. However, the *trans*-stereochemistry of the first isoprenyl unit is lost during the process [220]. The discovery of palladium cross-coupling reactions in the late seventies has opened new opportunities for building coenzyme Q_n. For example, the Stille coupling of geranyl- or farnesylstannanes **281**, previously prepared by Naruta [206], with bromoaryl **306** provides aryl isoprenoids **307-309**, which are the direct precursors of CoQ1 to CoQ3 [221]. The Kumada cross-coupling reaction brings an efficient solution to the problem of the S_N2/S_N2' selectivity. A striking example is the addition of benzyl Grignard reagent to farnesyl bromide (**259**) in the presence of a catalytic amount of Pd(0) that affords the α-adducts with a 93% selectivity. In the absence of palladium the reaction gives 10% of γ-alkylation along with 25-30% of Wurtz dimerization of the Grignard reagent [222]. This process is best illustrated by the efficient synthesis of menaquinone-7 (**286**) involving the Kumada reaction of Grignard reagent **310** with heptaprenyl bromide (**311**). Menaquinone-7 is obtained in 56% overall yield after reoxidation to naphthoquinone with CAN (Scheme **57**) [223].

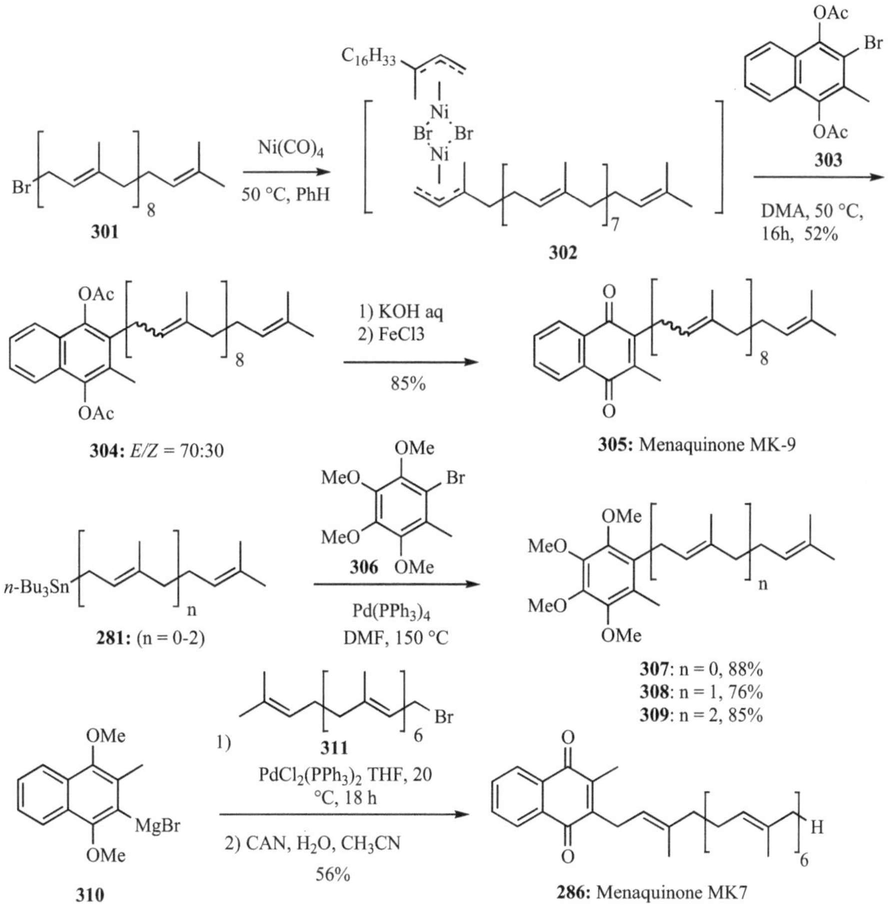

Scheme 57. Synthesis of polyprenylquinones by condensation of η^3-allylnickel coupling and Stille reaction.

The inversion of the two partners in the Stille coupling was explored for the synthesis of the 15-lipoxoxygenase inhibitor jaspaquinol (**314**). In this case, condensation of the aryl tributyltin derivative **312** with allylic chloride **313** gives the α-coupling product **314** in 92% yield after reductive debenzylation with LiAlH$_4$/NiCl$_2$. The high selectivity of the process can be tentatively rationalized by the complexation of the distal olefin with the palladium, as depicted in complex **315,** prohibiting the reductive elimination at the γ-carbon without prior decomplexation (Scheme **58**) [215].

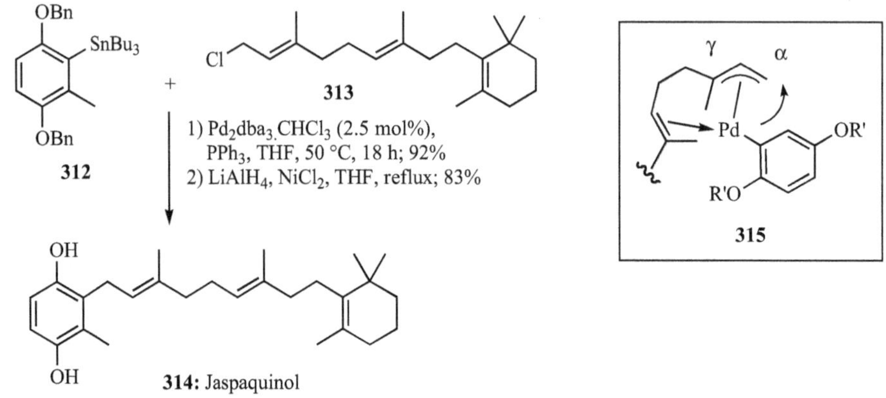

Scheme 58. Synthesis of jaspaquinol by Stille cross-coupling reaction.

Methods able to deliver coenzyme Q_n without the need to protect the quinone ring as hydroquinone ether, avoiding the lengthy protection/reoxidation sequence would reduce the overall number of steps, which is an important challenge to address when targeting an economically valuable dietary supplement such as ubiquinone. Lipshutz *et al.* have reported a very short synthesis of CoQ10 (**86**) using solanesol (**77**) as precursor of the decaprenyl side chain. The synthetic scheme involves the convergent nickel-catalyzed bond formation between the chloromethylated quinone **317** and the whole chain equipped with a terminal vinylalane function. To this aim, solanesyl chloride obtained by regioselective chlorination of solanesol with the Vilsmeier salt of DMF is alkylated with 1,3-dilithiopropyne to give alkyne **316**. The latter when subjected to a zirconium catalyzed carboalumination reaction furnishes stereoselectively the vinylalane **206** previously prepared by Negishi [148]. Condensation of this compound with the chloromethylquinone **317** in the presence of a catalytic amount of Ni(0) gives directly CoQ10 (**86**) (Scheme **59**) [28, 149, 224].

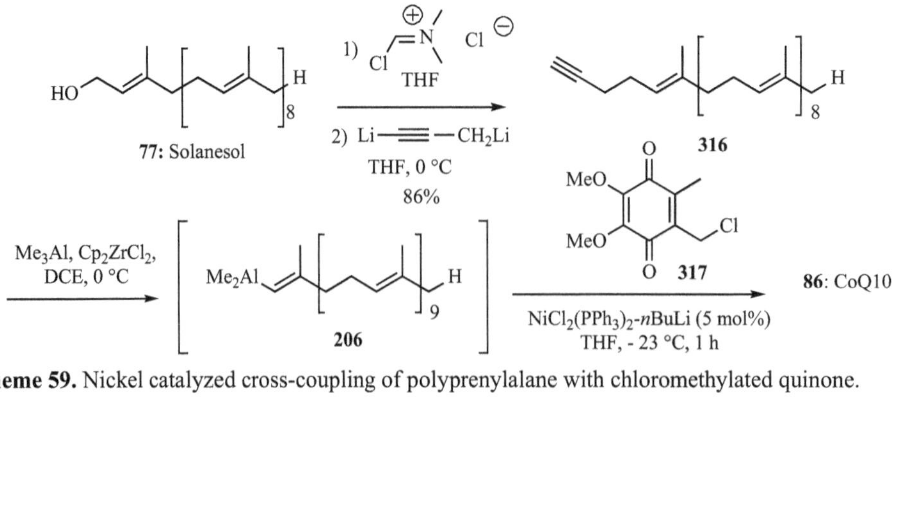

Scheme 59. Nickel catalyzed cross-coupling of polyprenylalane with chloromethylated quinone.

The rhodium catalyzed condensation of diene with phenol can be used to introduce a prenyl chain in the α-position of a phenol. For example, the condensation of β-springene (**319**) with hydroquinone (**318**) gives the adduct **320** in 89% yield. Although, the reaction affords a mixture of endo/exo Δ^3 double bond, the process is well adapted to formation of the benzopyran ring of vitamin E (Scheme **60**) [225].

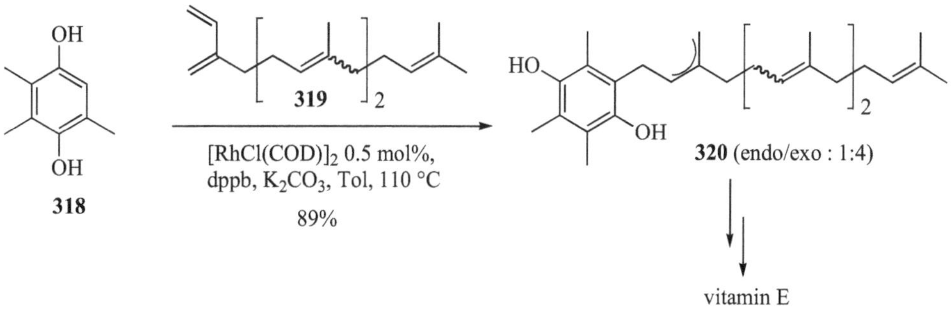

Scheme 60. Synthesis of vitamin E using rhodium(I)-catalyzed addition of hydroquinone to dienes.

Synthesis of Other Isoprenoid Substituted Carbocycles and Heterocycles

Direct Alkylation Reactions

Furanoterpenes and pyrroloterpenes constitute a small group of heterocyclic terpenes but with a large spectrum of biological activities. The 3-farnesyl pyrrole **322**, a precursor of nitropyrrolin A is simply obtained by direct alkylation of the 3-lithiopyrrole prepared from bromopyrrole **321** by lithium halide exchange reaction, with farnesyl bromide (**259**) [226]. Grignard-Schlosser cross-coupling reaction of allylic acetate with organomagnesium reagents provides polyprenyl alkylated derivatives with high α-selectivity as exemplified by the synthesis from acetate **323** of neotorreyol (**325**) a wood-oil constituent (Scheme **61**) [227, 228].

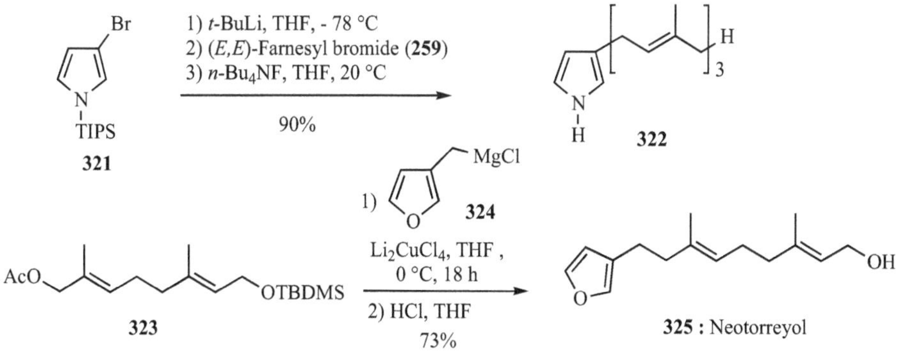

Scheme 61. Synthesis of prenylated heterocycles by direct alkylation reaction.

Aurachins are myxobacterial quinolone alkaloids bearing a simple or a functionalized farnesyl side chain. They block NADH oxidation in mammalian cells by the inhibition of complexes I and III of the mitochondrial respiratory chain. A straightforward synthesis of aurachins A (**328**) that displays an unusual furo[2,3-c]quinoline *N*-oxide core is described by a sequential process involving alkylation of ketone **326** with iodo-epoxide **327**, followed by quinoline and dihydrofuran ring formation [229].

The Shapiro reaction is one of the most powerful methods for the rapid assembly of trisubstituted olefins while addressing the crucial problem of the stereo-selectivity. Corey used it for a one-pot synthesis of polyisoprenoid compounds from easily available alkyl halides. For example, alkylation of the anion derived from acetone trisylhydrazone (**329**) with iodide **330** gives the hydrazone **331** which upon *in situ* re-deprotonation and rapid warming to effect the extrusion of N_2, metal exchange with lithium (2-thienyl)cyanocuprate furnishes the higher order cyanocuprate **332**. Alkylation of the latter with iodo-ketal **333** gives ketal **334** with an overall yield of 65% and a full control of the *E*-stereochemistry of the formed double bond (Scheme **62**) [230].

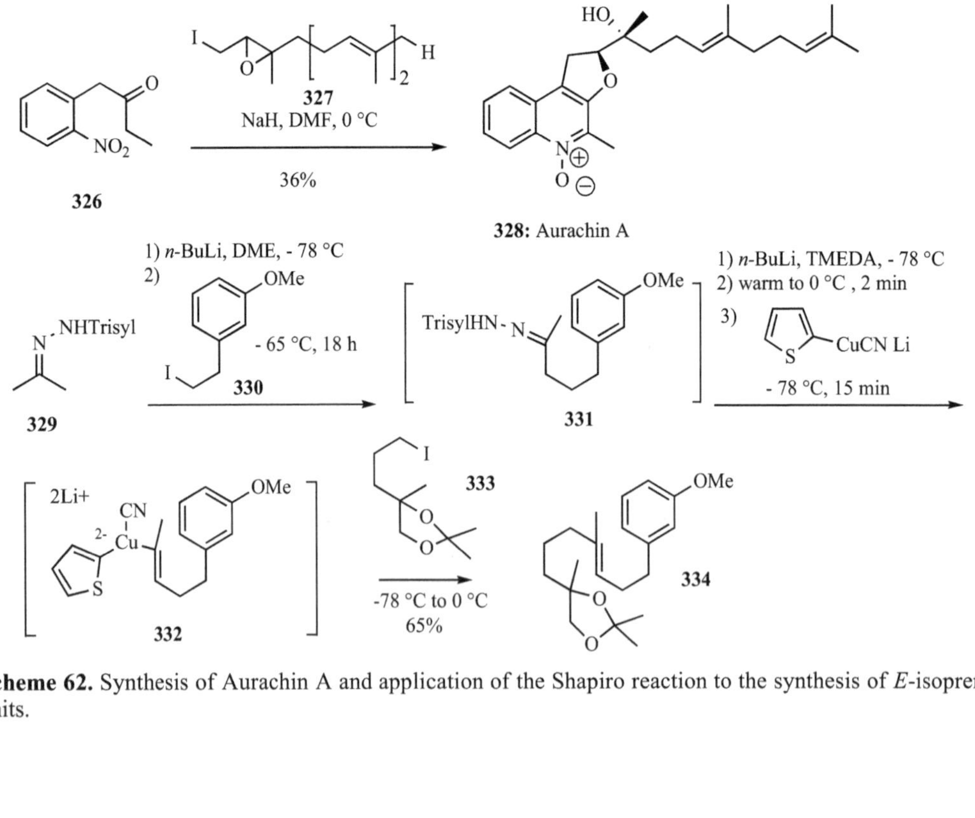

Scheme 62. Synthesis of Aurachin A and application of the Shapiro reaction to the synthesis of *E*-isoprene units.

Wittig Type Reactions

In some cases, it may be convenient to take advantage of a short carbon appendage, precursor of the future polyisoprenyl chain, previously introduced on the head-group of the target. This strategy is illustrated by the synthetic approach to anthroquinonol a promising anticancer agent. Initial Wittig reaction of lactol **335** with ethyl 2-(triphenylphosphanylidene)propanoate introduces a first *trans*-isoprenyl unit on the cyclohexyl ring. Conventional Biellmann-Ducep chain homologation with phenylgeranyl sulfide carbanion **59** completes the installation of the farnesyl chain [231]. Likewise, the unsaturated ester **340**, a simple precursor of tocopherol analogues is obtained with a 9:1 *E/Z* selectivity by HWE olefination of aldehyde **339** using sodium hydride as base (Scheme 63) [232].

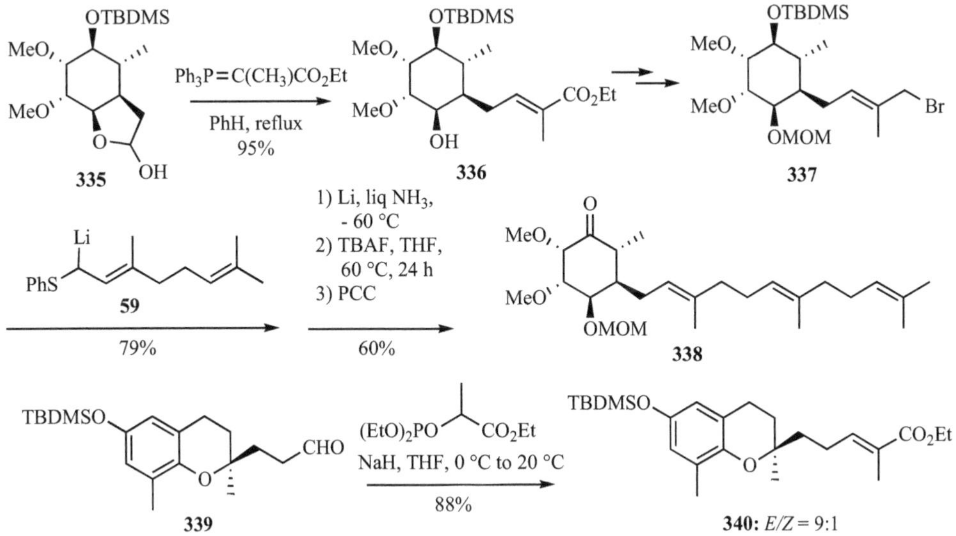

Scheme 63. Introduction of polyprenyl side chains by Wittig and HWE olefination reaction.

Sigmatropic Rearrangements

Sigmatropic rearrangements and especially Claisen rearrangements are well adapted to build polyisoprenyl units with high specificity. Thus, once a suitable precursor, usually an allylic alcohol, is fixed on the carbocycle the introduction of the polyisoprenyl chain is quite easy. The elaboration of the unusual side chain of the anticancer agent lehualide B (**347**) illustrates the synthetic potential the Claisen rearrangement. Functionalization of the γ-pyrone **341** is first achieved by condensation with 2-phenylseleno-isobutyraldehyde that gives the seleno-alcohol **342** in 78% yield. Sequential oxidation and syn-elimination of the resulting selenoxide provides allylic alcohol **343** setting the stage for a Claisen rearrangement with ethyl vinyl ether that delivers the corresponding unsaturated aldehyde **344** with a 10:1 *E/Z*-selectivity. Bromo(prop-1-en-2-yl)magnesium addition then affords the homologated allylic alcohol **345** ready for a new chain extension. The skipped diene of lehualide B is finally introduced in 50% yield by reductive cross-coupling reaction of 1,1-diispopropoxy-1-titanacycloprop-2-ene **346** with the allylic alcohol appendage of **345** [233] (Scheme **64**).

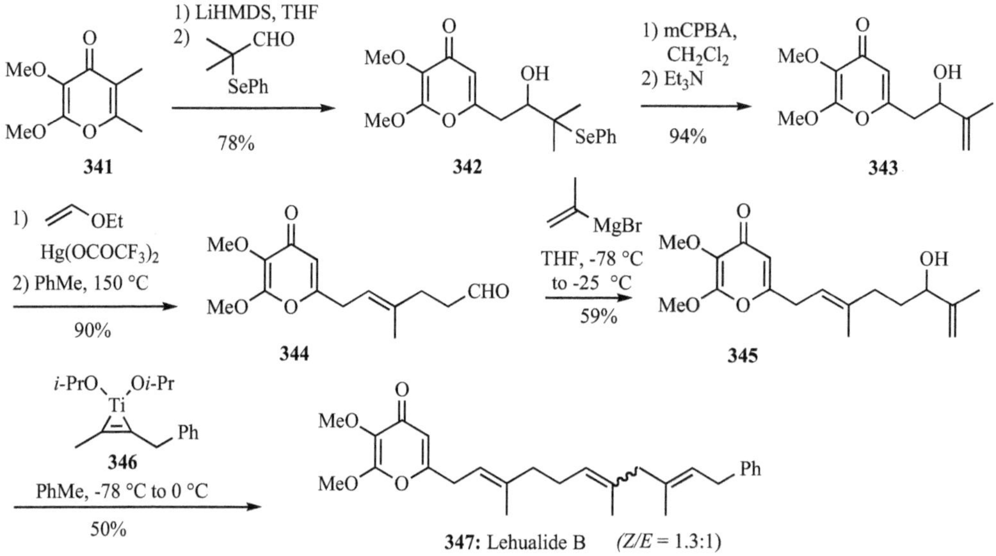

Scheme 64. Elaboration of the side chain of lehualide B by Claisen rearrangement.

Organometallic-Catalyzed Cross-Coupling Reactions and New Organometallic Processes

The introduction the ω-functionalized farnesyl side chain on the *trans*-decaline system of myrrhanol, an antitumor polypodane terpene, illustrates the versatility of palladium catalyzed cross-coupling reactions. Thus, conventional functionalization of geranyl acetate followed by installation of a terminal acetylenic group, open the way to a Negishi carboalumination reaction giving the *E*-vinyl iodide **349**. Suzuki-Miyaura cross-coupling with borane **350** then affords the myrrhanol advanced intermediate **351** (Scheme **65**) [234, 235]. The farsenyl side chains of iridal a monocyclic triterpenoidic extracted from *Iris germanica* L and anthroquinonol a potential anti-obesity drug are similarly introduced using a Suzuki-Miyaura cross coupling reaction. However, in both cases the two reactive partners are reversed, the (*E*)-1-alkenyl-9-BBN reagent brings the farnesyl appendage while the iodide group is fixed on the main carbocycle core [236, 237].

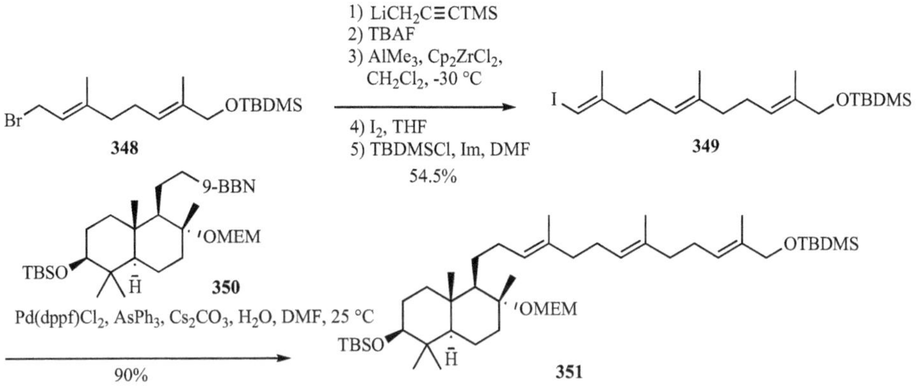

Scheme 65. Installation of a polyprenyl side chain by Negishi carboalumination and Suzuki cross-coupling reactions.

Negishi cross-coupling reaction of the allylzinc reagent **353** prepared according to the Knochel's protocol [238] offers an effective means for accessing farnesyl dibenzothiophene **354** (Scheme 66) [239].

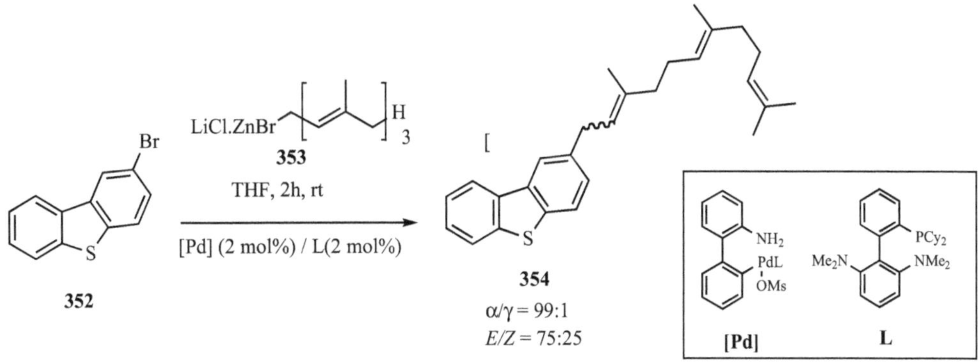

Scheme 66. Introduction of a polyprenyl side chains by sequential Negishi reaction, Suzuki cross-coupling.

The installation of the polyprenyl side-chain of many terpenes is achieved by palladium catalyzed rearrangement of tertiary allylic acetates according to a method previously reported for the rearrangement of linalyl acetate into geranylacetate For example the side chain of elegansidiol **356** can be elaborated by palladium-catalyzed [1,3]-sigmatropic rearrangement of the corresponding tertiary acetate [240, 164]. The decarboxylative alkenylation reaction recently reported by Baran *et al.* affords an expeditious access to α-tocotrienol. To this aim, the alkenylzinc reagent **358** available through lithium halogen exchange reaction and transmetalation with $ZnCl_2 \cdot LiCl$, is condensed with the redox-active ester **357** in the presence of nickel(II) catalyst to give after hydrolysis α-tocotrienol (**360**) in 50% yield with an excellent 20:1 *E/Z* selectivity (Scheme **67**) [241].

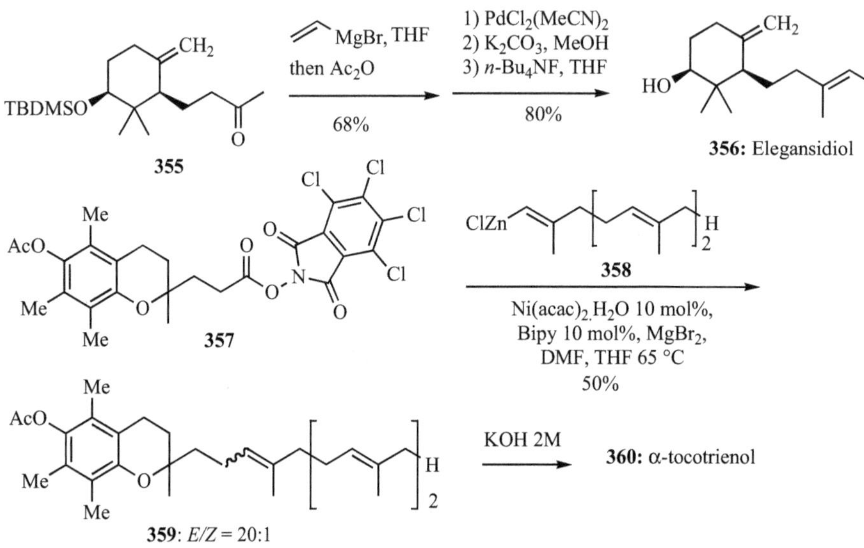

Scheme 67. Elaboration of a polyprenyl side chain by allylic isomerization reaction and nickel catalyzed decarboxylative alkenylation.

Beside palladium-catalyzed cross coupling reactions, new pattern of reactivity emerges using other metals. For example some ruthenium and iridium catalysts promote the transfer of hydrogen from primary alcohols to various π-unsaturated compounds. This reactivity is illustrated by the synthesis of 3-hydroxyl-oxindole **363** involving the condensation of the 3-hydoxypyrrolidinone **361** with myrcene (**362**) [242]. Similarly, *N*-nosyl vinylaziridine **364** reacts with geraniol (**26**) upon treatment with the chiral iridium catalyst **366**, to give amino-alcohol **365** with good diastereoselectivity and high enantiomeric excess. In both cases, the process is supposed to involve abstraction of the hydride from the alcohol to give the corresponding carbonyl compound and a η³-allylmetal complex that condense together to form the carbon-carbon bond addition product [243]

Hydrovinylation reactions using ethylene or propylene are economically promising methods for upgrading 1,3-dienes such as myrcene or β-farnesene [244]. Thus reaction of β-farnesene (**367**) with ethylene (25 psi) in the presence of the low-valent iron catalyst obtained upon reduction of iron (II) complex **370** with naphthalenide sodium gives tetraene **368** with 93.8% yield and 98%. Cross metathesis reaction of **368** with styrene using Grubbs-Hoveyda catalyst C711 takes place selectively on the terminal vinyl group of tetraene **368** to deliver aryl tetraene **369** in 20% yield (Scheme **68**) [245].

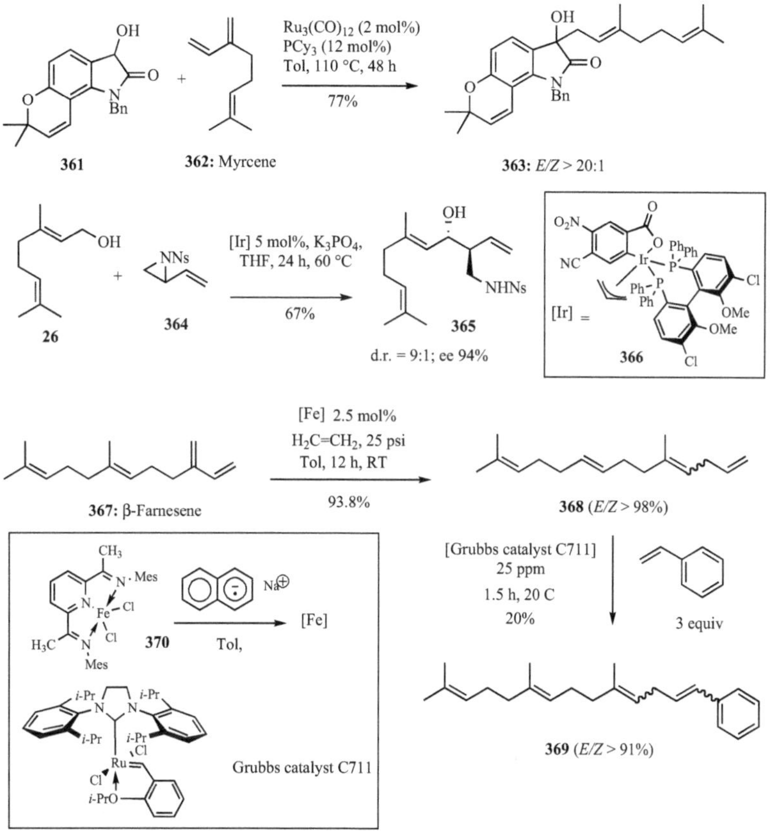

Scheme 68. reaction of η³-allyl iridium and ruthenium complexes with alcohols and synthesis of aryl-functionalized polyisoprenoids using hydrovinylation-cross metathesis sequential reactions.

POLYMERIZATION REACTIONS

Polyisoprenes occur naturally as *cis*-1,4-polyisoprenes (natural rubber) and as *trans*-1,4-polyisoprenes (gutta percha). Both isomers can also be prepared synthetically. Isoprene can be polymerized to 1,4-*cis*-poly(isoprene) (94% *cis*-1,4; 6% 3,4 structures) with lithium or alkylithium reagents while *trans* structures are mainly obtained by polymerization using Ziegler Natta catalyst or Lewis acids [246]. Alternative sources of natural rubber production including rubber-producing microorganisms, transgenic yeasts and bacteria are currently investigated [247]. Polyisoprene polymers with narrow polydispersity indices can be obtained by reversible addition-fragmentation chain transfer (RAFT) using a trithiocarbonate transfer agent and *t*-butyl peroxide as radical initiator. The majority of repeat units possess the 1,4-structure (as a *cis/trans* mixture) with the 1,2- and 3,4-isomers being present as minor components (*c.a.* 20%) [248].

The access to efficient metathesis catalysts in the early 2000s opened new opportunities for the syntheses of telechelic polyisoprene polymers. For example, ring-opening metathesis polymerization (ROP) of 1,5-dimethyl-1-5-cyclooctadiene (**371**) in the presence of 1 mol% of cis-1,4-diacetoxy-2-butene (**372**) as a chain transfer agent and 0.2 mol% of Grubbs II catalyst gives the triblock copolymer **373** in excellent yield [249].

There is a growing interest for drug-loaded polymer nanoparticles to treat severe diseases such as cancer or neurodegenerative disorders. Polyisoprene chains bring the hydrophobic character required to obtain amphiphilic conjugates of nucleoside analogues embedding self-assembling properties. Initially reported with squalene and regular polyprenols this concept was extend to polymeric isoprenoids chains [250, 251]. Such material can be grown directly from the chosen drug, using drug-initiated polymerization reactions. For example, nitroxide-mediated polymerization (NMP) of isoprene using the primer **375** prepared by grafting an alkoxyamine initiator group on the anticancer drug gemcitabine (**374**) provides the polymer prodrug **376** embedding self-assembling properties thanks to the polyisoprene appendage . Nanoparticles made from this material exhibit efficient anticancer activity *in vitro* on various cancer cell lines and *in vivo* on human pancreatic carcinoma-bearing mice, while suppressing the inherent toxicity of the parent drug (Scheme **69**) [252].

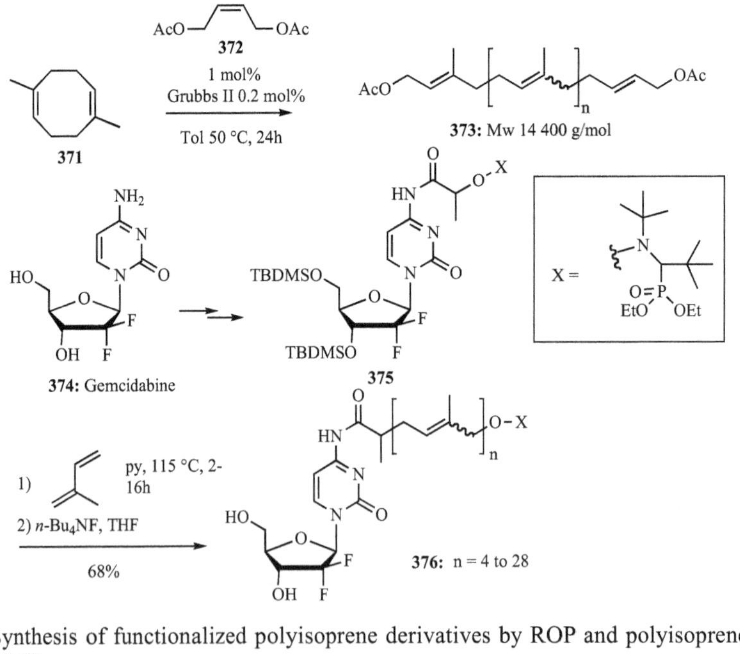

Scheme 69. Synthesis of functionalized polyisoprene derivatives by ROP and polyisoprene polymer drug conjugate by NMP.

CONCLUDING REMARKS

This overview of the available synthetic methods to prepare polyisoprenoid compounds highlights that the field is deeply dependent on progress to synthetize trisubstituted olefins in a stereocontrolled manner. Early efforts involving allylic couplings with heteroatom-stabilized carbanions derived from small polyprenols were attempts to work around this difficulty. Nowadays, organometallic cross-coupling reactions brought deep changes in the field, making the elaboration of the olefins central to the synthetic scheme. Nevertheless, truly iterative processes allowing efficient polyprenol chain elongation are rare. Among them the methods based on Tsuji-Trost reactions, sequential carboalumination/Negishi cross-coupling reactions, and the nickel-catalyzed coupling of 5-alkyl-2,3-dihydrofurans with Grignard reagents are the most efficient. It appears that recent trends in organic synthesis, such as cross-metathesis, C-H activation, alkylative M–H-type reactions, metal-catalyzed borylation, hydrovinylation reactions *etc.* are underexplored for polyisoprene synthesis. Although, development of terpenoid-production platforms in bacteria or yeast may compete with chemical processes, the complexity of the optimized expression of gene cluster will restraint the isoprenoid production in engineered bacteria to high commercial value compounds, leaving space to classical synthetic chemistry for other applications. Finally, the full-scale commercialization of specific prenyltransferase enzymes in a stabilized form would be a significant progress. It may be expected that new methods still in infancy will lead to fruitful future developments.

CONSENT FOR PUBLICATION

Not applicable.

ACKNOWLEGMENTS

Declared none

CONFLICT OF INTEREST

The author confirms that he has no conflict of interest to declare for this publication.

REFERENCES

[1]   Rohmer, M.A. Mevalonate-independent Route to Isopentenyl Diphosphate.*Comprehensive Natural Products Chemistry*; Cane, D., Ed.; Pergamon Press: New York, **1999**, pp. 45-68.
      [http://dx.doi.org/10.1016/B978-0-08-091283-7.00036-9]

[2]   Eisenreich, W.; Schwarz, M.; Cartayrade, A.; Arigoni, D.; Zenk, M.H.; Bacher, A. The deoxyxylulose phosphate pathway of terpenoid biosynthesis in plants and microorganisms. *Chem. Biol.,* **1998**, *5*(9), R221-R233.

[http://dx.doi.org/10.1016/S1074-5521(98)90002-3] [PMID: 9751645]

[3] Streiff, S.; Ribeiro, N.; Wu, Z.; Gumienna-Kontecka, E.; Elhabiri, M.; Albrecht-Gary, A-M.; Ourisson, G.; Nakatani, Y. "Primitive" membrane from polyprenyl phosphates and polyprenyl alcohols. *Chem. Biol.*, **2007**, *14*(3), 313-319.
[http://dx.doi.org/10.1016/j.chembiol.2006.11.017] [PMID: 17379146]

[4] Ciepichal, E.; Jemiola-Rzeminska, M.; Hertel, J.; Swiezewska, E.; Strzalka, K. Configuration of polyisoprenoids affects the permeability and thermotropic properties of phospholipid/polyisoprenoid model membranes. *Chem. Phys. Lipids*, **2011**, *164*(4), 300-306.
[http://dx.doi.org/10.1016/j.chemphyslip.2011.03.004] [PMID: 21440533]

[5] Semikolenov, V.A.; Ilyna, I.I.; Maksimovskaya, R.I. Linalool to geraniol/nerol isomerization catalyzed by $(RO)_3VO$ complexes: studies of kinetics and mechanism. *J. Mol. Catal. Chem.*, **2003**, *204–205*, 201-210.
[http://dx.doi.org/10.1016/S1381-1169(03)00299-1]

[6] Chen, W.; Viljoen, A.M. Geraniol — A review of a commercially important fragrance material. *S. Afr. J. Bot.*, **2010**, *76*, 643-651.
[http://dx.doi.org/10.1016/j.sajb.2010.05.008]

[7] Bråred Christensson, J.; Hagvall, L.; Karlberg, A-T. Fragrance Allergens, Overview with a Focus on Recent Developments and Understanding of Abiotic and Biotic Activation. *Cosmetics*, **2016**, *3*, 19.
[http://dx.doi.org/10.3390/cosmetics3020019]

[8] Pammi, M.; Liang, R.; Hicks, J.M.; Barrish, J.; Versalovic, J. Farnesol decreases biofilms of Staphylococcus epidermidis and exhibits synergy with nafcillin and vancomycin. *Pediatr. Res.*, **2011**, *70*(6), 578-583.
[http://dx.doi.org/10.1203/PDR.0b013e318232a984] [PMID: 21857375]

[9] Bonrath, W.; Beumer, R.; Medlock, J. A. Production of farnesol WO 2017046346 A1 2017 March;23

[10] Mori, T.; Sato, J.; Fukumoto, T.; Nakao, K.; Tamai, Y. Process for producing geranylgeraniol. EP 0711749 A1 1996 May;15

[11] Swiezewska, E.; Danikiewicz, W. Polyisoprenoids: structure, biosynthesis and function. *Prog. Lipid Res.*, **2005**, *44*(4), 235-258.
[http://dx.doi.org/10.1016/j.plipres.2005.05.002] [PMID: 16019076]

[12] Lee, Y.J.; Ishiwata, A.; Ito, Y. Synthesis of undecapeenyl pyrophosphate-linked glycans as donnor substrates for bacterial protein N-glycosidation. *Tetrahedron*, **2009**, *65*, 6310-6319.
[http://dx.doi.org/10.1016/j.tet.2009.06.032]

[13] Takajo, S.; Nagano, H.; Dannenmuller, O.; Ghosh, S.; Albrecht, A-M.; Nakatani, Y.; Ourisson, G. Membrane properties of sodium 2- and 6-(poly)prenyl-substituted polyprenyl phosphates. *New J. Chem.*, **2001**, *25*, 917-929.
[http://dx.doi.org/10.1039/b101802g]

[14] Naider, F.R.; Becker, J.M. Synthesis of prenylated peptides and peptide esters. *Biopolymers*, **1997**, *43*(1), 3-14.
[http://dx.doi.org/10.1002/(SICI)1097-0282(1997)43:1<3::AID-BIP2>3.0.CO;2-Z] [PMID: 9174408]

[15] Schäfer, M.; Brütting, C.; Meza-Canales, I.D.; Großkinsky, D.K.; Vankova, R.; Baldwin, I.T.; Meldau, S. The role of cis-zeatin-type cytokinins in plant growth regulation and mediating responses to environmental interactions. *J. Exp. Bot.*, **2015**, *66*(16), 4873-4884.
[http://dx.doi.org/10.1093/jxb/erv214] [PMID: 25998904]

[16] Rosales-García, T.; Jiménez-Martínez, C.; Cardador-Martínez, A.; Martín-del Campo, S.T.; Galicia-Luna, L.A.; Téllez-Medina, D.I.; Dávila-Ortiz, G. Squalene Extraction by supercritical fluids from traditionally puffed *Amaranthus hypochondriacus* seeds. *J. Food Qual.*, **2017**. Article ID 6879712, 8 pages

[17] Ghimire, G.P.; Thuan, N.H.; Koirala, N.; Sohng, J.K. Advances in biochemistry and microbial

production of squalene and its derivatives. *J. Microbiol. Biotechnol.*, **2016**, *26*(3), 441-451.
[http://dx.doi.org/10.4014/jmb.1510.10039] [PMID: 26643964]

[18] Xu, W.; Ma, X.; Wang, Y. Production of squalene by microbes: an update. *World J. Microbiol. Biotechnol.*, **2016**, *32*(12), 195.
[http://dx.doi.org/10.1007/s11274-016-2155-8] [PMID: 27730499]

[19] Riley, R.G.; Silverstein, R.M. Improved synthesis of 2-methyl-6-methylene-2,7-octadien-4-ol, a pheromone of Ips paraconfusus, and an alternative synthesis of the intermediate, 2-bromomethyl-l-3-butadiene. *J. Org. Chem.*, **1974**, *39*, 1957-1958.
[http://dx.doi.org/10.1021/jo00927a040]

[20] Pummerer, R.; Reindel, W. Uber die Oxyde des Isoprens und Butadiens. *Chem. Ber.*, **1933**, *66*, 335-339.
[http://dx.doi.org/10.1002/cber.19330660305]

[21] Yoo, S-E.; Lee, S-H.; Yi, K-Y.; Jeong, N. Synthesis of α-kainic acid and α-allokainic acid by Pd(0) mediated olefin insertion-carbonylation reaction. *Tetrahedron Lett.*, **1990**, *31*, 6877-6880.
[http://dx.doi.org/10.1016/S0040-4039(00)97195-8]

[22] Thomsen, D.S.; Schiøtt, B.; Jørgensen, K.A. Regioselective monoepoxidation of 1,3-dienes catalysed by transition-metal complexes. *J. Chem. Soc. Chem. Commun.*, **1992**, 1072-1074.
[http://dx.doi.org/10.1039/C39920001072]

[23] Shimizu, I.; Maruyama, T.; Makuta, T.; Yamamoto, A. Palladium-catalyzed reactions of alkenyloxiranes with carbon monoxide. *Tetrahedron Lett.*, **1993**, *34*, 2135-2138.
[http://dx.doi.org/10.1016/S0040-4039(00)60364-7]

[24] Socolsky, C.; Plietker, B. Total synthesis and absolute configuration assignment of MRSA active garcinol and isogarcinol. *Chemistry*, **2015**, *21*(7), 3053-3061.
[http://dx.doi.org/10.1002/chem.201406077] [PMID: 25537962]

[25] Mandal, A.K.; Schneekloth, J.S., Jr; Kuramochi, K.; Crews, C.M. Synthetic studies on amphidinolide B1. *Org. Lett.*, **2006**, *8*(3), 427-430.
[http://dx.doi.org/10.1021/ol052620g] [PMID: 16435851]

[26] Lambertin, F.; Wende, M.; Quirin, M-J.; Taran, M.; Delmond, B. New retinoid analogs from δ-pyronene, a natural synthon. *Eur. J. Org. Chem.*, **1999**, 1489-1494.
[http://dx.doi.org/10.1002/(SICI)1099-0690(199906)1999:6<1489::AID-EJOC1489>3.0.CO;2-1]

[27] Mornet, R.; Gouin, L. Synthèse de l'acétate du chloro-4 méthyl-2 ol-1 (*E*). Nouvelle voie d'accès à la *trans*-zéatine. *Tetrahedron Lett.*, **1977**, *18*, 167-168.
[http://dx.doi.org/10.1016/S0040-4039(01)92578-X]

[28] Lipshutz, B.H.; Bulow, G.; Fernandez-Lazaro, F.; Kim, S-K.; Lowe, R.; Mollard, P.; Stevens, K.L. A convergent cpproach to coenzyme Q. *J. Am. Chem. Soc.*, **1999**, *121*(50), 11664-11673.
[http://dx.doi.org/10.1021/ja992164p]

[29] Fox, D.T.; Poulter, C.D. Synthesis of (*E*)-4-hydroxydimethylallyl diphosphate. An intermediate in the methyl erythritol phosphate branch of the isoprenoid pathway. *J. Org. Chem.*, **2002**, *67*(14), 5009-5010.
[http://dx.doi.org/10.1021/jo0258453] [PMID: 12098326]

[30] Ward, J.L.; Beale, M.H. Synthesis of (2*E*)-4-hydroxy-3-methylbut-2-enyl diphosphate, a key intermediate in the biosynthesis of isoprenoids. *J. Chem. Soc., Perkin Trans. 1*, **2002**, 710-712.
[http://dx.doi.org/10.1039/b200712f]

[31] Giner, J-L. Convenient synthesis of (*E*)-4-hydroxy-3-methyl-2-butenyl pyrophosphate and its [4-^{13}C]-labeled form. *Tetrahedron Lett.*, **2002**, *43*, 5457-5459.
[http://dx.doi.org/10.1016/S0040-4039(02)01102-4]

[32] Watanabe, H.; Hatakeyama, S.; Tazumi, K.; Takano, S.; Masuda, S.; Okano, T.; Kobayashi, T.; Kubodera, N. Synthetic studies of vitamin D analogs. XXII. Synthesis and antiproliferation activity of

putative metabolites of 1α, 25-dihydroxy-22-oxavitamin D3. *Chem. Pharm. Bull. (Tokyo)*, **1996**, *44*(12), 2280-2286.
[http://dx.doi.org/10.1248/cpb.44.2280] [PMID: 8996858]

[33] Enders, D.; Schüßeler, T. First highly efficient asymmetric synthesis of the Hyrtios Erectus diketotriterpenoid. *Synthesis*, **2002**, 2280-2288.
[http://dx.doi.org/10.1055/s-2002-34949]

[34] Trost, B.M.; Machacek, M.R.; Tsui, H.C. Development of aliphatic alcohols as nucleophiles for palladium-catalyzed DYKAT reactions: total synthesis of (+)-hippospongic acid A. *J. Am. Chem. Soc.*, **2005**, *127*(19), 7014-7024.
[http://dx.doi.org/10.1021/ja050340q] [PMID: 15884945]

[35] Fringuelli, F.; Pizzo, F.; Germani, R. pH-controlled regioselectivity of epoxidation of Geraniol in Water. *Synlett*, **1991**, 474-476.
[http://dx.doi.org/10.1055/s-1991-20765]

[36] Demotie, A.; Fairlamb, I.J.S.; Radford, S.K. On the selective reduction of the distal olefin in geraniol and farnesol derivatives. *Tetrahedron Lett.*, **2003**, *44*, 4539-4542.
[http://dx.doi.org/10.1016/S0040-4039(03)01047-5]

[37] Umbriet, M.A.; Sharpless, K.B. Allylic oxidation of olefins by catalytic and stoichiometric selenium dioxide with tert-butyl hydroperoxide. *J. Am. Chem. Soc.*, **1977**, *99*, 5526-5528.
[http://dx.doi.org/10.1021/ja00458a072]

[38] Paz, J.L.; Rodrigues, J.A.R. Preparation of aromatic geraniol analogues *via* Cu(I)-mediated Grignard coupling. *J. Braz. Chem. Soc.*, **2003**, *14*, 975-981.
[http://dx.doi.org/10.1590/S0103-50532003000600014]

[39] Fairlamb, I.J.S.; Dickinson, J.M.; Pegg, M. Selenium dioxide E-methyl oxidation of suitably protected geranyl derivatives—synthesis of farnesyl mimics. *Tetrahedron Lett.*, **2001**, *42*, 2205-2208.
[http://dx.doi.org/10.1016/S0040-4039(01)00110-1]

[40] Li, Y.; Li, W.; Li, Y. Studies on Macrocyclic diterpenoids. Part 10. First total synthesis of (±)-isosarcophytol-A. *J. Chem. Soc., Perkin Trans 1*, **1993**, 2953-2956.
[http://dx.doi.org/10.1039/P19930002953]

[41] Adams, C.M.; Ghosh, I.; Kishi, Y. Validation of lanthanide chiral shift reagents for determination of absolute configuration: total synthesis of glisoprenin A. *Org. Lett.*, **2004**, *6*(25), 4723-4726.
[http://dx.doi.org/10.1021/ol048059o] [PMID: 15575670]

[42] Stork, G.; Gregson, M.; Grieco, P.A. A convenient route to *cis* and *trans*-trisubstituted olefins from geraniol and nerol. *Tetrahedron Lett.*, **1969**, *9*, 1391-1392.
[http://dx.doi.org/10.1016/S0040-4039(01)87895-3]

[43] Coates, R.M.; Ley, D.A. Cavender, P. L. Synthesis and carbon-^{13}C Nuclear Magnetic Resonance spectra of all-*trans*-geranylgeraniol and its nor analogues. *J. Org. Chem.*, **1978**, *43*, 4915-4922.
[http://dx.doi.org/10.1021/jo00420a003]

[44] Wise, M.L.; Pyun, H-J.; Hems, G.; Assink, B.; Coates, R.M.; Croteau, R.B. Stereochemical disposition of the geminal dimethyl groups in the enzymatic cyclization of geranyl diphosphate to (+)-bornyl diphosphate by recombinant (+)-bornyl diphosphate synthase from Salvia officinalis. *Tetrahedron*, **2001**, *57*, 5327-5334.
[http://dx.doi.org/10.1016/S0040-4020(01)00451-3]

[45] Barrero, A.F.; Quíllez del Moral, J.F.; Herrador, M.M.; Sánchez, E.M.; Arteaga, J.F. Regio- and enantioselective functionalization of acyclic polyprenoids. *J. Mex. Chem. Soc.*, **2006**, *50*, 149-156.

[46] Masaki, Y.; Hashimoto, K.; Sakuma, K.; Kaji, K. A facile functionalization of the isopropylidene terminus of isoprenoids. Application to the synthesis of terminal *trans* allylic alcohols. *Tetrahedron Lett.*, **1978**, *19*, 4539-4542.
[http://dx.doi.org/10.1016/S0040-4039(01)95272-4]

[47] Masaki, Y. Hashimoto, K.; Sakuma, K.; Kaji, K. Facile regio- and stereo-specific allylic oxidation of gem-dimethyl olefins *via* addition of benzenesulphenyl chloride. Synthesis of allylic oxygenated terpenes. *J. Chem. Soc. Perkin,* **1984**, *1*, 1289-1295.
[http://dx.doi.org/10.1039/p19840001289]

[48] Bulliard, M.; Balme, G.; Goré, J. Chloration allylique d'olefines de type isoprenique a l'aide du chlorure de sulfuryle. *Tetrahedron Lett.,* **1989**, *30*, 5767-5770.
[http://dx.doi.org/10.1016/S0040-4039(00)76192-2]

[49] Torii, S.; Tanaka, H.; Tada, N.; Nagao, S.; Sasaoka, M. Ene-type chlorination of olefins with dichlorine monoxide. *Chem. Lett.,* **1984**, *13*, 877-880.
[http://dx.doi.org/10.1246/cl.1984.877]

[50] Sato, W.; Ikeda, N.; Yamamoto, H. An efficient double chlorination of olefins by *tert*-butyl hypochlorite. *Chem. Lett.,* **1982**, *11*, 141-144.
[http://dx.doi.org/10.1246/cl.1982.141]

[51] Hori, T.; Sharpless, K.B. Selenium-Catalyzed Nonradical Chlorination of Olefins with N-Chlorosuccinimide. *J. Org. Chem.,* **1979**, *44*, 4204-4208.
[http://dx.doi.org/10.1021/jo01337a046]

[52] Torii, S.; Uneyama, K.; Nakai, T.; Yasuda, T. An electrochemical chlorinative ene-type reaction of isoprenoids. *Tetrahedron Lett.,* **1981**, *22*, 2291-2294.
[http://dx.doi.org/10.1016/S0040-4039(01)92913-2]

[53] Mignani, G.; Grass, J-P.; Chabardes, P.; Morel, D. Convenient synthesis of isoprenoid chlorides by a direct chlorination process. *Tetrahedron Lett.,* **1992**, *33*, 495-498.
[http://dx.doi.org/10.1016/S0040-4039(00)93978-9]

[54] Hegde, S.G.; Vogel, M.K.; Saddler, J.; Hrinyo, T.; Rockwell, N.; Haynes, R.; Oliver, M.; Wolinsky, J. The reaction of hypochlorous acid with olefins. A convenient synthesis of allylic chlorides. *Tetrahedron Lett.,* **1980**, *21*, 441-444.
[http://dx.doi.org/10.1016/S0040-4039(00)71427-4]

[55] Inoue, S.; Iwase, N.; Miyamoto, O.; Sato, K. Regio- and stereoselective oxidation of gem-dimethyl olefins *via* [2,3]-sigmatropic rearrangement of allyl amine oxides. *Chem. Lett.,* **1986**, 2035-2038.
[http://dx.doi.org/10.1246/cl.1986.2035]

[56] Barrero, A.F.; Quílez Del Moral, J.F.; Herrador, M.M.; Cortés, M.; Arteaga, P.; Catalán, J.V.; Sanchez, E.M.; Arteaga, J.F. Solid-phase selenium-catalyzed selective allylic chlorination of polyprenoids: facile syntheses of biologically active terpenoids. *J. Org. Chem.,* **2006**, *71*(15), 5811-5814.
[http://dx.doi.org/10.1021/jo060760d] [PMID: 16839173]

[57] Gnanadesikan, V.; Corey, E.J. A strategy for position-selective epoxidation of polyprenols. *J. Am. Chem. Soc.,* **2008**, *130*(25), 8089-8093.
[http://dx.doi.org/10.1021/ja801899v] [PMID: 18494468]

[58] Vidari, G.; Dapiaggi, A.; Zanoni, G.; Garlaschelli, L. Asymmetric dihydroxylation of geranyl, neryl and *trans, trans*-farnesyl acetates. *Tetrahedron Lett.,* **1993**, *34*, 6485-6488.
[http://dx.doi.org/10.1016/0040-4039(93)85077-A]

[59] Corey, E.J.; Noe, M.C.; Shieh, W-C. A short and convergent enantioselective synthesis of (3S)-2,3-oxidosqualene. *Tetrahedron Lett.,* **1993**, *34*, 5995-5998.
[http://dx.doi.org/10.1016/S0040-4039(00)61710-0]

[60] Xu, D.; Park, C.Y.; Sharpless, K.B. Study of the regio- and enantioselectivity of the reactions of osmium tetroxide with allylic alcohols and allylic sulfonamides. *Tetrahedron Lett.,* **1994**, *35*, 2495-2498.
[http://dx.doi.org/10.1016/S0040-4039(00)77153-X]

[61] Corey, E.J.; Noe, M.C.; Lin, S.A. Mechanistically designed bis-cinchona alkaloid ligand allows

position- and enantioselective dihydroxylation of farnesol and other oligoprenyl derivatives at the terminal isopropylidene unit. *Tetrahedron Lett.,* **1995**, *36*, 8741-8744.
[http://dx.doi.org/10.1016/0040-4039(95)01920-D]

[62] Corey, E.J.; Zhang, J. Highly effective transition structure designed catalyst for the enantio- and position-selective dihydroxylation of polyisoprenoids. *Org. Lett.,* **2001**, *3*(20), 3211-3214.
[http://dx.doi.org/10.1021/ol016577i] [PMID: 11574033]

[63] Kodama, M.; Minami, H.; Mima, Y.; Fukuyama, Y. Convenient synthesis of chiral epoxyisoprenoids by yeast reduction. *Tetrahedron Lett.,* **1990**, *31*, 4025-4026.
[http://dx.doi.org/10.1016/S0040-4039(00)94489-7]

[64] Fourneron, J.D.; Archelas, A.; Furtoss, R. Microbial Transformations 12. Regiospecific and asymmetric oxidation of the remote double bond of geraniol. *J. Org. Chem.,* **1989**, *54*, 4686-4689.
[http://dx.doi.org/10.1021/jo00280a043]

[65] Katzenellenbogen, J.A.; Crumrine, A.L. Selective γ alkylation of dienolate anions derived from α–β-unsaturated acids. Applications to the synthesis of isoprenoid olefins. *J. Am. Chem. Soc.,* **1976**, *98*, 4925-4935.
[http://dx.doi.org/10.1021/ja00432a038]

[66] Pitzele, B.S.; Baran, J.S.; Steinman, D.H. γ-Alkylation of α,β-unsaturated acids: A technique of isoprenoid homologation. *Tetrahedron,* **1976**, *32*, 1347-1351.
[http://dx.doi.org/10.1016/0040-4020(76)85008-9]

[67] Freise, M. Verfahren zur Herstellung von *E,E-* und *E,Z-*Dimethyl-10-methylen-dodeca-2-6,11-trienalgemischen un deren Verwendung als Aroma- oder Riechstoffe. German Patent DE 41 27888 A1 1993 February 25;

[68] Altman, L.J.; Ash, L. Marson, S. A new, highly stereoselective synthesis of all trans geranygeraniol. *Synthesis,* **1974**, 129-131.
[http://dx.doi.org/10.1055/s-1974-23262]

[69] Sato, K.; Inoue, S.; Onishi, A.; Uchida, N; Minowa, N. Stereoselective synthesis of solanesol and all-trans-decaprenol. *J. Chem. Soc. Perkin Trans. 1,* **1981**, 761-769.
[http://dx.doi.org/10.1039/P19810000761]

[70] Tsuji, J.; Shimizu, I.; Minami, I.; Ohashi, Y.; Sugiura, T.; Takahashi, K. Allylic carbonates. efficient allylating agents of carbonucleophiles in palladium-catalyzed reactions under neutral conditions. *J. Org. Chem.,* **1985**, *50*, 1523-1529.
[http://dx.doi.org/10.1021/jo00209a032]

[71] Yan, N.; Liu, Y.; Gong, D.; Du, Y.; Zhang, H.; Zhang, Z. Solanesol: a review of its resources, derivatives, bioactivities, medicinal applications, and biosynthesis. *Phytochem. Rev.,* **2015**, *14*, 403-417.
[http://dx.doi.org/10.1007/s11101-015-9393-5]

[72] Masaki, Y.; Hashimoto, K.; Kaji, K. A novel terminal functionalization of farnesol and related polyisoprenoids. Application to the synthesis of solanesol. *Tetrahedron Lett.,* **1978**, *19*, 5123-5126.
[http://dx.doi.org/10.1016/S0040-4039(01)85828-7]

[73] Yu, X-J.; Zhang, H.; Xiong, F-J.; Chen, X-X.; Chen, F-E. An improved convergent strategy for the synthesis of oligoprenols. *Helv. Chim. Acta,* **2008**, *91*, 1967-1974.
[http://dx.doi.org/10.1002/hlca.200890211]

[74] Mohri, M.; Kinoshita, H.; Inomata, K.; Kotake, H.; Takagaki, H.; Yamazaki, K. Palladium-catalyzed regio- and stereoselective reduction of allylic compounds with LiBHEt$_3$: Application to the synthesis of co-enzyme Q10. *Chem. Lett.,* **1986**, 1177-1180.
[http://dx.doi.org/10.1246/cl.1986.1177]

[75] Trost, B.M.; Dong, G.; Vance, J.A. A diosphenol-based strategy for the total synthesis of (-) -terpestacin. *J. Am. Chem. Soc.,* **2007**, *129*(15), 4540-4541.

[http://dx.doi.org/10.1021/ja070571s] [PMID: 17343388]

[76] Ji, M.; Choi, H.; Park, M.; Kee, M.; Jeong, Y.C.; Koo, S. A highly efficient chain-extension process in the systematic syntheses of carotenoid natural products. *Angew. Chem. Int. Ed. Engl.,* **2001**, *40*(19), 3627-3629.
[http://dx.doi.org/10.1002/1521-3773(20011001)40:19<3627::AID-ANIE3627>3.0.CO;2-E] [PMID: 11592202]

[77] Kuk, J.; Kim, B.S.; Jung, H.; Choi, S.; Park, J-Y.; Koo, S. General preparation and controlled cyclization of acyclic terpenoids. *J. Org. Chem.,* **2008**, *73*(5), 1991-1994.
[http://dx.doi.org/10.1021/jo702303a] [PMID: 18247490]

[78] van Tamelen, E.E.; Curphey, T.J. The selective *in vitro* oxidation of the terminal double bonds in squalene. *Tetrahedron Lett.,* **1962**, *3*, 121-124.
[http://dx.doi.org/10.1016/S0040-4039(00)71112-9]

[79] Terao, S.; Kato, K.; Shiraishi, M.; Morimoto, H. Synthesis of ubiquinones. Elongation of the heptaprenyl side-chain in ubiquinone-7. *J. Chem. Soc. Perkin Trans. 1,* **1978**, 1101-1108.
[http://dx.doi.org/10.1039/P19780001101]

[80] Mehta, D.; Mohan, P.; Shastri, M.; Reid, T. Method of preparation of stereospecific quinone derivative. *EP 2 868 658 A1,* **2015**, *10*, 605-610.

[81] Yu, X.; Wang, S.; Chen, F. Solid-phase synthesis of solanesol. *J. Comb. Chem.,* **2008**, *10*(4), 605-610.
[http://dx.doi.org/10.1021/cc800069t] [PMID: 18558751]

[82] Chang, Y-F.; Liu, C-Y.; Guo, C-W.; Wang, Y-C.; Fang, J-M.; Cheng, W-C. Solid-phase organic synthesis of polyisoprenoid alcohols with traceless sulfone linker. *J. Org. Chem.,* **2008**, *73*(18), 7197-7203.
[http://dx.doi.org/10.1021/jo8010182] [PMID: 18707172]

[83] Hutchinson, D.A.; Beck, K.R.; Benkeser, R.A.; Grutzner, J.B. Concerning the structure of allylic Grignard reagents. *J. Am. Chem. Soc.,* **1973**, *95*(21), 7075-7082.
[http://dx.doi.org/10.1021/ja00802a031]

[84] Yanagisawa, A.; Noritake, Y.; Nomura, N.; Yamamoto, H. Superiority of Phosphate Ester as Leaving Group for Organocopper Reactions. Highly S_N2' -, (*E*)-, and antiselective alkylation of allylic alcohol derivatives. *Synlett,* **1991**, 251-253.
[http://dx.doi.org/10.1055/s-1991-20696]

[85] Julia, M.; Verpeaux, J-N. Synthèse à l'aide de sulfones-XXVI. Synthèse d'alcools allyliques et de polyprenols par attachement d'un synthon prénol en position 4 E. *Tetrahedron,* **1983**, *39*, 3289-3291.
[http://dx.doi.org/10.1016/S0040-4020(01)91578-9]

[86] Yanagisawa, A.; Nobura, N.; Yamamoto, H. Iron-catalyzed Kharasch-type reaction between Grignard reagents and allylic phosphates. Highly S_N2 selective cross-coupling process. *Synlett,* **1991**, 513-514.
[http://dx.doi.org/10.1055/s-1991-20783]

[87] Ellwart, M.; Makarov, I.S.; Achrainer, F.; Zipse, H.; Knochel, P. Regioselective transition-metal-free allyl–allyl cross-couplings. *Angew. Chem. Int. Ed. Engl.,* **2016**, *55*(35), 10502-10506.
[http://dx.doi.org/10.1002/anie.201603923] [PMID: 27430745]

[88] Cardillo, G.; Contento, M.; Sandri, S. Synthesis of compounds containing the isoprene unit. A new stereospecific synthesis of the geranyl and farnesyl skeleton. *Tetrahedron Lett.,* **1974**, *15*, 2215-2216.
[http://dx.doi.org/10.1016/S0040-4039(01)93180-6]

[89] Corey, E.J.; Hamanaka, E. A New Synthetic approach to medium-size carbocyclic systems. *J. Am. Chem. Soc.,* **1964**, *86*(8), 1641-1642.
[http://dx.doi.org/10.1021/ja01062a041]

[90] Sato, K.; Inoue, S.; Ota, S.; Fujita, Y. Reactions of π-allylic nickel(II) bromide with organic halides. A novel synthesis of monoterpenoid compounds. *J. Org. Chem.,* **1972**, *37*(3), 462-466.
[http://dx.doi.org/10.1021/jo00968a029]

[91] Yanagisawa, A.; Hibino, H.; Habaue, S.; Hisada, Y.; Yamamoto, H. Highly Selective Homocoupling Reaction of Allylic Halides Using Barium Metal. *J. Org. Chem.*, **1992**, *57*(24), 6386-6387.
[http://dx.doi.org/10.1021/jo00050a006]

[92] Yanagisawa, A.; Habaue, S.; Yamamoto, H. Direct insertion of alkali (alkaline-earth) metals into allylic carbon-halogen bonds avoiding stereorandomization. *J. Am. Chem. Soc.*, **1991**, *113*(23), 8955-8956.
[http://dx.doi.org/10.1021/ja00023a058]

[93] Corey, E.J.; Shieh, W-C. A simple synthetic process for the elaboration of oligoprenols by stereospecific coupling of di-, tri-, or oligoisoprenoid units. *Tetrahedron Lett.*, **1992**, *33*, 6435-6438.
[http://dx.doi.org/10.1016/S0040-4039(00)79008-3]

[94] Ishihara, K.; Ishibashi, H.; Yamamoto, H. Enantio- and diastereoselective stepwise cyclization of polyprenoids induced by chiral and achiral LBAs. A new entry to (-)-ambrox, (+)-podocarpa-8,11,-3-triene diterpenoids, and (-)-tetracyclic polyprenoid of sedimentary origin. *J. Am. Chem. Soc.*, **2002**, *124*(14), 3647-3655.
[http://dx.doi.org/10.1021/ja0124865] [PMID: 11929254]

[95] Araki, S.; Butsugan, Y. Transition-metal-catalysed Grignard Reaction of Secondary Allylic Phosphates. *J. Chem. Soc. Perkin Trans. 1*, **1984**, 969-972.
[http://dx.doi.org/10.1039/p19840000969]

[96] Kichisaburo, H.; Yutaka, O.; Yukio, N.; Yoshihiko, H.; Takashi, B.; Shizumasa, K. Process for the preparation of terpenes. EP 0 414 106 A2 1990 August 14;

[97] Masaki, Y.; Sakuma, K.; Kaji, K. Regio- and stereoselective desulphurizative γ-substitution of α-substituted p-methylallyl suphoxides and sulphones with lithium dialkylcuprates providing trisubstituted olefins. *J. Chem. Soc. Perkin Trans. 1*, **1985**, 1171-1173.
[http://dx.doi.org/10.1039/P19850001171]

[98] Calo, V.; Lopez, L.; Pesce, G. Regio reversed nucleophilic substitution of 2-(arlyoxy) benzothiazole by allylic Grignard reagents. A regioselective synthesis of 1,5-dienes. *J. Chem. Soc. Perkin Trans. 1*, **1988**, 1301-1304.
[http://dx.doi.org/10.1039/p19880001301]

[99] Wang, F.; Jiang, X.; Hu, L.; Dong, S.; Wu, X.; Bai, H.; Zhang, Y.; Stöckigt, J.; Zhao, Y. A Novel and convenient method for the synthesis of Ubiquinone-10. *Lett. Org. Chem.*, **2006**, *3*, 610-612.
[http://dx.doi.org/10.2174/157017806778559464]

[100] Yanagisawa, A.; Nomura, N.; Noritake, Y.; Yamamoto, H. Highly SN2'-, (E)-, and antiselective alkylation of allylic phosphates. Facile synthesis of coenzyme Q10. *Synthesis*, **1991**, 1130-1136.
[http://dx.doi.org/10.1055/s-1991-28404]

[101] West, D.D. Synthesis of coenzyme Q10 ubiquinone. U.S. Patent 6,506,915 B1 2003 January 14;

[102] Kawashima, M.; Sato, T.; Fujisawa, T. Regio- and stereoselective ring opening of ω-alkenyl lactone using organocopper reagents. *Bull. Chem. Soc. Jpn.*, **1988**, *61*, 3255-3264.
[http://dx.doi.org/10.1246/bcsj.61.3255]

[103] Keinan, E.; Eren, D. Total Synthesis of Linear Polyprenoids. 3. Syntheses of ubiquinones via palladium-catalyzed oligomerization of monoterpene monomers. *J. Am. Chem. Soc.*, **1988**, *110*(13), 4356-4362.
[http://dx.doi.org/10.1021/ja00221a040]

[104] Bouzbouz, S.; Kirschleger, B. Synthesis of pure trans, trans,trans–geranylgeraniol. *Synlett*, **1994**, 763-764.
[http://dx.doi.org/10.1055/s-1994-23002]

[105] Yanagisawa, A.; Noritake, Y.; Yamamoto, H. Selective 1,5-diene synthesis. A radical approach. *Chem. Lett.*, **1988**, *17*, 1899-1902.
[http://dx.doi.org/10.1246/cl.1988.1899]

[106] Désaubry, L.; Nakatani, Y.; Ourisson, G. Toward higher polyprenols under 'prebiotic' conditions. *Tetrahedron Lett.*, **2003**, *44*, 6959-6961.
[http://dx.doi.org/10.1016/S0040-4039(03)01624-1]

[107] Banthorpe, D.V. Branch. S. A. A convenient enzymatic preparation of specifically labelled geraniol. *J. Labelled Comp. Radiopharm.*, **1988**, *25*, 913-920.
[http://dx.doi.org/10.1002/jlcr.2580250811]

[108] Koyama, T.; Ogura, K. Synthesis and Absolute Configuration of 4-Methyl Juvenile Hormone I (4-MeJH I) by a biogenetic approach: A Combination of enzymatic synthesis and biotransformation. *J. Am. Chem. Soc.*, **1987**, *109*(9), 2853-2854.
[http://dx.doi.org/10.1021/ja00243a064]

[109] Nagaki, M.; Takaya, A.; Maki, Y.; Ishibashi, J.; Kato, Y.; Nishino, T.; Koyama, T. One-pot syntheses of the sex pheromone homologs of a codling moth, Laspeyresia promonella L. *J. Mol. Cat. B-Enzymatic*, **2000**, *10*, 517-522.
[http://dx.doi.org/10.1016/S1381-1177(00)00094-1]

[110] Ohnuma, S-i.; Koyama, T.; Ogura, K. Enzymatic synthesis of glycinoprenols. *Tetrahedron Lett.*, **1991**, *32*, 241-242.
[http://dx.doi.org/10.1016/0040-4039(91)80865-4]

[111] Fujiwara, S.; Yamanaka, A.; Yamada, Y.; Hirooka, K.; Higashibata, H.; Fukuda, W.; Nakayama, J.; Imanaka, T.; Fukusaki, E. Efficient synthesis of trans-polyisoprene compounds using two thermostable enzymes in an organic-aqueous dual-liquid phase system. *Biochem. Biophys. Res. Commun.*, **2008**, *365*(1), 118-123.
[http://dx.doi.org/10.1016/j.bbrc.2007.10.133] [PMID: 17976371]

[112] Julien, B.; Burlingame, R. Method for production of isoprenoid compounds. US Patent, 0242658 A1 2014 August 28;

[113] Mirata, M.A. Microbial cell factories: Towards an industrial production of isoprenoids. *ChemCatChem*, **2014**, *6*, 955-957.
[http://dx.doi.org/10.1002/cctc.201300833]

[114] Li, Y.; Wang, G. Strategies of isoprenoids production in engineered bacteria. *J. Appl. Microbiol.*, **2016**, *121*(4), 932-940.
[http://dx.doi.org/10.1111/jam.13237] [PMID: 27428054]

[115] Radetich, B.; Corey, E.J. A general stereocontrolled, convergent synthesis of oligoprenols that parallels the biosynthetic pathway. *J. Am. Chem. Soc.*, **2002**, *124*(11), 2430-2431.
[http://dx.doi.org/10.1021/ja0127537] [PMID: 11890779]

[116] Radetich, B.; Corey, E.J. A study of oligoprenyl coupling reactions with allylic stannanes. *Org. Lett.*, **2002**, *4*(20), 3463-3464.
[http://dx.doi.org/10.1021/ol026568p] [PMID: 12323044]

[117] Sato, K.; Inoue, S.; Ota, S. The synthesis of isoprenoid ketones. *J. Org. Chem.*, **1970**, *35*(3), 565-566.
[http://dx.doi.org/10.1021/jo00828a005]

[118] Roe, S.J.; Oldfield, M.F.; Geach, N.; Baxter, A. A convergent stereocontrolled synthesis of [3-^{14}C] solanesol. *J. Labelled Comp. Radiopharm.*, **2013**, *56*(9-10), 485-491.
[http://dx.doi.org/10.1002/jlcr.3083] [PMID: 24285526]

[119] Citron, C.A.; Rabe, P.; Barra, L.; Nakano, C.; Hoshino, T.; Dickschat, J.S. Synthesis of isotopically labelled oligoprenyl diphosphates and their application in mechanistic investigations of terpene cyclases. *Eur. J. Org. Chem.*, **2014**, 7684-7691.
[http://dx.doi.org/10.1002/ejoc.201403002]

[120] Walter, W.M. Synthesis of geranylgeraniol-2-^{14}C. *J. Labelled Compd.*, **1967**, *3*, 54-56.
[http://dx.doi.org/10.1002/jlcr.2590030110]

[121] Yu, J.S.; Kleckley, T.S.; Wiemer, D.F. Synthesis of farnesol isomers *via* a modified Wittig procedure. *Org. Lett.,* **2005**, *7*(22), 4803-4806.
[http://dx.doi.org/10.1021/ol0513239] [PMID: 16235893]

[122] Klinge, S.; Demuth, M. An improved procedure for the preparation of all-*trans*-geranylgeraniol. *Synlett,* **1993**, 783-784.
[http://dx.doi.org/10.1055/s-1993-22608]

[123] Surendra, K.; Rajendar, G.; Corey, E.J. Useful catalytic enantioselective cationic double annulation reactions initiated at an internal π-bond: method and applications. *J. Am. Chem. Soc.,* **2014**, *136*(2), 642-645.
[http://dx.doi.org/10.1021/ja4125093] [PMID: 24359428]

[124] Kotoku, N.; Fujioka, S.; Nakata, C.; Yamada, M.; Sumii, Y.; Kawachi, T.; Arai, M.; Kobayashi, M. Concise synthesis and structure-activity relationship of furospinosulin-1, a hypoxia-selective growth inhibitor from marine sponge. *Tetrahedron,* **2011**, *67*, 6673-6678.
[http://dx.doi.org/10.1016/j.tet.2011.05.009]

[125] Netscher, T. Preparation of trialkyl-substituted olefins by ruthenium catalyzed cross-metathesis. *J. Organomet. Chem.,* **2006**, *691*, 5155-5162.
[http://dx.doi.org/10.1016/j.jorganchem.2006.09.030]

[126] Zhang, W.; Tian, F.; Zhang, Y. Method for synthesizing coenzyme Q10 by double decomposition of alkene. Chinese Patent CN101139274A 2007 September 27;

[127] Trost, B.M.; Dong, G.; Vance, J.A. Cyclic 1,2-diketones as core building blocks: a strategy for the total synthesis of (-)-terpestacin. *Chem. -Eur. J.,* **2010**, *16*(21), 6265-6277.
[http://dx.doi.org/10.1002/chem.200903356] [PMID: 20411537]

[128] McGrath, N.A.; Lee, C.A.; Araki, H.; Brichacek, M.; Njardarson, J.T. An efficient substrate-controlled approach towards hypoestoxide, a member of a family of diterpenoid natural products with an inside-out [9.3.1]bicyclic core. *Angew. Chem. Int. Ed. Engl.,* **2008**, *47*(49), 9450-9453.
[http://dx.doi.org/10.1002/anie.200804237] [PMID: 18979485]

[129] Zakarian, J.E.; El-Azizi, Y.; Collins, S.K. Exploiting quadrupolar interactions in the synthesis of the macrocyclic portion of longithorone C. *Org. Lett.,* **2008**, *10*(14), 2927-2930.
[http://dx.doi.org/10.1021/ol800821f] [PMID: 18572944]

[130] LeNoble, W.J. The configurations of some substituted β-haloacrylic acids. *J. Am. Chem. Soc.,* **1961**, *83*(18), 3897-3899.
[http://dx.doi.org/10.1021/ja01479a036]

[131] Posner, G.H.; Ting, J-S.; Lentz, C.M. A Mechanistic and synthetic study of organocopper substitution reactions with some homoallylic and cyclopropylcarbinyl substrates, application to isoprenoid synthesis. *Tetrahedron,* **1976**, *32*, 2281-2287.
[http://dx.doi.org/10.1016/0040-4020(76)88002-7]

[132] Svatoš, A.; Urbanová, K.; Valterová, I. The first synthesis of geranyllinalool enantiomers. *Collect. Czech. Chem. Commun.,* **2002**, *67*, 83-90.
[http://dx.doi.org/10.1135/cccc20020083]

[133] Sum, F.W.; Weiler, L. Synthesis of isoprenoid natural products from β-keto esters. *Tetrahedron,* **1981**, *37* Suppl. 1, 303-317.
[http://dx.doi.org/10.1016/0040-4020(81)85068-5]

[134] Eis, K.; Schmalz, H-G. Synthesis of (*E,E,E*)-(1,2,3,4-^{13}C4)-Geranylgeraniol. *Synthesis,* **1997**, 202-206.
[http://dx.doi.org/10.1055/s-1997-1154]

[135] Maynor, M.; Scott, S.A.; Rickert, E.L.; Gibbs, R.A. Synthesis and evaluation of 3- and 7-substituted geranylgeranyl pyrophosphate analogs. *Bioorg. Med. Chem. Lett.,* **2008**, *18*(6), 1889-1892.
[http://dx.doi.org/10.1016/j.bmcl.2008.02.014] [PMID: 18321704]

[136] Gibbs, R.A.; Krishnan, U.A. Pd(0)-Catalyzed route to 13-methylidenefarnesyl diphosphate. *Tetrahedron Lett.*, **1994**, *35*, 2509-2512.
[http://dx.doi.org/10.1016/S0040-4039(00)77157-7]

[137] Mu, Y.Q.; Gibbs, R.A. Coupling of isoprenoid triflates with organoboron nucleophiles: Synthesis of all-*trans*-geranylgeraniol. *Tetrahedron Lett.*, **1995**, *36*, 5669-5672.
[http://dx.doi.org/10.1016/00404-0399(50)11193-]

[138] Xie, H.; Shao, Y.; Becker, J.M.; Naider, F.; Gibbs, R.A. Synthesis and biological evaluation of the geometric farnesylated analogues of the a-factor mating peptide of Saccharomyces cerevisiae. *J. Org. Chem.*, **2000**, *65*(25), 8552-8563.
[http://dx.doi.org/10.1021/jo000942m] [PMID: 11112575]

[139] Mu, Y.; Eubanks, L.M.; Poulter, C.D.; Gibbs, R.A. Coupling of isoprenoid triflates with organoboron nucleophiles: synthesis and biological evaluation of geranylgeranyl diphosphate analogues. *Bioorg. Med. Chem.*, **2002**, *10*(5), 1207-1219.
[http://dx.doi.org/10.1016/S0968-0896(01)00390-X] [PMID: 11886785]

[140] Kobayashi, S.; Takei, H.; Mukaiyama, T. The stereospecific preparation of trisubstituted olefins. *Chem. Lett.*, **1973**, *2*, 1097-1100.
[http://dx.doi.org/10.1246/cl.1973.1097]

[141] Kobayashi, S.; Mukaiyama, T. The stereospecifique preparation of methyl geranate and synthetic precursors of C18- and C17-Juvenile hormones. *Chem. Lett.*, **1974**, *3*, 1425-1428.
[http://dx.doi.org/10.1246/cl.1974.1425]

[142] Buchanan, G.S.; Cole, K.P.; Li, G.; Tang, Y.; You, L-F.; Hsung, R.P. Constructing the architecturally distinctive ABD-tricycle of phomactin A through an intramolecular oxa-[3+3] annulation strategy. *Tetrahedron*, **2011**, *67*(52), 10105-10118.
[http://dx.doi.org/10.1016/j.tet.2011.09.111] [PMID: 23750054]

[143] Manabe, A.; Ohfune, Y.; Shinada, T. LStereoselective total syntheses of insect juvenile hormones JH 0 and JH I. *Synlett*, **2012**, *23*, 1213-1216.
[http://dx.doi.org/10.1055/s-0031-1290803]

[144] Totsuka, Y.; Ueda, S.; Kuzuyama, T.; Shinada, T. Facile synthesis of deuterium-labelled geranylgeraniols. *Bull. Chem. Soc. Jpn.*, **2015**, *88*, 575-577.
[http://dx.doi.org/10.1246/bcsj.20140384]

[145] Miyaura, N.; Ishiyama, T.; Sasaki, H.; Ishikawa, M.; Satoh, M.; Suzuki, A. Palladium-catalyzed inter- and intramolecular cross-coupling reactions of B-Alkyl-9-borabicyclo[3.3.1]nonane derivatives with 1-halo-1-alkenes or haloarenes. Syntheses of functionalized alkenes, arenes, and cycloalkenes *via* a hydroboration-coupling sequence. *J. Am. Chem. Soc.*, **1989**, *111*(1), 314-321.
[http://dx.doi.org/10.1021/ja00183a048]

[146] Surendra, K.; Corey, E.J. Rapid and enantioselective synthetic approaches to germanicol and other pentacyclic triterpenes. *J. Am. Chem. Soc.*, **2008**, *130*(27), 8865-8869.
[http://dx.doi.org/10.1021/ja802730a] [PMID: 18597440]

[147] Rand, C.L.; Van Horn, D.E.; Moore, M.W.; Negishi, E-i. A versatile and selective route to difunctional trisubstituted (*E*)-alkene synthons *via* zirconium-catalyzed carboalumination of alkynes. *J. Org. Chem.*, **1981**, *46*(20), 4096-4097.
[http://dx.doi.org/10.1021/jo00333a041]

[148] Negishi, E.; Liou, S-Y.; Xu, C.; Huo, S. A novel, highly selective, and general methodology for the synthesis of 1,5-diene-containing oligoisoprenoids of all possible geometrical combinations exemplified by an iterative and convergent synthesis of coenzyme Q(10). *Org. Lett.*, **2002**, *4*(2), 261-264.
[http://dx.doi.org/10.1021/ol010263d] [PMID: 11796065]

[149] Lipshutz, B.H.; Mollard, P.; Pfeiffer, S.S.; Chrisman, W. A short, highly efficient synthesis of

coenzyme Q_{10}. *J. Am. Chem. Soc.*, **2002**, *124*(48), 14282-14283.
[http://dx.doi.org/10.1021/ja021015v] [PMID: 12452683]

[150] Wang, G.; Negishi, E.I. $AlCl_3$-promoted facile *E*-to-*Z* isomerization route to (*Z*)-2-methyl-1-buten-1, 4-ylidene synthons for highly efficient and selective (*Z*)-isoprenoid synthesis. *Eur. J. Org. Chem.*, **2009**, *2009*(11), 1679-1682.
[http://dx.doi.org/10.1002/ejoc.200801188] [PMID: 24307863]

[151] Wadman, S.; Whitby, R.; Yeates, C.; Kocienski, P.; Cooper, K. An Efficient and Stereoselective Synthesis of homoallylic alcohols *via* nickel-catalysed coupling of 5-alkyl-2,3-dihydrofurans with Grignard reagents. *J. Chem. Soc. Chem. Commun.*, **1987**, 241-243.
[http://dx.doi.org/10.1039/c39870000241]

[152] Kocienski, P.; Wadman, S.A. Highly stereoselective and iterative approach to isoprenoid chains: Synthesis of homogeraniol, homofarnesol, and homogeranylgeraniol. *J. Org. Chem.*, **1989**, *54*(5), 1215-1217.
[http://dx.doi.org/10.1021/jo00266a047]

[153] Temple, K. J.; Wright, E. N.; Fierke, C. A.; Gibbs, R. A. Synthesis of non-natural, frame-shifted isoprenoid diphosphate analogues. *Org. Lett,* **2016**, *18*, 6038-6041.
[http://dx.doi.org/10.1021/acs.orglett.6b02977]

[154] Naruta, Y.; Maruyama, K. Highly regioselective addition of allylstannanes to vinyl epoxides by Lewis acid mediation. *Chem. Lett.*, **1987**, *16*, 963-966.
[http://dx.doi.org/10.1246/cl.1987.963]

[155] Mikami, K.; Kishi, N.; Nakai, T.; Fujita, Y. New sigmatropic sequences based on the [2,3] Wittig rearrangement of bis(allylic) ethers. A general approach to regiocontrolled C-C bond formation of allylic moieties leading to unsaturated carbonyl compounds. *Tetrahedron,* **1986**, *42*, 2911-2918.
[http://dx.doi.org/10.1016/S0040-4020(01)90580-0]

[156] Barrero, A.F.; Altarejos, J.; Alvarez-Manzaneda, E.J.; Ramos, J.M.; Salido, S. Synthesis of (±)-ambrox from (E)-nerolidol and β-ionone *via* allylic alcohol [2,3] sigmatropic rearrangement. *J. Org. Chem.*, **1996**, *61*(6), 2215-2218.
[http://dx.doi.org/10.1021/jo951908o]

[157] Terao, S.; Shiraishi, M.; Kato, K.; Ohkawa, S.; Ashida, Y.; Maki, Y. Quinones. Part 2. General synthetic routes to quinone derivatives with modified polyprenyl side chains and the inhibitory effects of these quinones on the generation of the slow reacting substance of anaphylaxis (SRS-A). *J. Chem. Soc., Perkin Trans. 1,* **1982**, 2909-2920.
[http://dx.doi.org/10.1039/p19820002909]

[158] Johnson, W.S.; Werthemann, L.; Bartlett, W.R.; Brocksom, T.J.; Li, T-t.; Faulkner, D.J.; Petersen, M.R. A simple steresoselective version of the Claisen rearrangement leading to *trans*-trisubstituted olefinic bonds. Synthesis of squalene. *J. Am. Chem. Soc.*, **1970**, *92*(3), 741-743.
[http://dx.doi.org/10.1021/ja00706a074]

[159] Takayanagi, H.; Sugiyama, S.; Morinaka, Y. A ketal Claisen rearrangement for α-ketol isoprene unit elongation: application to a practical synthesis of sarcophytol A intermediate. *J. Chem. Soc. Perkin Trans I,* **1995**, 751-756.
[http://dx.doi.org/10.1039/P19950000751]

[160] Baeckström, P.; Li, L. Syntheses of all-*trans* acyclic isoprenoid pheromone components. *Tetrahedron,* **1991**, *47*, 6533-6538.
[http://dx.doi.org/10.1016/S0040-4020(01)86580-7]

[161] Honda, K.; Tabuchi, M.; Kurokawa, H.; Asami, M.; Inoue, S. Stereocontrolled synthesis of acyclic terpenoids *via* N-ylide [2,3]rearrangement of ammonium salts with the stereodefined isoprene unit. *J. Chem. Soc. Perkin Trans. 1,* **2002**, 1387-1396.
[http://dx.doi.org/10.1039/b201304e]

[162] Masaki, Y.; Sakuma, K.; Kaji, K. Regio-and stereoselective terminal allylic carboxymethylation of

gem-dimethyl olefins. Synthesis of biologically important linear degraded terpenoids. *Chem. Pharm. Bull. (Tokyo),* **1985**, *33*, 1930-1940.
[http://dx.doi.org/10.1248/cpb.33.1930]

[163] Grigorieva, N.Y.; Avrutov, I.M.; Semenovsky, A.V. Novel approach to the stereoselective synthesis of polyprenols *via* directed aldol condensation. Preparation of heptaprenols ωtttcccOH and ωtttcctOH. *Tetrahedron Lett.,* **1983**, *24*, 5531-5534.
[http://dx.doi.org/10.1016/S0040-4039(00)94132-7]

[164] Bakkestuen, A.K.; Gundersen, L-L.; Petersen, D.; Utenova, B.T.; Vik, A. Synthesis and antimycobacterial activity of agelasine E and analogs. *Org. Biomol. Chem.,* **2005**, *3*(6), 1025-1033.
[http://dx.doi.org/10.1039/b417471b] [PMID: 15750645]

[165] Inoue, M.; Araki, H.; Yokota, K. Method for preparation of carboxylic acid prenyl esters and phenols by rearrangement of allyl alcohols using bis(b-diketonato)dioxomolybdenum complex as catalyst. Jpn. Kokai, JP 2017071557 A 2017 April 13;

[166] Inoue, M.; Araki, H.; Yokota, K.; Katsuhiro, S. Method for producing prenols and prenyl esters. Jpn. Kokai, JP 2016190790 A 2016 November 10;

[167] Barrero, A.F.; Alvarez-Manzaneda, E.J.; Chahboun, R.; Coral Páiz, M. A new enantiospecific route toward monocarbocyclic terpenoids: Synthesis of (-)-caparrapi oxide. *Tetrahedron Lett.,* **1998**, *39*, 9543-9544.
[http://dx.doi.org/10.1016/S0040-4039(98)02119-4]

[168] Yuasa, Y.; Yuasa, Y. Stereoselective synthesis of (2*E*, 6*E*, 10*E*)-geranylgeraniol from geranyllinalyl acetate *via* palladium-catalyzed amination. *Synth. Commun.,* **2006**, *36*, 1671-1677.
[http://dx.doi.org/10.1080/00397910600616610]

[169] Sreekumar, C.; Darst, K.P.; Clark Still, W. A direct synthesis of Z-trisubstituted allylic alcohols *via* the Wittig reaction. *J. Org. Chem.,* **1980**, *45*(21), 4260-4262.
[http://dx.doi.org/10.1021/jo01309a051]

[170] Wu, B.; Woodward, R.; Wen, L.; Wang, X.; Zhao, G.; Wang, P.G. Synthesis of a comprehensive polyprenol library for the evaluation of bacterial enzyme lipid substrate specificity. *Eur. J. Org. Chem.,* **2013**, *2013*(36), 8162-8173.
[http://dx.doi.org/10.1002/ejoc.201301089] [PMID: 24511260]

[171] Hesek, D.; Lee, M.; Zajícek, J.; Fisher, J. F.; Mobashery, S. Synthesis and NMR characterization of (Z,Z,Z,Z,E,E,ω)-heptaprenol. *J. Am. Chem. Soc.,* **2012**, *134*(33), 13881-13888.
[http://dx.doi.org/10.1021/ja306184m]

[172] Sato, K.; Miyamoto, O.; Inoue, S.; Furusawa, F.; Matsuhashi, Y. General method of stereospecific synthesis of natural polyprenols. Synthesis of betualprenol-6,-7,-8, and -9. *Chem. Lett.,* **1984**, *13*, 1105-1108.
[http://dx.doi.org/10.1246/cl.1984.1105]

[173] Sato, K.; Miyamoto, O.; Inoue, S. Stereoselective synthesis of a cisoid C10 isoprenoid building block and some all-cis-polyprenols. *Chem. Lett.,* **1983**, *12*, 725-728.
[http://dx.doi.org/10.1246/cl.1983.725]

[174] van Tamelen, E.E. Bioorganic chemistry: Sterols and acyclic terpene terminal epoxides. *Acc. Chem. Res.,* **1968**, *1*, 111-120.
[http://dx.doi.org/10.1021/ar50004a003]

[175] Hauptfleisch, R.; Franck, B. Stereoselectivive syntheses of 1,24-dihydroxy squalene 2,3;22,23-dioxides by double sharpless epoxidation. *Tetrahedron Lett.,* **1997**, *38*, 383-386.
[http://dx.doi.org/10.1016/S0040-4039(96)02332-5]

[176] Ceruti, M.; Viola, F.; Dosio, F.; Cattel, L.; Bouvier-Navé, P.; Ugliengo, P. Stereospecific synthesis of squalenoid epoxide vinyl ethers as Inhibitors of 2,3-oxidosqualene cyclase. *J. Chem. Soc. Perkin Trans. 1,* **1988**, 461-469.

[http://dx.doi.org/10.1039/P19880000461]

[177] Abad, J-L.; Casas, J.; Sánchez-Baeza, F.; Messeguer, A. Dioxidosqualenes: Characterization and activity as inhibitors of 2,3-oxidosqualene-lanosterol cyclase. *J. Org. Chem.*, **1993**, *58*(15), 3991-3997.
[http://dx.doi.org/10.1021/jo00067a036]

[178] Crispino, G.A.; Sharpless, K.B. Asymmetric dihydroxylation of squalene. *Tetrahedron Lett.*, **1992**, *33*, 4273-4274.
[http://dx.doi.org/10.1016/S0040-4039(00)74236-5]

[179] Sobot, D.; Mura, S.; Rouquette, M.; Vukosavljevic, B.; Cayre, F.; Buchy, E.; Pieters, G.; Garcia-Argote, S.; Windbergs, M.; Desmaële, D.; Couvreur, P. Circulating Lipoproteins: A Trojan Horse guiding squalenoylated drugs to LDL-accumulating cancers cells. *Mol. Ther.*, **2017**, *25*(7), 1596-1605.
[http://dx.doi.org/10.1016/j.ymthe.2017.05.016] [PMID: 28606375]

[180] Buchy, E.; Vukosavljevic, B.; Windbergs, M.; Sobot, D.; Dejean, C.; Mura, S.; Couvreur, P.; Desmaële, D. Synthesis of a deuterated probe for the confocal Raman microscopy imaging of squalenoyl nanomedicines. *Beilstein J. Org. Chem.*, **2016**, *12*, 1127-1135.
[http://dx.doi.org/10.3762/bjoc.12.109] [PMID: 27559365]

[181] Cornforth, J.W.; Cornforth, R.H.; Mathew, K.K. A Stereoselective synthesis of squalene. *J. Chem. Soc.*, **1959**, 2539-2547.
[http://dx.doi.org/10.1039/jr9590002539]

[182] Bhalerao, U.T.; Rapoport, H. Stereospecific synthesis of 2,7-dimethyl-trans,trans-2,6-octadiene-1, 8-dial, a tail-to-tail all-trans bifunctional isoprenoid synthetic unit. Convenient synthesis of squalene. *J. Am. Chem. Soc.*, **1971**, *93*(20), 5311-5313.
[http://dx.doi.org/10.1021/ja00749a087]

[183] Biellmann, J.F.; Ducep, J.B. Synthèse du squalène par couplage queue à queue. *Tetrahedron Lett.*, **1969**, *10*, 3707-3710.
[http://dx.doi.org/10.1016/S0040-4039(01)88493-8]

[184] Biellmann, J.F.; Ducep, J.B. Synthese du squalène et d'analogues. *Tetrahedron*, **1971**, *27*, 5861-5872.
[http://dx.doi.org/10.1016/S0040-4020(01)91751-X]

[185] Mohri, M.; Kinoshita, H.; Inomata, K.; Kotake, H. Palladium-catalyzed regio- and stereoselective desulfonylation of allylic sulfones with LiBHEt$_3$. Application to the synthesis of squalene. *Chem. Lett.*, **1985**, *14*, 451-454.
[http://dx.doi.org/10.1246/cl.1985.451]

[186] Kim, H.J.; Su, L.; Jung, H.; Koo, S. Selective deoxygenation of allylic alcohol: stereocontrolled synthesis of lavandulol. *Org. Lett.*, **2011**, *13*(10), 2682-2685.
[http://dx.doi.org/10.1021/ol200779y] [PMID: 21510630]

[187] Faulkner, D.J.; Petersen, M.R. Application of the claisen rearrangement to the synthesis of trans trisubstituted olefinic bonds. Synthesis of squalene and insect juvenile hormone. *J. Am. Chem. Soc.*, **1973**, *95*(2), 553-563.
[http://dx.doi.org/10.1021/ja00783a040] [PMID: 4687674]

[188] Werthemann, L.; Johnson, W.S. Application of the chloro ketal Claisen reaction to the total synthesis of squalene. *Proc. Natl. Acad. Sci. USA*, **1970**, *67*(3), 1465-1467.
[http://dx.doi.org/10.1073/pnas.67.3.1465] [PMID: 5274470]

[189] Lindel, T.; Franck, B. Synthesis and biomimetic rearrangement of a chiral diterpene dioxide. *Tetrahedron Lett.*, **1995**, *36*, 9465-9468.
[http://dx.doi.org/10.1016/0040-4039(95)02066-7]

[190] Rodríguez-López, J.; Pinacho Crisóstomo, F.; Ortega, N.; López-Rodríguez, M.; Martín, V.S.; Martín, T. Epoxide-opening cascades triggered by a Nicholas reaction: total synthesis of teurilene. *Angew. Chem. Int. Ed. Engl.*, **2013**, *52*(13), 3659-3662.
[http://dx.doi.org/10.1002/anie.201209159] [PMID: 23436322]

[191] Mc Murry, J.E.; Silvestri, M. A Simplified method for the titanium(II)-induced coupling of allylic and benzylic alcohols. *J. Org. Chem.*, **1975**, *40*(18), 2687-2688.
[http://dx.doi.org/10.1021/jo00906a027]

[192] Sharpless, K.B.; Hanzlik, R.P.; van Tamelen, E.E. A one-step Synthesis of 1,5-dienes involving reductive coupling of allyl alcohols. *J. Am. Chem. Soc.*, **1968**, *90*(1), 209-210.
[http://dx.doi.org/10.1021/ja01003a037]

[193] Momose, D-I.; Iguchi, K.; Sugiyama, T.; Yamada, Y. reaction of organic halides with chlorotris (triphenylphosphine)cobalt (I). *Chem. Pharm. Bull. (Tokyo)*, **1984**, *32*, 1840-1853.
[http://dx.doi.org/10.1248/cpb.32.1840]

[194] Sasaoka, S-i.; Yamamoto, T.; Kinoshita, H. Palladium catalyzed coupling of allylic acetate with zinc. *Chem. Lett.*, **1985**, *14*, 315-318.
[http://dx.doi.org/10.1246/cl.1985.315]

[195] Masuyama, Y.; Otake, K.; Kurusu, Y. Hexacarbonylmolybdenum(0) catalyzed reductive coupling of allylic acetates. *Bull. Chem. Soc. Jpn.*, **1987**, *60*, 1527-1528.
[http://dx.doi.org/10.1246/bcsj.60.1527]

[196] Kitagawa, Y.; Oshima, K.; Yamamoto, H.; Nozaki, H. A new stereospecific synthesis of 1.5-dienes. *Tetrahedron Lett.*, **1975**, *16*, 1859-1862.
[http://dx.doi.org/10.1016/S0040-4039(00)75277-4]

[197] Yanagisawa, A.; Hibino, H.; Habaue, S.; Hisada, Y.; Yasue, K.; Yamamoto, H. regio- and stereoselective synthesis of 1,5-dienes using allylic barium reagents. *Bull. Chem. Soc. Jpn.*, **1995**, *68*, 1263-1268.
[http://dx.doi.org/10.1246/bcsj.68.1263]

[198] Barrero, A.F.; Herrador, M.M.; del Moral, J.F.; Arteaga, P.; Arteaga, J.F.; Diéguez, H.R.; Sánchez, E.M. Mild TiIII- and Mn/ZrIV-catalytic reductive coupling of allylic halides: efficient synthesis of symmetric terpenes. *J. Org. Chem.*, **2007**, *72*(8), 2988-2995.
[http://dx.doi.org/10.1021/jo062630a] [PMID: 17375959]

[199] Corey, E.J.; Noe, M.C.; Shieh, W-C. A short and convergent enantioselective synthesis of 2,3-oxidosqualene. *Tetrahedron Lett.*, **1993**, *34*, 5995-5998.
[http://dx.doi.org/10.1016/S0040-4039(00)61710-0]

[200] Hernández-Camacho, J.D.; Bernier, M.; López-Lluch, G.; Navas, P. Coenzyme Q_{10} Supplementation in Aging and Disease. *Front. Physiol.*, **2018**, *9*, 44.
[http://dx.doi.org/10.3389/fphys.2018.00044] [PMID: 29459830]

[201] Fieser, L.F.; Campbell, W.P.; Fry, E.M.; Gates, M.D., Jr Synthetic approach to vitamin K1. *J. Am. Chem. Soc.*, **1939**, *61*(9), 2559-2559.

[202] Duralski, A.A.; Watts, A. Synthesis of isotopically labelled ubiquinones. *Tetrahedron Lett.*, **1992**, *33*, 4983-4984.
[http://dx.doi.org/10.1016/S0040-4039(00)61251-0]

[203] Suhara, Y.; Murakami, A.; Kamao, M.; Mimatsu, S.; Nakagawa, K.; Tsugawa, N.; Okano, T. Efficient synthesis and biological evaluation of omega-oxygenated analogues of vitamin K2: study of modification and structure-activity relationship of vitamin K2 metabolites. *Bioorg. Med. Chem. Lett.*, **2007**, *17*(6), 1622-1625.
[http://dx.doi.org/10.1016/j.bmcl.2006.12.082] [PMID: 17239598]

[204] Srinivasa Rao, A.; Bhavani, R.; Basava Raju, D. A new method of synthesis of coenzyme Q10 from isolated solanesol from tobacco waste. *Int. J. Pharma Sci.*, **2014**, *6*, 499-502.

[205] Hamamura, K.; Yamatsu, I.; Minami, N.; Yamagishi, Y.; Inai, Y.; Kijima, S.; Nakamura, T. Synthesis of [3'-^{14}C] coenzyme Q_{10}. *J. Labelled Comp. Radiopharm.*, **2002**, *45*, 823-829.
[http://dx.doi.org/10.1002/jlcr.588]

[206] Naruta, Y. Regio- and stereoselective synthesis of coenzymes Qn (n = 2-10), Vitamin K, and related polyprenylquinones. *J. Org. Chem.,* **1980**, *45*(21), 4097-4104.
[http://dx.doi.org/10.1021/jo01309a006]

[207] Baj, A.; Wałejlo, P.; Kutner, A.; Kaczmarek, L.; Morzycki, J.W.; Witkowski, S. Convergent synthesis of menaquinone-7 (MK-7). *Org. Process Res. Dev.,* **2016**, *20*, 1026-1033.
[http://dx.doi.org/10.1021/acs.oprd.6b00037]

[208] Zheng, Y-F.; Lin, J-D.; Li, C-P.; Li, J-H. Friedel-Crafts allylation of 2-(benzyloxy)-3,4-5-trimethoxytoluene catalyzed by a metal trifluoromethanesulfonate salt in synthesis of coenzyme Q10. *J. Chem. Res.,* **2007**, *12*, 686-688.
[http://dx.doi.org/10.3184/030823407X270338]

[209] Min, J-H.; Lee, J-S.; Yang, J-D.; Koo, S. The Friedel-Crafts allylation of a prenyl group stabilized by a sulfone moiety: expeditious syntheses of ubiquinones and menaquinones. *J. Org. Chem.,* **2003**, *68*(20), 7925-7927.
[http://dx.doi.org/10.1021/jo0350155] [PMID: 14510583]

[210] Keinan, E.; Eren, D. Total synthesis of linear polyprenols. 2. Improved preparation of the aromatic nucleus of ubiquinone. *J. Org. Chem.,* **1987**, *52*(17), 3872-3875.
[http://dx.doi.org/10.1021/jo00226a028]

[211] Skattebøl, L.; Aukrust, I. R.; Sandberg, M. Process for the preparation of vitamin K2. WO2011/117324 A2, September 29, **2011**.

[212] Roy, M.; Upare, A. A.; Chavan, A. A.; Karnalkar, D. R. Novel intermediates, process for their preparation and process for the preparation of COQ_{10} employing the said novel intermediates. WO 2007/129269A2, November 15, **2007**

[213] Yu, X-J.; Dai, H-F.; Chen, F-E. Synthetic Studies on Coenzyme Q_{10}. *Helv. Chim. Acta,* **2007**, *90*, 967-971.
[http://dx.doi.org/10.1002/hlca.200790099]

[214] Mechelke, M.F.; Wiemer, D.F. Synthesis of farnesol analogues through Cu(I)-mediated displacements of allylic THP ethers by Grignard reagents. *J. Org. Chem.,* **1999**, *64*(13), 4821-4829.
[http://dx.doi.org/10.1021/jo990161p] [PMID: 11674556]

[215] Demotie, A.; Fairlamb, I.J.S.; Lu, F-J.; Shaw, N.J.; Spencer, P.A.; Southgate, J. Synthesis of jaspaquinol and effect on viability of normal and malignant bladder epithelial cell lines. *Bioorg. Med. Chem. Lett.,* **2004**, *14*(11), 2883-2887.
[http://dx.doi.org/10.1016/j.bmcl.2004.03.055] [PMID: 15125952]

[216] Dai, H-F.; Chen, F.E.; Yu, X.J. Synthetic studies on coenzyme Q_{10}. An efficient and improved synthesis of coenzyme Q_{10} via the C5+C45 approach. *Helv. Chim. Acta,* **2006**, *89*, 1317-1321.
[http://dx.doi.org/10.1002/hlca.200690130]

[217] Odejinmi, S.I.; Wiemer, D.F. Application of benzyl protecting groups in the synthesis of prenylated aromatic compounds. *Tetrahedron Lett.,* **2005**, *46*, 3871-3874.
[http://dx.doi.org/10.1016/j.tetlet.2005.03.187]

[218] Garcías, X.; Ballester, P.; Capó, M.; Saá, J.M. $^2\Delta$-Stereocontrolled Entry to (*E*)- or (2)-prenyl aromatics and quinones. Synthesis of menaquinone-4. *J. Org. Chem.,* **1994**, *59*(19), 5093-5096.
[http://dx.doi.org/10.1021/jo00096a069]

[219] Rüttimann, A.; Lorenz, P. Ein neuer synthetischer Zugang zu Ubichinonen. *Helv. Chim. Acta,* **1990**, *73*, 790-796.
[http://dx.doi.org/10.1002/hlca.19900730404]

[220] Sato, K.; Inoue, S.; Saito, K. A New synthesis of vitamin K via p-allylnickel intermediates. *J. Chem. Soc. Perkin,* **1973**, 2289-2293.
[http://dx.doi.org/10.1039/P19730002289]

[221] Jung, Y-S.; Joe, B-Y.; Seong, C-M.; Park, N-S. Synthesis of Ubiquinones utilizing Pd(0) catalyzed Stille coupling. *Bull. Korean Chem. Soc.*, **2000**, *21*, 463-464.

[222] Rosales, V.; Zambrano, J.L.; Demuth, M. Regioselective palladium-catalyzed alkylation of allylic halides with benzylic grignard reagents. Two-step synthesis of abietane terpenes and tetracyclic polyprenoid compounds. *J. Org. Chem.*, **2002**, *67*(4), 1167-1170.
[http://dx.doi.org/10.1021/jo010786z] [PMID: 11846658]

[223] Skattebøl, L.; Aukrust, I. R. Process for the preparation of vitamin K2. WO 2010/035000A1. April 01, **2010**.

[224] Lipshutz, B.H.; Lower, A.; Berl, V.; Schein, K.; Wetterich, F. An improved synthesis of the "miracle nutrient" coenzyme Q_{10}. *Org. Lett.*, **2005**, *7*(19), 4095-4097.
[http://dx.doi.org/10.1021/ol051329y] [PMID: 16146360]

[225] Bienaymé, H.; Ancel, J-E.; Meilland, P.; Simonato, J-P. Rhodium(I)-catalyzed addition of phenols to dienes. A new convergent synthesis of vitamin E. *Tetrahedron Lett.*, **2000**, *41*, 3339-3343.
[http://dx.doi.org/10.1016/S0040-4039(00)00381-6]

[226] Schmidt, J.; Stark, C. B. W. Synthetic endeavors toward 2□nitro-4-alkylpyrroles in the context of the total synthesis of heronapyrrole C and preparation of a carboxylate natural product analogue. *J. Org. Chem.*, **2014**, *79*(5), 1920-1928.
[http://dx.doi.org/10.1021/jo402240g]

[227] Boukouvalas, J.; Albert, V. Synthesis of the hypoxic signaling inhibitor furospongolide. *Synlett*, **2011**, 2541-2544.
[http://dx.doi.org/10.1055/s-0030-1260329]

[228] Fouquet, G.; Schlosser, M. Improved carbon-carbon linking by controlled copper catalysis. *Angew. Chem. Int. Ed. Engl.*, **1974**, *13*, 82-83.
[http://dx.doi.org/10.1002/anie.197400821]

[229] Hattori, H.; Yokoshima, S.; Fukuyama, T. Total Syntheses of Aurachins A and B. *Angew. Chem. Int. Ed. Engl.*, **2017**, *56*(24), 6980-6983.
[http://dx.doi.org/10.1002/anie.201702204] [PMID: 28471077]

[230] Corey, E.J.; Lee, J.; Roberts, B.E. The application of the Shapiro reaction to the stereoselective synthesis of E-trisubstituted olefins for cation-olefin cyclization by three component coupling. *Tetrahedron Lett.*, **1997**, *38*, 8915-8918.
[http://dx.doi.org/10.1016/S0040-4039(97)10424-5]

[231] Sulake, R.S.; Chen, C. Total synthesis of (+)-antroquinonol and (+)-antroquinonol D. *Org. Lett.*, **2015**, *17*(5), 1138-1141.
[http://dx.doi.org/10.1021/acs.orglett.5b00046] [PMID: 25679542]

[232] Liu, X.; Gujarathi, S.; Zhang, X.; Shao, L.; Boerma, M.; Compadre, C.M.; Crooks, P.A.; Hauer-Jensen, M.; Zhou, D.; Zheng, G. Synthesis of (2R,8′ S,3′ E)-δ-tocodienol, a tocoflexol family member designed to have a superior pharmacokinetic profile compared to δ-tocotrienol. *Tetrahedron*, **2016**, *72*(27-28), 4001-4006.
[http://dx.doi.org/10.1016/j.tet.2016.05.028] [PMID: 27773949]

[233] Jeso, V.; Micalizio, G. C. Total synthesis of Lehualide B by allylic alcohol-alkyne reductive cross-coupling. *J. Am. Chem. Soc.*, **2010**, *132*(33), 11422-11424.
[http://dx.doi.org/10.1021/ja104782u]

[234] Domingo, V.; Silva, L.; Diéguez, H.R.; Arteaga, J.F.; Quílez del Moral, J.F.; Barrero, A.F. Enantioselective total synthesis of the potent anti-inflammatory (+)-myrrhanol A. *J. Org. Chem.*, **2009**, *74*(16), 6151-6156.
[http://dx.doi.org/10.1021/jo901011m] [PMID: 19575536]

[235] Domingo, V.; Lorenzo, L.; Quilez del Moral, J.F.; Barrero, A.F. First synthesis of (+)-myrrhanol C, an anti-prostate cancer lead. *Org. Biomol. Chem.*, **2013**, *11*(4), 559-562.

[http://dx.doi.org/10.1039/C2OB26947C] [PMID: 23147747]

[236] Corbu, A.; Aquino, M.; Pratap, T.V.; Retailleau, P.; Arseniyadis, S. Enantioselective synthesis of iridal, the parent molecule of the iridal triterpenoid class. *Org. Lett.,* **2008,** *10*(9), 1787-1790.
[http://dx.doi.org/10.1021/ol8005425] [PMID: 18396888]

[237] Chen, C.-p.; Shivaji, S. R.; Huang, Z.-L. Use of antroquinonol for treating obesity and process for preparation of antroquinonol. WO 2016/037566A, March 17, **2016.**

[238] Krasovskiy, A.; Malakhov, V.; Gavryushin, A.; Knochel, P. Efficient synthesis of functionalized organozinc compounds by the direct insertion of zinc into organic iodides and bromides. *Angew. Chem. Int. Ed. Engl.,* **2006,** *45*(36), 6040-6044.
[http://dx.doi.org/10.1002/anie.200601450] [PMID: 16900548]

[239] Yang, Y.; Mustard, T.J.L.; Cheong, P.H-Y.; Buchwald, S.L. Palladium-catalyzed completely linear-selective Negishi cross-coupling of allylzinc halides with aryl and vinyl electrophiles. *Angew. Chem. Int. Ed. Engl.,* **2013,** *52*(52), 14098-14102.
[http://dx.doi.org/10.1002/anie.201308585] [PMID: 24353232]

[240] Barrero, A.F.; Cuerva, J.M.; Alvarez-Manzaneda, E.J.; Oltra, J.E.; Chahboun, R. First synthesis of achilleol A using titanium(III) chemistry. *Tetrahedron Lett.,* **2002,** *43*, 2793-2796.
[http://dx.doi.org/10.1016/S0040-4039(02)00358-1]

[241] Edwards, J.T.; Merchant, R.R.; McClymont, K.S.; Knouse, K.W.; Qin, T.; Malins, L.R.; Vokits, B.; Shaw, S.A.; Bao, D-H.; Wei, F-L.; Zhou, T.; Eastgate, M.D.; Baran, P.S. Decarboxylative alkenylation. *Nature,* **2017,** *545*(7653), 213-218.
[http://dx.doi.org/10.1038/nature22307] [PMID: 28424520]

[242] Chen, T-Y.; Krische, M.J. Regioselective ruthenium catalyzed hydrohydroxyalkylation of dienes with 3-hydroxy-2-oxindoles: prenylation, geranylation, and beyond. *Org. Lett.,* **2013,** *15*(12), 2994-2997.
[http://dx.doi.org/10.1021/ol401184k] [PMID: 23721207]

[243] Wang, G.; Franke, J.; Ngo, C.Q.; Krische, M.J. Diastereo- and enantioselective iridium catalyzed coupling of vinyl aziridines with alcohols: Site-selective modification of unprotected diols and synthesis of substituted piperidines. *J. Am. Chem. Soc.,* **2015,** *137*(24), 7915-7920.
[http://dx.doi.org/10.1021/jacs.5b04404] [PMID: 26074091]

[244] McNeill, E.; Ritter, T. 1,4-Functionalization of 1,3-dienes with low-valent iron catalysts. *Acc. Chem. Res.,* **2015,** *48*(8), 2330-2343.
[http://dx.doi.org/10.1021/acs.accounts.5b00050] [PMID: 26214092]

[245] Johns, A. M.; Pederson, R. L.; Kiser, R. C.; Nickel, A. Cross metathesis of poly-branched poly-olefins. US 2016/0107980 2016 April 21;

[246] Elias, H.G., Ed. *Macromolecules: Volume 1: Chemical Structures and Syntheses*; Wiley VCH, **2005.**

[247] Mooibroek, H.; Cornish, K. Alternative sources of natural rubber. *Appl. Microbiol. Biotechnol.,* **2000,** *53*(4), 355-365.
[http://dx.doi.org/10.1007/s002530051627] [PMID: 10803889]

[248] Germack, D.S.; Wooley, K.L. Isoprene polymerization *via* reversible addition fragmentation chain transfer polymerization. *J. Polym. Sci. A Polym. Chem.,* **2007,** *45*, 4100-4108.
[http://dx.doi.org/10.1002/pola.22226]

[249] Thomas, R.M.; Grubbs, R.H. Synthesis of telechelic polyisoprene *via* ring-opening metathesis polymerization in the presence of chain transfer agent. *Macromolecules,* **2010,** *43*(8), 3705-3709.
[http://dx.doi.org/10.1021/ma902749q]

[250] Desmaële, D.; Gref, R.; Couvreur, P. Squalenoylation: a generic platform for nanoparticular drug delivery. *J. Control. Release,* **2012,** *161*(2), 609-618.
[http://dx.doi.org/10.1016/j.jconrel.2011.07.038] [PMID: 21840355]

[251] Maksimenko, A.; Mougin, J.; Mura, S.; Sliwinski, E.; Lepeltier, E.; Bourgaux, C.; Lepêtre, S.; Zouhiri,

F.; Desmaële, D.; Couvreur, P. Polyisoprenoyl gemcitabine conjugates self assemble as nanoparticles, useful for cancer therapy. *Cancer Lett.,* **2013**, *334*(2), 346-353.
[http://dx.doi.org/10.1016/j.canlet.2012.08.023] [PMID: 22935679]

[252] Harrisson, S.; Nicolas, J.; Maksimenko, A.; Bui, D.T.; Mougin, J.; Couvreur, P. Nanoparticles with *in vivo* anticancer activity from polymer prodrug amphiphiles prepared by living radical polymerization. *Angew. Chem. Int. Ed. Engl.,* **2013**, *52*(6), 1678-1682.
[http://dx.doi.org/10.1002/anie.201207297] [PMID: 23255475]

CHAPTER 2

Monosubstituted Ferrocene-Containing Thermotropic Liquid Crystals

Irina Carlescu[1,*], Daniela Apreutesei Wilson[2,*], Gabriela Lisa[1], Nicolae Hurduc[1] and Dan Scutaru[1]

[1] *Faculty of Chemical Engineering and Environmental Protection, "Gheorghe Asachi" Technical University of Iasi, 73 Prof. dr.docent Dimitrie Mangeron street; 700050 Iași, România*

[2] *Systems Chemistry, Radboud University, Heyendaalseweg 135, 6525 AJ Nijmegen, The Netherlands*

Abstract: This chapter describes the influence of micromolecular structure on self-assembly macroscopic properties of liquid crystalline ordered systems and focuses on structure-properties relationships of some thermotropic monosubstituted ferrocenomesogens obtained in our research group. The structural features evaluated are the connectivity between ferrocene and mesogenic units, the number of aromatic rings and the types of linking and ending groups responsible for mesomorphic behavior. It is shown that systems based on monosubstituted ferrocene derivatives exhibit unique liquid crystalline properties which may be used for designing materials with specific functions and properties. The aim of this paper is to present essentially the work carried out in our laboratory over a period of 10 years therefore; this chapter will focus on the main conclusions drawn on our own synthesized ferrocenomesogens.

Keywords: Anisotropic molecular structure, Azobenzene, Azo-aromatic compounds, Calamitic, Cholesteryl esters, Cholesteric mesophases, Differential scanning calorimetry (DSC), Ferrocene, Ferrocenomesogens, Flexible aliphatic chains, Metallomesogens, Molecular modelling, Multifunctional materials, Nematic, Optical polarized microscopy (OPM), Schiff Bases, Smectic, Structure - mesomorphic properties relationship, Thermotropic liquid crystals, Thermal degradation.

---

[*] **Corresponding authors Irina Carlescu:** "Gheorghe Asachi" Technical University of Iasi, Faculty of Chemical Engineering and Environmental Protection, 73 Prof. dr. Docent Dimitrie Mangeron street; 700050 Iași, România; Tel: +40 - 232 278683; Fax: +40 - 232 271311; E-mail: icarlescu@ch.tuiasi.ro
Daniela Apreutesei Wilson: Radboud Univ Nijmegen, Inst Mol & Mat, Heyendaalseweg 135, NL-6525 AJ Nijmegen, The Netherlands; Tel: +31 (024) 365 3421; E-mail: d.wilson@science.ru.nl

Atta-ur-Rahman (Ed.)
All rights reserved-© 2018 Bentham Science Publishers

INTRODUCTION

Thermotropic liquid crystalline materials are omnipresent in many areas of modern life as they found wide applications in communications and consumer electronics: electro-optical flat panel display, mobile phones, watches or notebook computers. In addition to optoelectronics, liquid crystals are used as well in medicine or nanomedicine and material testing for thermography or in biotechnology, food science or pharmacology [1, 2]. Further applications are expected in the future. The main goal of modern liquid crystal research is to design new functional materials that incorporate units with specific properties (optical, redox, magnetic, chemical stability, good solubility) that combine with those of liquid crystals (ordering, anisotropy).

Introduction of ferrocene as a component in thermotropic liquid crystals (metallomesogens) is based on its remarkable properties, useful for multifunctional materials [3]. The physical properties of liquid crystals containing metals are based on the large polarizable electron density of metal that increases the birefringence of the mesogens. There are excellent available reviews that describe in detail metallomesogens with ferrocene moieties carried out up to 2017 [4 - 18]. In this type of molecules, the combination of different structural units contributes to the obtaining of special physical properties that became very important when materials with potential applicability are targeted. For practical uses, the compound does not have to only contain the necessary molecular structure required to promote mesomorphic properties at a certain temperature, but also an appropriate combination of physical properties. The structure - mesomorphic properties relationship has been described in detail in many studies [19 - 20].

Factors that contribute to the molecular structure are various and include core unit, linking groups, terminal groups, lateral groups, the length of flexible chain *etc*. The core units include rigid fragments based on polarizable and planar aromatic rings which are connected one to another directly or through linking groups, *e.g.* carboxyl, azo, azomethine or conformationally flexible methylene groups. Usually the mesogenic cores contain a terminal alkyl, alkyloxy or acyloxy group.

Although each structural unit influences the mesomorphic behavior, this is not enough; the magnitude of the dipole moment of the whole molecule, which determines the intermolecular forces, plays a very important role. Consequently, the appearance of mesophase depends on several cumulative factors: dipole – dipole attractions induced by permanent dipoles of the molecule, the overall polarizability and electronic conjugation effects, the shape and the size of the molecule.

The compounds synthesized in our group contain monosubstituted ferrocenyl units attached to rigid mesogenic moieties (MU) (consisting from two to four aromatic rings connected by esteric, azo or imino linkages) or cholesteryl units. The compounds include flexible aliphatic chains situated at the end or inside the structures.

The introduction of a metal atom into ordered systems allows the combination of liquid crystalline properties (fluidity) with its own attributes (color, electronic density, paramagnetism and polarisability) and leads to different geometries that cannot be met by organic compounds [3]. Ferrocene has several valuable properties that underly its vast uses: high reactivity towards electrophiles, although is a non-benzoid aromatic structure [21], good solubility in organic solvents, photochemically/thermally and especially chemically high stability due to the presence of iron that has the same electronic structure to krypton, with reversible change of the valence state, low toxicity and antimicrobial activity [22].

Although this bulky unit introduces a "kink" in the linear shape of monosubstituted ferrocenomesogens which may produce steric repulsions with the neighboring molecules as well as favorable molecular packing, nevertheless it induces the anisotropy of intermolecular interactions and promote the anisotropic geometry, which stabilizes liquid crystalline phases. Likewise, ferrocene provides a certain degree of structural rigidity, required in liquid crystals. Additionally, because of the very small rotation energy barrier, the sandwich structure of ferrocene can freely rotate about the C_5 axis and exists in the form of two conformations (Fig. **1**). While in solution at 98 K, the cyclopentadienyl rings possess an eclipsed geometry; in crystalline structures, the rings are rotated about 9 degrees into staggered conformation, identified at higher temperatures [23].

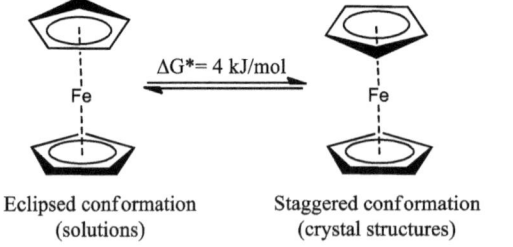

Fig. (1). Rotation of cyclopentadienyl rings in solution and in crystalline structures.

This phenomenon induces some degree of flexibility to ferrocenyl derivatives, which ensures lower transition temperatures, if compared to other transitions in metal-containing liquid crystals.

Following the systematic investigations of Loubser *et al.* on monosubstituted ferrocenomesogens [24, 25], our goal was to design elongated ferrocene derivatives, in order to vary the structural parameters that may promote and stabilize the liquid crystalline properties. According to the first X-ray crystal structure reported in 1993, two monosubstituted ferrocenomesogens molecules adopt a parallel arrangement, forming an extended S-shape association with the ferrocenyl-phenyl units in the middle and the end of each molecule orientated in opposite directions [26]. Such intermolecular association evidences that the presence of ferrocenyl group in terminal position favors the mesomorphism. In accordance with guidelines proposed by Loubser *et al.*, two main criteria are necessary in order to induce mesophase: 1) a long linear shape and 2) a large length – depth (l/d) ratio of a single molecule. To meet these criteria, it was established that a minimum of three aromatic rings in the mesogenic core are enough to stabilize a nematic phase. The insertion of at least one benzene ring between the ferrocenyl unit and the carbonyloxy linking group leads to a decrease of clearing temperature with the increase of thermal and photochemical stabilities, by attenuating the withdrawal effect of electron density from the ferrocenyl group. However, the mesogenic moiety determines the temperature domain where the liquid crystalline phase occurs and also the type of mesophase. Thus, if the number of aromatic ring increases from three to four, the nematic behavior is emphasized but often affects the thermal stability. In order to balance the rigidity with the flexibility of the molecule, terminal flexible substituents are needed.

Generally, the presence of cholesterol induces chiral mesophases, in which molecules are organized into helical structures, the pitch length of which is a function of the chemical structure, the nature and the concentration of the impurities in the system, temperature and external electric field. The molecular ordering of cholesteric liquid crystals induces particular optical properties (selective reflection of circularly polarized light, transmission of light, thermochromism, circular dichroism and electro-optic effect) that can be changed with applied electrical fields. On the whole, this makes cholesteryl derivates valuable candidates for applications in modern LC technology [27].

The presence of photo-controllable unit, such as an azo-aromatic mesogenic group in chiral/achiral mesophases allows an excellent control of mesophase color, with application in detecting, conversion and reproduction of color information [28 - 30].

Ferrocenomesogens based materials reported so far show different liquid crystalline mesophases such as nematic (N), smectic (S), twist-grain boundary (TGBA), blue (BP), columnar or cubic. However, the most common mesophases associated with monosubstituted ferrocene derivatives are the nematic, smectic A

(S_A) and chiral nematic (N*) or chiral smectic C (S_C*) respectively for derivatives with chiral center such as cholesterol (cholesteric phase, Ch). Otherwise liquid crystals presenting chiral smectic C phase gained increasing attention because of their utility in flat screen technology.

The purpose of this chapter is to provide a comprehensive survey of the results we obtained on liquid crystals containing ferrocene units since 2005, with emphasis on structure – mesomorphic properties – chemical thermostability relation. This work focuses on monosubstituted ferrocenomesogens, where ferrocene is rigidly or flexible connected to the calamitic mesogenic units, which contain mainly azobenzene or cholesterol units into its structure.

Ferrocenomesogens studied in our research group fit in the structural types shown in Fig. (2).

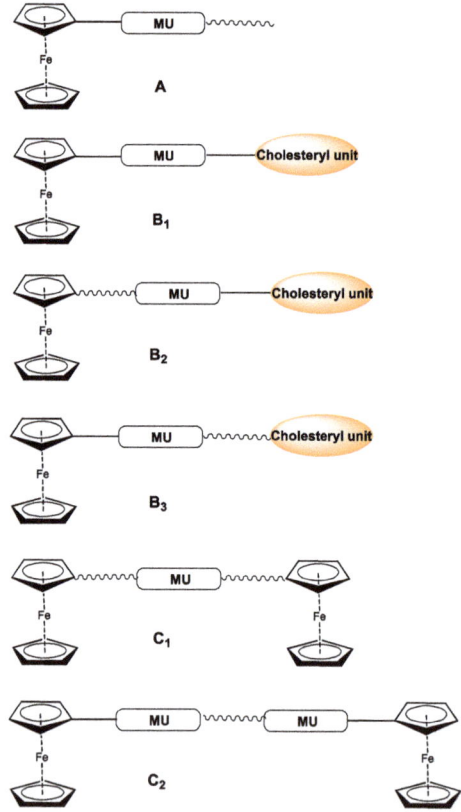

Fig. (2). General structures of ferrocene derivatives (MU= mesogenic unit).

In order to meet the above-mentioned criteria regarding the appearance of liquid crystalline behavior, based on the length/diameter ratio of the molecules, ferrocene derivatives were designed to incorporate systematically small structural changes. The relationship between structure, liquid crystalline properties and thermostability was investigated by comparing several parameters as the influence of ferrocene, mesogenic unit, presence of cholesterol, length and position of the flexible chain, influence of the bond/linking type of connecting groups and number of benzenic units. The introduction of azo and imine linking groups was meant to increase the polarities of the moieties, while maintaining the linear pattern, whereas the nature of the terminal substituents had the purpose to balance rigidity with flexibility of the molecule.

The synthetic strategies for obtaining ferrocene derivatives were based mainly on the synthesis of different structural units followed by a convergent assembly of the target molecules in a final step. The compounds were thus obtained either through reactions of esterification of the ferrocenyl or phenyl units with the mesogenic block, using DCC/DMAP method or through reactions of condensation of the ferrocenyl or phenyl amine with aromatic aldehydes containing the flexible terminal chain or cholesterol. The ferrocenyl units were obtained by the arylation of ferrocene with the corresponding diazonium salts, in conditions of phase transfer catalysis.

The confirmation of the synthesized structures was performed using spectroscopic methods (^{1}H-RMN, ^{13}C-RMN, FTIR, MS) as well as elemental analysis. The liquid crystalline properties were investigated by: optical polarized microscopy (OPM), differential scanning calorimetry (DSC), X-ray powder diffraction (XRD). Primary identification of mesophases was performed by polarizing microscopy. While DSC and optical microscopy analysis indicated the types of mesophases from the thermal effects of phase transitions, X-ray technique provided information on the type of molecular packing into the mesophase. Some aspects regarding the mechanism of thermal degradation of ferrocenomesogens were obtained by using a TG/FTIR/MS system.

Conformational theoretical studies were performed to get insight into the 3D structure of the synthesized ferrocenomesogens. In order to find out if monosubstituted ferrocenomesogens meet the linearity condition and present a suitable length – depth (l/d) ratio, molecular simulations were performed, by minimization of the individual molecules with the HYPERCHEM 4.5 program (Hypercube Inc.) or quantum DMol 3 module. The initial molecular conformation of the simulated products was optimized using a MM+ field force, when the total potential energy of the single molecule was obtained. In order to find out the real value for minimum energy (not a local minimum) the obtained conformation was followed by a molecular dynamic cycle and re-minimized. The criterion of energy convergence was to obtain a residual root-mean-square force in the simulated system of less than 0.05 kJ/ mol ·A°. Minimization was performed using the steepest-descent and conjugate-gradient algorithm described by Fletcher and Reeves. The systems consisting from several molecules were optimized with molecular mechanic Forcite module (Materials Studio 4.0).

LIQUID CRYSTALLINE FERROCENES WITH FERROCENE RIGIDLY CONNECTED TO MESOGENIC UNIT AND FLEXIBLE ALKYL CHAIN AS TERMINAL GROUP

The designing of calamitic (rod-like) ferrocene liquid crystals with a broad mesomorphic domain depends on carefully selection of structural elements such as: linear shape or anisotropic molecular structure, rigidity, high length/depth or optimal rigid/flexible ratios. The liquid crystalline properties are determined not only by the number of aromatic rings or the nature of linking groups in the rigid core, but also by the length of the flexible chain in the terminal group (compounds of **A** type).

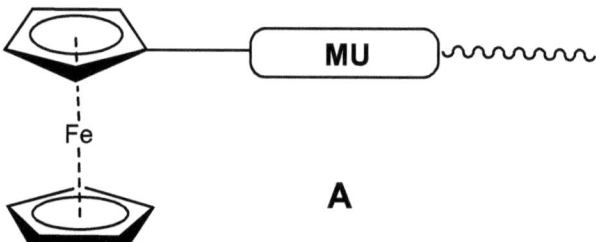

A

When mesogenic unit rigidly connected to ferrocene contains only two aromatic rings (compounds of **1** type), no liquid crystalline properties are present whatever the nature of terminal flexible group [31].

1

Compound	R	Transitions
1a	$-OC_{18}H_{37}$	K 112 I
1b	$-OC(O)C_{17}H_{35}$	K 115 I

K – crystalline, I – isotropic

Even though all these molecules have a linear shape, still the length – depth (l/d) ratio is too small to facilitate the appearance of the mesophase.

The increase in the length of the mesogenic unit by insertion of a third aromatic ring caused the increase of the anisotropic shape, which favored the appearance of mesophases (compounds **2** and **3** type) [32, 33]. The presence of two esteric linkages increases the molecular polarizability and consequently the intermolecular anisotropic forces of the molecules that stabilize the liquid crystalline phase, toward a single ester group, which display only monotropic phase (compound **2b**). However, given that the alkoxy group is stereochemically equivalent to a methylene unit, this contributes to the decreasing of the melting point as well. Permanent removal of the flexible terminal chain results in a disappearing of the mesogenic behavior.

2

Compound	R	Transitions*
2a	-OC$_{18}$H$_{37}$	K [135 N] 136 I
2b	-OC(O)C$_{17}$H$_{35}$	K 128 N 145 I
2c	-	K 194 I

*transition in square brackets refers to monotropic phase

Although the position of azo linking group is changed (in compound **3b**), the liquid crystalline behavior remains almost the same with compound **2a**, with slightly increase of melting point for compound **3b**.

3

Compound	R	Transitions
3a	-OC$_9$H$_{19}$	K$_1$ 155 K$_2$ 161 N 173 I
3b	-OC$_{18}$H$_{37}$	K [138 N] 138 I
3c	-OC(O)C$_7$H$_{15}$	K$_1$ 136 K$_2$ 181 N 191 I
3d	-OC(O)C$_9$H$_{19}$	K$_1$ 89 K$_2$ [175 N] 174 I
3e	-OC(O)C$_{17}$H$_{35}$	K 147 I
3f	-OC(O)-cis-9-C$_{17}$H$_{33}$	K 143 I
3g	-OC(O)-trans-9-C$_{17}$H$_{33}$	K 154 I

However, the mesomorphic behavior disappears when R is replaced by -OC(O)C$_{17}$H$_{35}$ (compound **3e**), meaning that the direct connection of the azobenzene unit to ferrocene moiety and the presence of two esteric linking groups restrains the favorable intermolecular packing and prevents the appearance of a mesophase.

A further study was performed on compounds of **3** type, where the number of carbon atoms on flexible terminal chain was varied from 6 to 10 for alkyloxy chain and from 4 to 9 for acyloxy ones [33]. The results showed enantiotropic behavior only for compounds where R = -OC$_9$H$_{19}$ and -OC(O)C$_7$H$_{15}$. On both DSC thermograms, endothermic peaks corresponding to crystalline – crystalline transitions are observed on heating, before the appearance of crystalline – liquid crystal peaks, probably because of molecular rearrangements (Fig. **3**). The same phenomenon appears on cooling as well, after liquid crystalline – crystalline transition, the thermal effect of the exothermic peaks being much smaller.

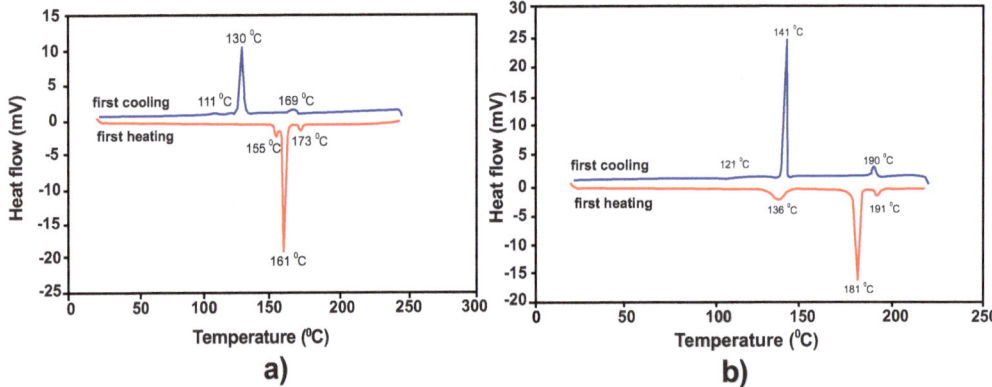

Fig. (**3**). DSC thermograms: a) compound **3a** and b) compound **3c**.

All other compounds exhibited only a monotropic nematic behavior, as a consequence of a better packing in the crystalline phase (reflected in the presence of polymorphism phenomenon) and unfavorable rigid/flexible ratio. The nematic domain on cooling is relatively short and increases with the number of carbon atoms from the terminal chain up to 8 and starts to drop with chains longer than 9 (Fig. **4**).

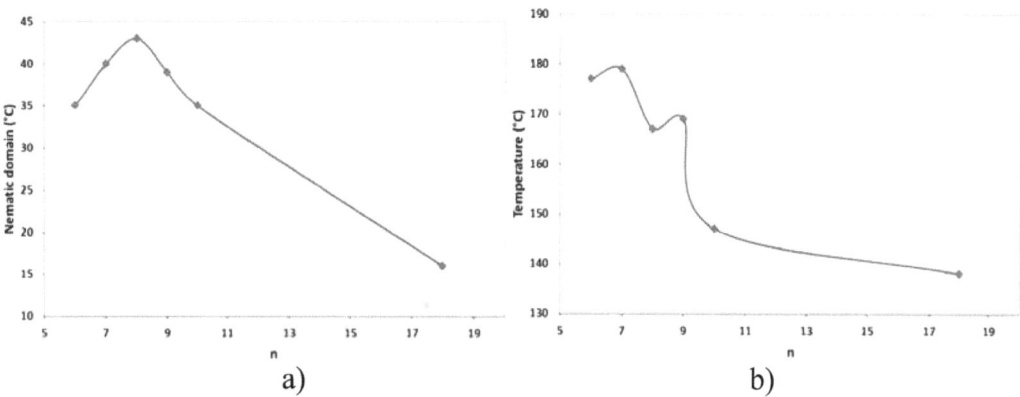

Fig. (4). Variation of a) nematic domain on cooling and b) isotropic-nematic temperature transition with the number of carbon atoms on the flexible chain.

Although ferrocene plays an important role in promoting liquid crystalline properties, the presence of three-dimensional bulky unit decreases the molecular interactions in between layers by steric repulsion, inducing just an orientation ordering and nematic phase appearance respectively. For this reason, most of compounds show predominantly only a monotropic nematic phase. In such homologous series, the liquid crystalline properties are induced by van der Waals dispersion forces of the hydrocarbonate chains. Although the using of a long alkyl chain generally reduces the steric repulsion of ferrocenyl unit, the last compound of the series with 18 carbon atoms shows an unstable monotropic phase with a very short nematic domain.

As seen from Fig. (5) the monosubstituted cyclopentadienyl and benzene rings are slightly rotated one to another, with a certain angle that holds the planarity, while the cyclopentadienyl rings from ferrocene are eclipsed. The azo unit maintains the planarity of the two aromatic cycles and the esteric linkage rotates the plane of the third aromatic cycle substituted with the alkyl chain. Overall, the molecular shape is linear, with a longer axis than another, which fulfills the condition for mesophase occurrence.

Therefore, the measured molecular parameters as length, diameter, dipole moment and molecular asymmetry coefficient by molecular modelling studies were in agreement with the mesomorphic behavior. The presence of two ester linking groups increases the polarity of the molecules, with a slight increase in the isotropization temperature.

The nematic mesophase has been evidenced with polarized optical microscope by Schlieren textures and characteristic nematic droplets. The nematic droplets appear on cooling from the isotropic phase (Fig. **6a**) joining into nematic thread-like texture (Fig. **6b**) and Schlieren texture respectively, (Fig. **6c**) as the temperature decreases, where fourfold curved dark brushes are evidenced.

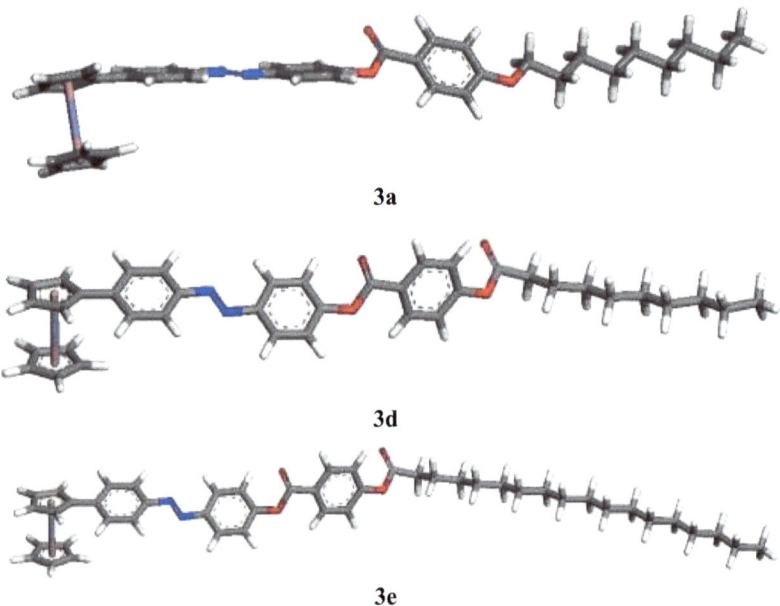

Fig. (5). Geometry of the compound of **3** type.

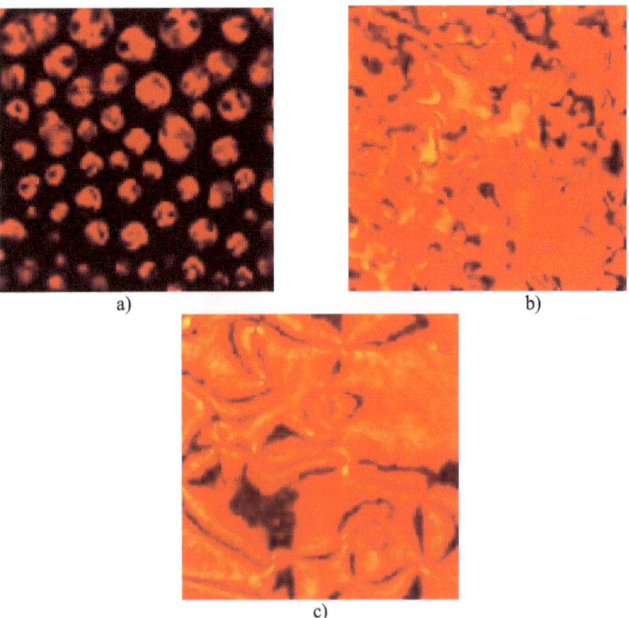

Fig. (6). Optical photomicrographs of compounds **3**: a) **3a**, nematic droplets; b) **3b**, nematic thread-like texture and c) **3c**, characteristic nematic Schlieren textures.

The ordering of monosubstituted ferrocenomesogens into stable mesophases is the consequence of some molecular association between mesogens [34]. In the case of **3** type compound, four theoretical stacking arrangements were considered: parallel (head to head or head to tail) and antiparallel (rectangle form and with perpendicular molecular axis) (Fig.7) [33]. From these, the most stable arrangement is the parallel head to head one, which allows the insertion of tails of other molecules in the free space formed for optimum packaging (Fig. 7a).

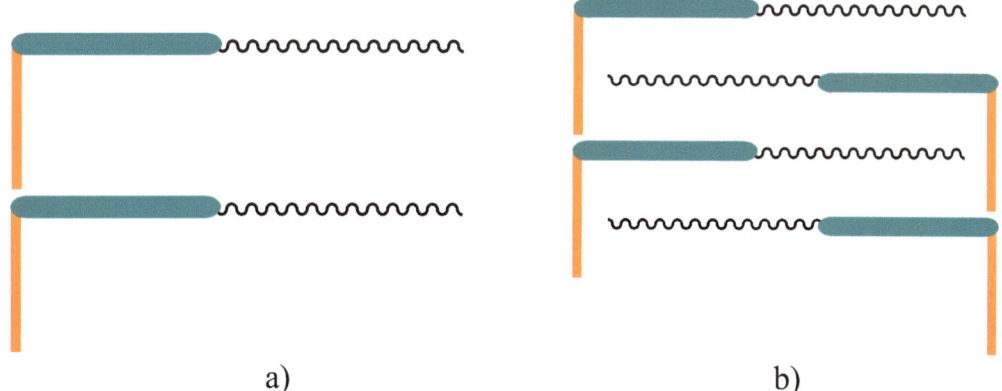

a)   b)

Fig. (7). Parallel and intercalated packing of monosubstituted ferrocenomesogens in nematic mesophase.

No liquid crystalline properties were found when compound **3d** contains a double bond in the middle of the flexible terminal chain of 17 carbon atoms.

The molecular modeling data predicted the absence of mesomorphism due to a change into the overall geometry of the ferrocene derivatives responsible for both lower values of the length/diameter ratio as well as larger ellipsoidal contours (Fig. **8**).

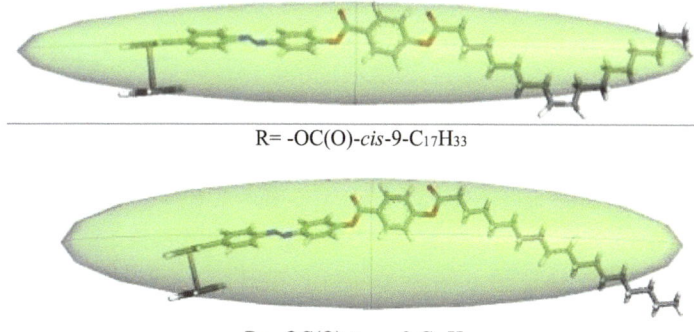

R= -OC(O)-*cis*-9-$C_{17}H_{33}$

R= -OC(O)-*trans*-9-$C_{17}H_{33}$

Fig. (8). Ellipsoidal models for compounds **3f** and **3g** containing a double bond.

Thermogravimetric analyses showed that ferrocene derivatives containing ether terminal linking group are more stable, if compared with the ones with ester group.

The removal of azo linking group (compounds of **4** type) maintains the same mesomorphic behavior to that of compound **2a** and **2b,** but the liquid crystalline domain is considerable reduced (Fig. **9a**).

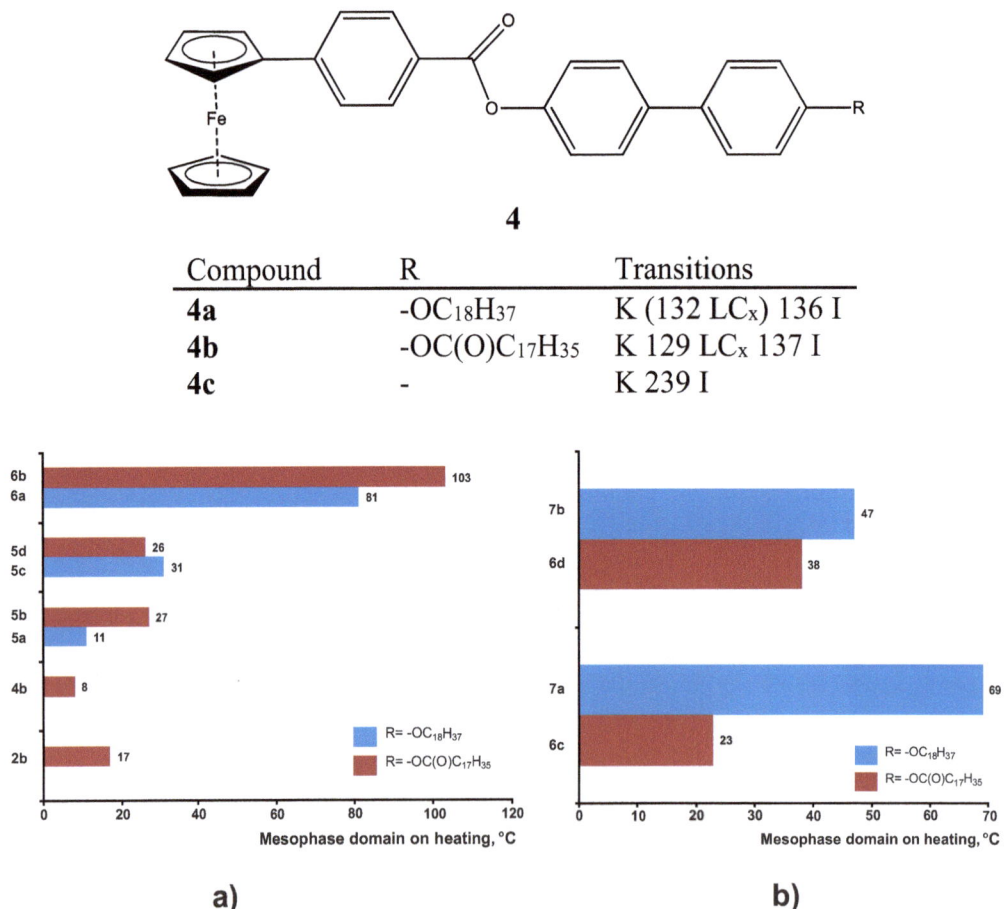

Compound	R	Transitions
4a	$-OC_{18}H_{37}$	K (132 LC_x) 136 I
4b	$-OC(O)C_{17}H_{35}$	K 129 LC_x 137 I
4c	-	K 239 I

Fig. (9). Comparison between mesophases' domains: a) compounds **2, 4-6** type; b) compounds **6** and **7**.

Compounds from **4** category displayed a nematic mesophase with surface disclination lines and characteristic Schlieren texture on cooling, which develops from nematic droplets (Fig. **10**). When heating on a slow rate (1°C/min), the compound **4b** presented a ribbon like texture [32].

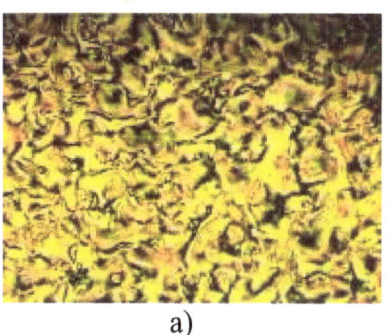 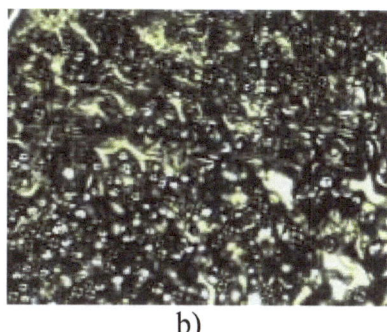

Fig. (10). Optical photomicrographs of compounds **4**: a) **4a**, characteristic nematic Schlieren texture; b) **4b**, nematic droplet texture.

The role of ferrocenyl unit in the rigid core on mesophase existence and stability was investigated by removing the bulky ferrocene unit from compounds of **2** and **4** type. As a result, benzoate diester derivatives exhibited enantiotropic behavior however on shorter ranges (of around 3°C), when compared with monosubstituted ferrocene analogous, when the molecular weights of ferrocenomesogens are much larger [35]. Again, the presence of two ester groups causes the change of the dipole moment in the extended rigid core.

Replacement of the azo moiety with an azomethine linking group was investigated in compounds **5** [31]. In contrast to compounds **2a** and **2b**, the first two analogous of compounds of **5** (**5a** and **5b**) showed a better stabilization of the liquid crystalline phases, with polymorphic behavior. Analysis of the molecular geometry explains the favorable mesomorphic behavior based on the presence of repulsion effects between the hydrogen atom of the azomethine group and the aromatic system, larger intermolecular interactions and decrease in the tendency of a compact packing.

Additional aromatic ring in compounds **1** leads to larger rigid/flexible ratio and consequently enantiotropic liquid crystal behavior over broader ranges. Compounds **5** differ by the number and the orientation of the ester linking groups. While the presence of the second ester group led to a slight increase in the clearing temperatures, the change in the ester orientation leads to almost comparable clearing temperatures, confirming thus the similarity of the physical intermolecular interactions.

[Structure of compound 5: ferrocene–C6H4–X–C6H4–N=CH–C6H4–R]

5

Compound	X	R	Transitions
5a	-C(O)O-	-OC$_{18}$H$_{37}$	K$_1$ 80 K$_2$ 133 LC$_x$ 144 I
5b	-C(O)O-	-OC(O)C$_{17}$H$_{35}$	K$_1$ 114 K$_2$ 126 LC$_x$ 153 I
5c	-OC(O)-	-OC$_{18}$H$_{37}$	K$_1$ 97 K$_2$ 114 LC$_x$ 145 I
5d	-OC(O)-	-OC(O)C$_{17}$H$_{35}$	K$_1$ 66 K$_2$ 129 LC$_x$ 155 I

The identified phase textures are varied, from nematic marble texture (Fig. **11a**) to smectic C Schlieren texture exhibiting fourfold singularities, defects of strength (Fig. **11d**). In the case of compound **5d**, interesting textures were noted on cooling, which were assigned as spiral smectic instabilities (Fig. **11c**). Besides nematic Schlieren and smectic textures, some dendritic growth like aggregates were separated from isotropic melt, which on further cooling transform into a mosaic like texture (Fig. **11b**) [35].

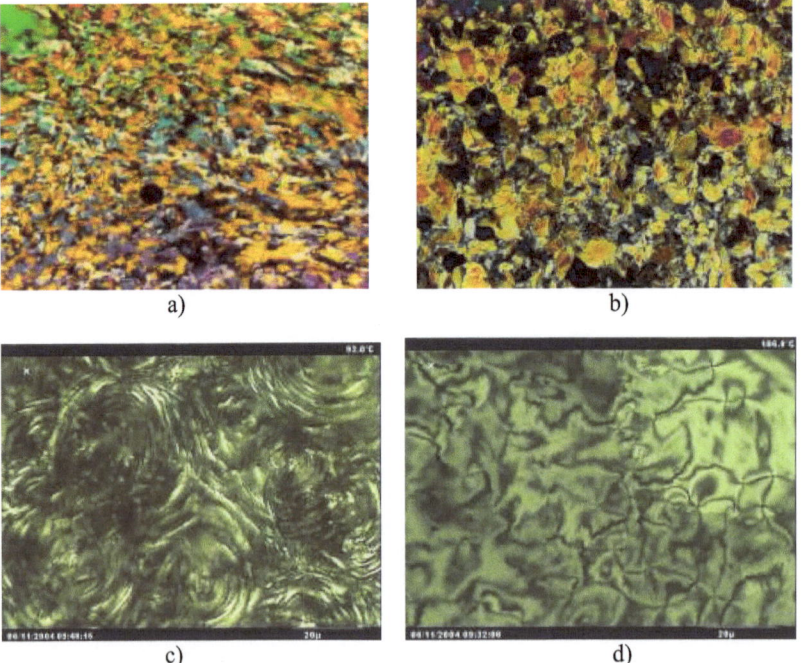

Fig. (11). Optical photomicrographs of compounds of **5** type: a) **5a**, nematic marble texture; b) **5b**, mosaic texture; c) and d) **5d**, smectic X textures.

Compared to compounds **5**, compounds **6** contain an additional azo-aromatic moiety, which emphasizes the anisotropic shape, conveys more polarity and hence increases polarizability [36]. As a result, compounds **6a** and **6b** displays organized mesophases over ranges larger with 70°C and 76°C, respectively, than compounds **5a** and **5b**. The different packing arrangement in the crystalline phase induces in this case polymorphic behaviors.

6

Compound	X	Y	R	Transitions
6a	-N=N-	-N=CH-	-OC$_{18}$H$_{37}$	K$_1$ 113 K$_2$ 166 N 247 I
6b	-N=N-	-N=CH-	-OC(O)C$_{17}$H$_{35}$	K$_1$ 117 K$_2$ 139 K$_3$ 151 N 254 I
6c	-CH=N-	-N=N-	-OC$_{18}$H$_{37}$	K$_1$ 116 K$_2$ 142 N 165 I
6d	-CH=N-	-N=N-	-OC(O)C$_{17}$H$_{35}$	K$_1$ 97 K$_2$ 130 N 168 I
6e	-CH=N-	-N=N-	-OH	K 300 I

The presence of the second ester linkage between the rigid core and flexible terminal chain (compounds **6b** and **6d**) stabilizes better the mesophase domains, if compared with compounds containing one ester linkage (compounds **6a** and **6c**). These compounds showed nematic and mosaic textures by OPM investigations (Fig. 12).

In order to evaluate the influence of the flexible unit upon liquid crystalline properties, the terminal flexible chain R in compounds **6** was replaced by a hydroxyl group, (compound **6e**). This structural change resulted in larger molecular rigidity and caused the disappearance of the mesomorphic properties. The formation of intermolecular hydrogen bonds is a possible explanation for the increased value of the melting point. By removing p-ferocenyl-phenyl moiety from compounds **6a** and **6b**, the liquid crystalline properties maintained, but the mesophase ranges of the obtained compounds are much smaller (with 74°C and 78°C). The position of the linking groups in the structure plays an important role in stabilization of the mesophase. Thus, the inversion of the position of the azo unit with the imine ones was expected to change the electronic structure of the ferrocene compounds and lead to increase in the melting point of about 24°C (**6c** towards **6a**) and 21°C (**6d** towards **6b**).

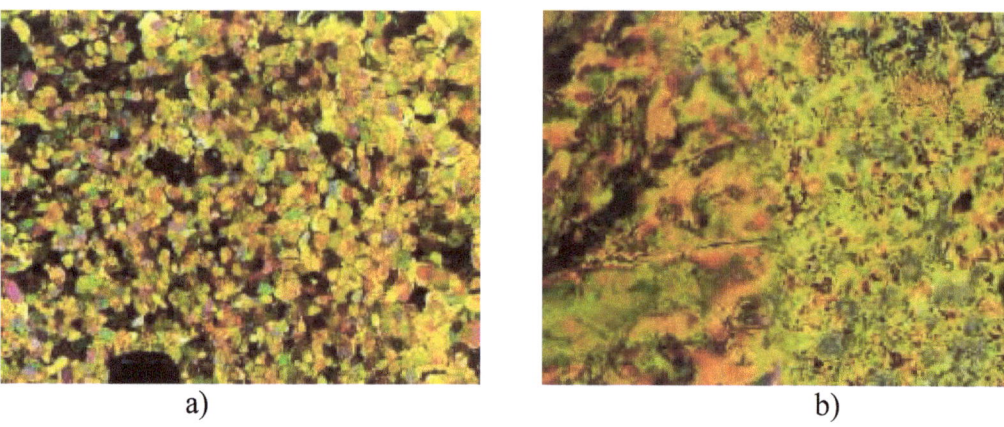

Fig. (12). Typical optical photomicrographs of compounds of **6** type: a) **6c**, mosaic texture; b) **6d**, coexistence of nematic marble, Schlieren and fingerprint textures.

Replacement of the ester linkage with an azo one (compounds of **7** type) increases the liquid crystalline domains, if compared with those of compound **6c** and **6d**, however with considerably higher melting points [36]. When trying to obtain compounds containing shorter terminal flexible chains (with 4 to 10 carbon atoms) partial thermal degradation into mesophase before isotropisation was observed.

7

Compound	R	Transitions
7a	-OC$_{18}$H$_{37}$	K$_1$ 122 K$_2$ 194 S$_x$ 211 N 263 I
7b	-OC(O)C$_{17}$H$_{35}$	K$_1$ 141 K$_2$ 197 N 244 I

The compounds of **7** type show a rich polymorphism evidenced on heating and cooling cycles of DSC thermograms (**Fig. 13**).

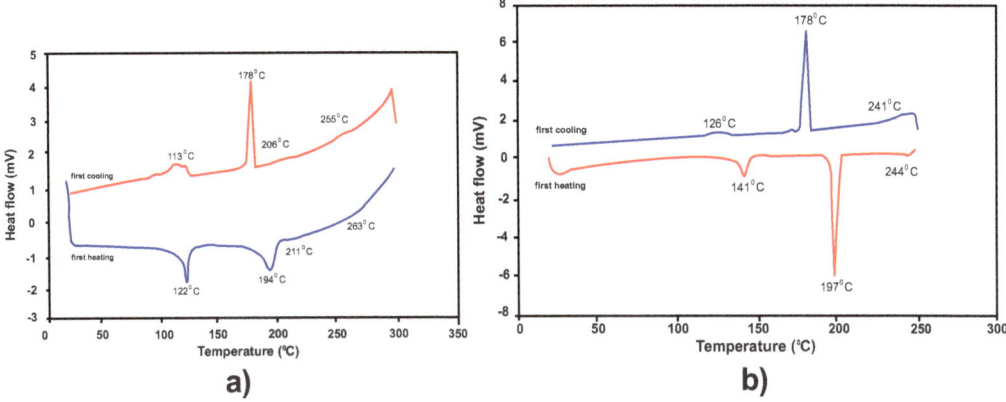

Fig. (13). DSC thermograms of a) **7a** and b) **7b**.

While compound **7a** presents a stable smectic A phase with characteristic focal conic texture on heating and cooling cycles (Fig. **14a**), compound **7b** presents smectic A phase only on cooling, from nematic phase (Fig. **14b** and **14c**). The presence of esteric linkage between alkyl chain and the rest of the molecule rotates the flexible plane against the rigid one, which induces a slightly decrease of intermolecular interactions and influences the liquid crystalline properties.

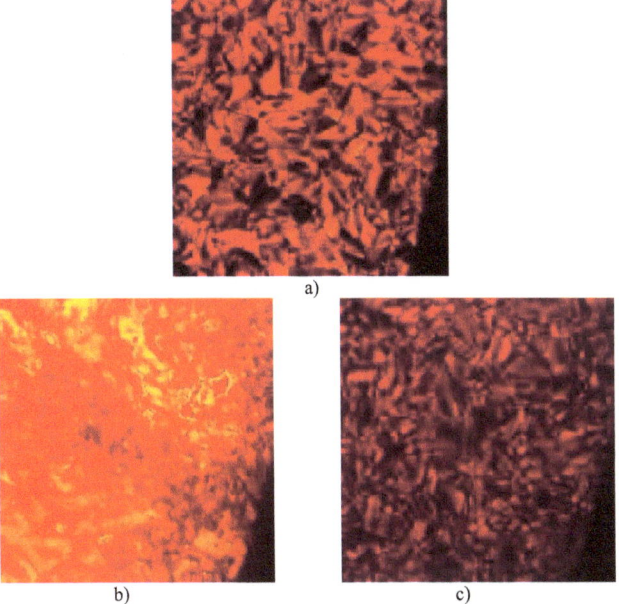

Fig. (14). Typical optical photomicrographs of compounds of **7** type: a) **7a**, smectic texture; b) and c) **7b**, nematic and smectic texture.

The presence of two azo linking groups maintains the extended conjugation in the rigid core and moreover enhances the anisotropic shape of the molecules, in conjunction with a strong increase of lateral interactions which favor tight packing and increase of the isotropization temperatures (Fig. **15**).

Fig. (15). Geometry of the compounds 7.

When comparing the mesophase domains between **6** and **7** analogous compounds (Fig. **9b**), it is noticeable that there is a larger difference in compounds with alkyloxy chain on the end (**7a** and **6c**) as against the ones with acyloxy chain (**7b** and **6d**). This may have the meaning that the molecules associate better in parallel stacking arrangements when the ester linkage is missing, although better packaging delays the isotropization and favors the beginning of degradation.

So far, among the ferrocenomesogens of **1-7** type, compound **6b** presents the optimal length/width ratio in conjunction with a better strength of intermolecular association due to both π-π and dipolar interactions and favorable steric factors that ensure long range ordering phenomenon.

Thermal investigations of ferrocenomesogens **1-7** indicated that the thermal decomposition takes place in two stages of degradation, with different mass loss. For imine derivatives of **1** and **5** type, the initial temperatures of the first stage are higher than 400°C, which proves a good thermal stability. However, the highest thermostability belongs to iminic derivative **1** containing two ester bonds into its structure. For this compound, the temperature for which the degradation rate is maximum was found between 580°C and 805°C and the activation energy from first stage was larger than that corresponding to analogous derivatives **5** [37]. Compounds from class **7** containing four aromatic rings into its structure presented the lowest thermostability, probably because the presence of an azo junction induced a lower stability if compared with the ester or imine ones.

MONOSUBSTITUTED LIQUID CRYSTALLINE FERROCENES CONTAINING A CHOLESTERYL GROUP

Since Reinitzer discovery of the mesomorphic behavior in cholesteryl benzoate in 1888, a large number of cholesteryl containing liquid crystals have been reported [38 - 42]. These compounds show liquid crystalline properties as a consequence of the pro-mesogenic character of the cholesteryl unit, although cholesterol itself doesn't form liquid crystalline textures [43]. This bulky non-planar asymmetrical group is rigid and elongated and meets the rod like main structural criterions.

In addition to these features, its ability to interact favorably with the hydrophobic aliphatic chains is added, which permits parallel compact molecular packing in solid state. The distinct feature of cholesteric liquid crystals is their ability to organize into twisted supramolecular structure that changes the optical properties when an electric field is applied. The phenomenon is especially useful in modern liquid crystals technology; in this respect, chiral liquid crystals containing ferrocene may provide fast response on switching, high birefringence and the presence of physical colors.

Ferrocene Derivatives with Ferrocene Rigidly Connected to the Cholesteryl Unit

As a function of the connective possibilities between ferrocenyl and cholesteryl units in molecules, various changes in molecule polarity and molecular interactions may take place, with important consequences on mesophase transitions. For example, the direct connection of ferrocene to rigid mesogenic unit leads to the increase of melting points and decrease in thermal stability. As consequence, ferrocene derivatives with cholesteryl unit rigidly connected to mesogenic unit (B_1) show liquid crystalline properties but with higher melting points, if compared with the analogous with flexible terminal chain end (compound **2**) instead of cholesteryl group.

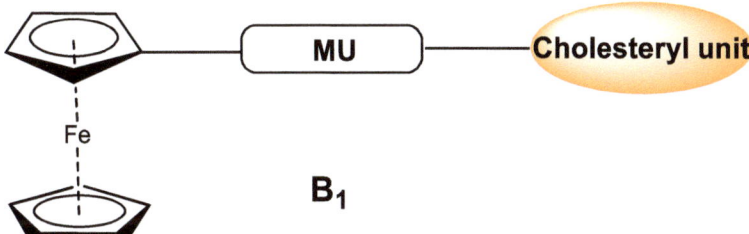

This is caused by compact packing of molecules in solid state as a consequence of the strong interactions between cholesteryl units, which increases significantly the melting points, sometimes beyond the thermal stability range.

The presence of only two aromatic rings into the structure and the replacement of azo with an imine linking group (compound **9a**) decreases the transition temperatures and promotes the appearance of smectic A phase, followed by TGBA and cholesteric phases. The liquid crystalline behavior is due to increase lateral interactions between molecules and consequently ordering into smectic layers.

The appearance of smectic phases in ferrocene derivatives containing only two aromatic rings is quite unusual and may be explained only by the presence of some molecular associations, favored by hydrogen bonding interactions.

Introduction of a third aromatic ring next to ferrocene (compound **9c**) increases the transition temperatures because of closer packing of molecules in the crystal state, while changing the orientation of ester group between the two aromatic rings leads to loss of liquid crystalline properties.

8

Compound	n	Transitions
8a	0	K_1 194 K_2 208 Ch (d)*
8b	1	K 226 SA 261 Ch (d)

Ch – cholesteric phase
*decomposition process

9

Compound	n	X	Transitions
9a	0	COO	K 198 SmA 228 TGBA 230 Ch (d)
9b	1	COO	K 229 Ch (d)
9c	1	OC(O)	K 230 I

Since detection by DSC of the characteristic peaks for liquid-liquid crystal transition upon cooling was difficult due to thermal decomposition of the structures, the influence of various structural factors upon degradation process was studied by thermal stability comparative analysis. The influence of the linking groups was investigated by comparing thermal stability of ferrocenomesogens of **8** and **9** type having similar length of the mesogenic block, but different types of linking groups (ester or imine) that connect the ferrocenyl and cholesteryl units. The influence of the ferrocene unit was elucidated by comparing the thermal stability of compounds with similar length of mesogenic groups but without ferrocene. The obtained data studies indicated a complex degradation mechanism which takes place in two or several stages of degradation.

The thermostability of these compounds varies as a function of thermal degradation atmosphere (air and or nitrogen) and depends on several structural factors such as the connection type and the number of linking groups.

Thermal investigations indicated a lower stability for compounds **8a** and **9a**, where the ester linkage is adjacent to the ferrocenyl unit (decomposition begins at 306°C for **8a** and 315°C for **9a**, in nitrogen atmosphere), opposed to the ones with a phenyl group next to ferrocene, when the compounds starting to lose weight above these values [44]. This behavior was attributed to the electron withdrawing effect of the carbonylic group next to ferrocene expected to induce a destabilization of the π bond between the iron atom and the cyclopentadienylic cycles, with consequences on thermal degradation. In the case when a phenyl ring is directly linked to ferrocene, the thermal degradation starts at the connecting groups between the aromatic rings (compounds **8b**, **9b**, **9c**). Thermostability decreased with increasing the number of the connecting groups and aromatic rings. This phenomenon was explained by the easy breaking of the linkages azo, imine or ester with releasing of small molecules such as N_2, CO_2, CO or CH_2O. Particularly, the azo - junction induces a lower stability than the ester one [45].

At the same time, the presence of cholesterol has favorable effects on thermostabiliy, as indicated in the comparative studies of compounds of **8** and **9** type with their homologous derivatives without cholesteryl units, where the thermal stability domain is about 10°C lower.

Ferrocene Derivatives with Ferrocene Flexible Connected to Cholesteryl Unit

The increase of the distance between the ferrocenyl and cholesteryl mesogenic units by inserting a flexible chain (structure B_2) diminishes the negative influence of repulsive steric effects of the ferrocene on mesomorphic properties. The presence of flexible chain balances as well the rigidity of the molecules, which is reflected on lower temperature transitions.

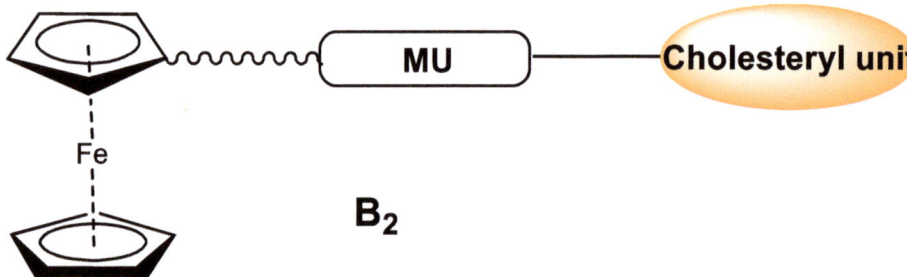

The mesomorphic properties of B_2 type compounds show a polymorphic behavior, as a result of different adopted conformational shapes, which are allowed because of internal degrees of freedom of molecules granted by the presence of a flexible chain between ferrocene and the mesogenic unit [48]. Compounds **10a** and **10b** show cholesteric mesophase on a broad temperature domain; nevertheless the degradation processes appear before clearing. The increase of the number of methylene group (compound **10b**) in the flexible chain seems to induce stronger interactions between molecules in the solid state reflected by higher melting point. The high dipole moment values of these compounds correlated with high melting temperatures suggests strong interactions and compact packing in structures. Addition to the second cyclopentadienyl ring of ferrocene of an 18 carbon atoms flexible acyl chain (compound **10c**) results in the decreasing of both melting and clearing temperatures. Such behavior is the result of the increased flexibility of the structure, determined by the increase of the flexible: rigid ratio, which thus changes the interactions between adjacent molecules.

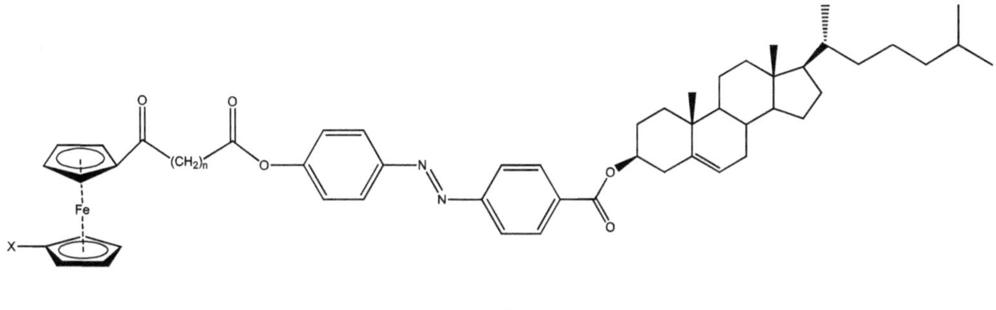

10

Compound	n	X	Transitions
10a	2	H	K$_1$ 63 K$_2$ 163 Ch 249 (d)
10b	3	H	K$_1$ 85 K$_2$ 181 K$_3$ 193 Ch 265 (d)
10c	2	-C(O)C$_{17}$H$_{35}$	K 98 Ch 175 I

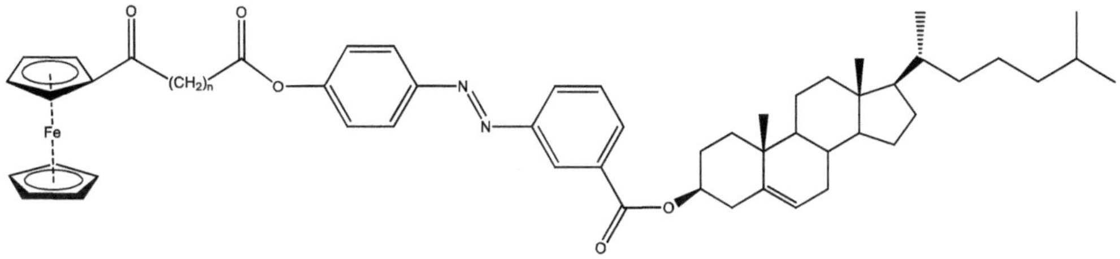

11

No liquid crystalline properties were observed in the absence of azobenzene rigid group or when the cholesteryl unit is connected in *meta* position of the phenylene ring (compound **11**).

In this case, the presence of a bend in the molecular structure destabilizes the organization into liquid crystalline structures, if the larger diameter of the molecule is considered.

The removal of the carbonyl link next to ferrocene leads to a slight increase of the flexible chain length, from 2.54 to 2.97 Å and has as result a decrease in the melting point (compound **12a**) [46].

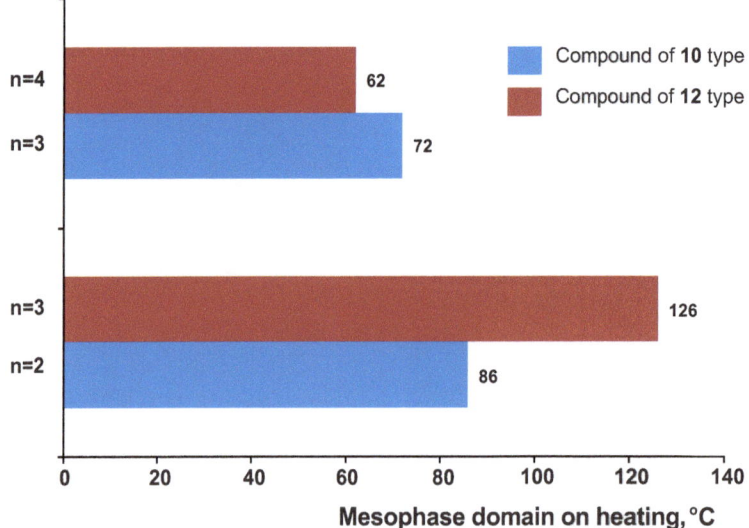

12

Compound	n	Transitions
12a	3	K_1 149 K_2 161 Ch 287 (d)
12b	4	K_1 127 K_2 139 Ch 201 I

Besides this, the mesophase domain of compound **12a** is increased with about 54°C, if compared with the analogous compound with the same number of carbon atoms, compound **10b** (Fig. **16**).

Fig. (16). Comparison between mesophases' domains of compounds of **10** and **12** type.

As in the case of compounds class **10**, the liquid crystalline behavior is suppressed in the absence of azobenzene rigid moiety or in the presence of a bend in the molecular structure determined by the substitution in the *meta* position of the benzene ring.

When the cholesteryl unit is removed from compounds of **10** and **12** type the disappearance of liquid crystalline properties occurs, partly because the cholesterol is the most important promoter of liquid crystalline behavior and partly because the molecules are too short to generate a mesophase. The influence of ferrocene unit upon the mesomorphic properties was studied by comparing the transitions of compounds of type **B₂** with some analogous compounds with the same core length, but where ferrocene was replaced by phenyl. It was found that both types of compounds posses similar high clearing points, showing that ferrocenyl unit doesn't contribute to the decrease of mesophase stability through steric repulsions with neighboring molecules.

Considering the high transition temperatures, and particularly the isotropic transition temperatures of these compounds, a systematic study of thermal stability was performed [44]. In most cases, the first phase of thermal degradation started above 310°C. Compounds from class **10** showed a lower thermal stability, if compared to analogous compounds **12**, where the carbonyl group next to ferrocene was reduced to a methylene one. In contrast to phenyl analogous, the presence of ferrocenyl unit stabilizes towards thermal degradation. The nature of the functional groups and their connecting position between the aromatic ring of the molecule influence as well the thermal stability, especially in the case of *meta* substitution [47, 48].

Ferrocene Derivatives with Cholesterol Flexible Connected to Mesogenic Unit

Another study focused on liquid crystalline properties of compounds of type **B₃**, where ferrocenyl unit is rigidly connected to the mesogenic part while the cholesteryl moiety is flexible attached to the structure by a chain containing ten methylene groups [49].

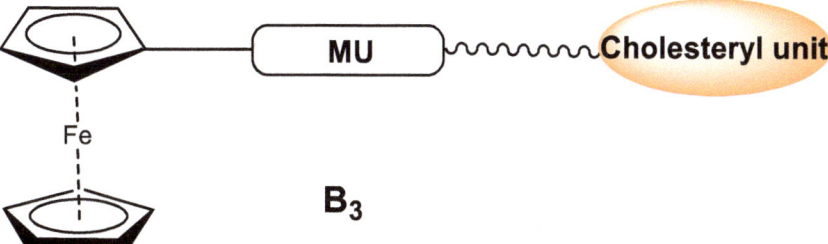

It was found that separation of cholesterol from mesogenic unit through a long flexible spacer contributes to the increase of lateral interactions between molecules, with smectic ordering as dominant phase towards the chiral nematic one.

Compared to previous ferrocenomesogens of type **B₂**, the mesophase of compound **13** appears at lower temperatures, below thermal degradation. The separation of cholesteryl and aromatic mesogenic units by a flexible spacer of ten methylene unit causes the formation of layered phases at lower temperatures.

13

Transitions
K (154 S$_A$) 152 Ch 179 I

Conventional or chiral smectic A phases prevailed over chiral nematic ones, as a consequence of the intense lateral interactions between the rigid parts.

The mesomorphic behavior of compound **13** on cooling was more complex, since a pseudo focal conic texture (SmA) was observed after the cholesteric ones, which was followed by an unidentified higher ordered smectic-like phase [49].

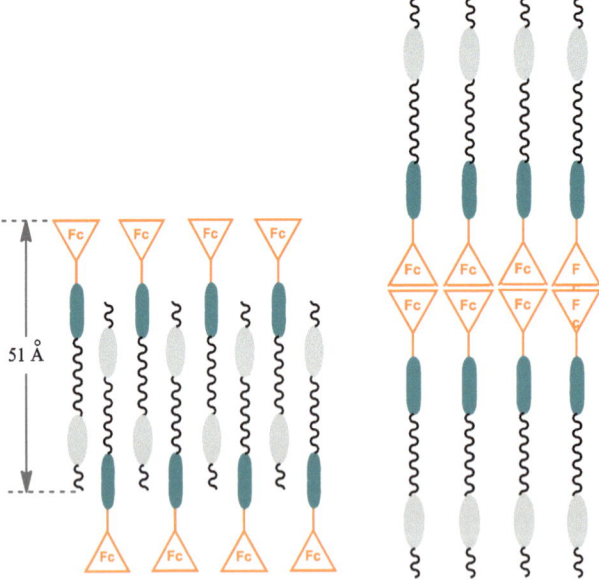

Fig. (17). Two-dimensional proposed structure models for the smectic phase of compound **13**.

The existence of smectic phase was investigated by X-ray diffraction (XRD), the diffractograms showing high intensity at small-angles. The molecular modeling data correlated with XRD allowed the establishing of a possible ordering model of compound **13** in the smectic phase (Fig. **17**).

Because of microphase separation, steric requirements and anisotropic polarizabilities, the aromatic groups are placed at the interfaces of the layers, while the aliphatic moieties containing the bulky cholesteryl unit are interdigitated in the centre of the layers. These represent opposite arrangements to the one proposed by Loubser *et al.*, although it was particular for nematic textures [26].

Changing the esteric connection in the rigid mesogen to a more rigid imine one (**14a**) modifies the self-organization process. These structural modifications results in a slightly different packing behavior in the smectic A phase towards a monolayer organization which subsequently leads to significantly changes of the isotropization temperatures, with a difference of 35°C, if compared with compound **13**.

14

Compound	X	Transitions
14a	-	K (176 S$_A$*) 177 Ch 214 I
14b	-C(O)O	K 179 Ch 220 I

Although the structural changes are minor (about the same length of the mesogenic unit, consisting of three aromatic rings and the same flexible spacer), the behavior of these compounds is different, both in terms of mesophase type (ordering mode) or the presence of mesophase and the transition temperatures as well. Compound **14a** differs from **13** only by the nature of the linking groups between aromatic rings. This structural change affects the nature of interactions between molecules by modifying the dipole moment and the geometry, which explains the organization into smectic mesophase. Their calculated dipole moment has high values, which are induced by strong lateral interactions between molecules and contributes to their organization into smectic structures. Also, the higher melting point of compound **14a** is probably due to a very compact packing of molecules into crystalline phase, when hydrogen bonding between imine groups of two neighborhood molecules can be established.

Adding one more ester linking group into the structure (compound **14b**) increases even further the isotropization temperature, with complete suppression on cooling of the smectic phase. The presence only of cholesteric phase is explained by the total dipole moment value of the molecule.

Compound **15** containing only two benzene rings doesn't show liquid crystalline properties, the sample is very viscous and ordering into the mesophase is hampered.

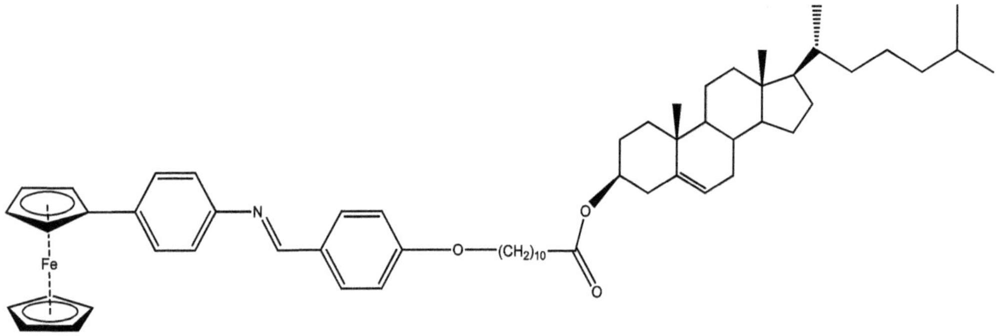

15

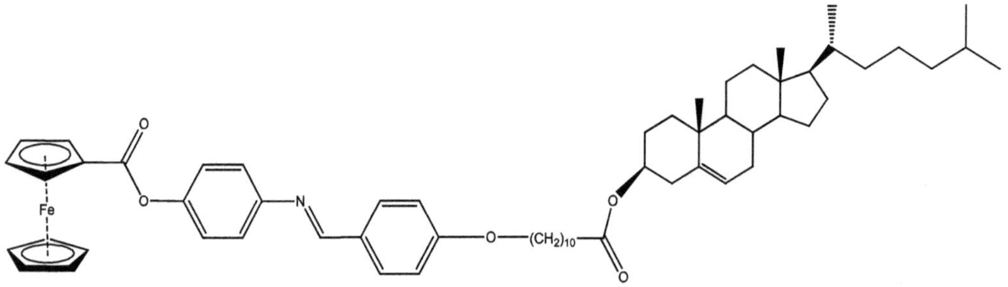

16

Transitions
K (119 LC) 141 I

Rapidly cooled at room temperature, compound **15** tends to order but is freezing suddenly on touching, proving that mesophase is not thermodynamic stable while the crystalline solid state is more stable (Fig. 18).

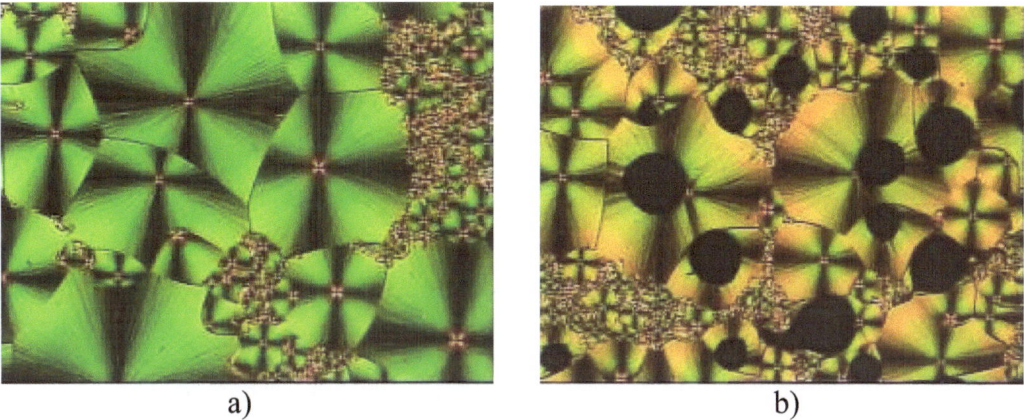

Fig. (18). Optical photomicrograph of compound **15**, rapidly cooled at room temperature.

Consequently, the obtained order is a virtual state of the material, in extreme conditions.

Adding an extra ester linkage next to ferrocene (compound **16**) allows the mesophase stabilization only on cooling, over a very narrow temperature domain (6.5 °C).

Symmetrical Derivatives with Two Monosubstituted Ferrocenyl Units

In most cases, symmetrical diferrocene derivatives with flexible unit as side part do not show liquid crystalline properties. The presence of two ferocenyl units does not have a favorable effect on monosubstituted ferrocenomesogens of type C_1.

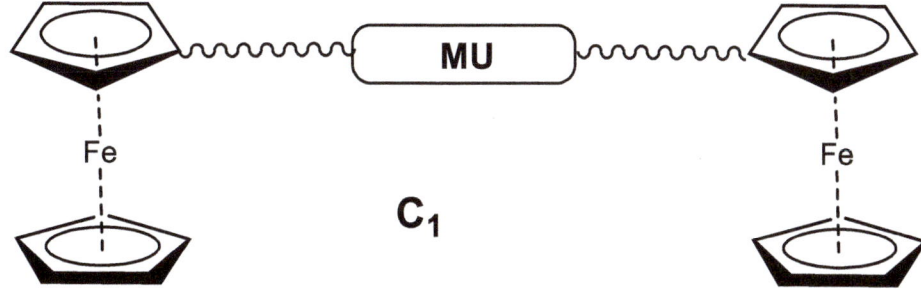

Therefore, the lack of mesomorphism for dimeric compounds **17** and **18** is probably due to the short length of the mesogenic unit and weak interactions which don't allow the ordering into liquid crystalline structures.

17

18

Even after the addition of two mesogenic units symmetrically inserted between a flexible central chain (structure **C₂**), the rigid/flexible ratio is not good enough to promote mesomorphism.

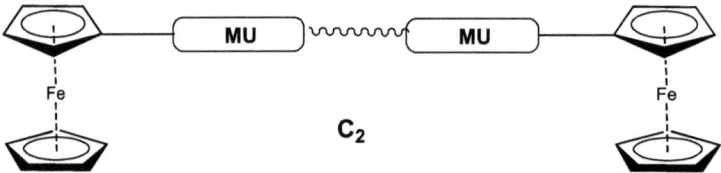

C₂

However, while the compounds **19** and **20** with two aromatic rings on mesogenic unit do not show liquid crystalline ordering, the introduction of a third aromatic unit favors the mesomorphism (compound **21**).

19

20

21

Nevertheless, the mesophase of compound **21** is unstable since was observed only on polarized light microscopy. The thermochemical and thermophysical properties of compounds **17÷21** indicated a lower stability, if compared to previous ferrocene derivatives, since the increase in the number of aromatic rings induces a decrease in thermostability. However, when considering the position of the flexible spacer into the molecular structure in compounds **C1** and **C2**, the presence of the flexible part in the centre of the structure induces a better thermostability (compounds of **C2** type), than for compounds with flexible chain on both sides of the mesogenic unit [50].

Using a combination of available experimental database, molecular design software and artificial intelligence (neural networks and genetic algorithms), a method for the design of ideal molecular liquid crystalline structures was established [50]. The assignment of a quantitative relation between structure and thermal stability was possible by taking into account the molecular descriptors (input data) as molecular weight, polarization of the molecule; several structural parameters: the number of aromatic rings, number of ferrocene units, number of cholesteryl units, number and type of bonds in the molecules (>C=O, -N=C<, -N=N-) and melting temperature respectively. The theoretical prediction of thermostability of ferrocene derivatives indicated that compounds with a melting temperature smaller than the decomposition temperature have a better stability, if compared to analogous derivatives. The optimized neural models using specific algorithms thus enable an accurate evaluation of thermal stability and above all allow the prediction of parameters of the liquid crystalline state of ferrocenomesogens.

CONCLUSIONS

In order to understand the correlation between molecular structure and liquid crystalline properties, our goal was to explore a large variety of mesogenic units and ferrocenomesogens. Study of monosubstituted ferrocene derivatives evidenced the importance of molecular geometry in appearance of liquid crystalline properties. We found out that the introduction of ferrocenyl unit in pro-mesogenic structures can stabilize the mesophases and additionally may bring particular properties related to the arrangement of the molecules into mesophase. The modification of structural parameters in monosubstituted ferrocenomesogens influences the mesophase domain, transition temperatures and thermal stability. Hence it was found that the optimal structures require a high length/width ratio of molecules. To fulfill this criterion, the elongation of mesogenic units to three and four benzene rings was performed. An important issue targeted the thermal stability of synthesized liquid crystals that may affect their practical applications, especially in those fields in which high temperature processing is required. Our results indicated that thermostability depends mainly on the nature and the number of the linking groups attached directly to ferrocenyl unit or from mesogenic units. With all these arguments, we hope that our results in the field of ferrocenomesogens will contribute further for designing of promising materials with novel properties.

CONSENT FOR PUBLICATION

Not applicable.

CONFLICT OF INTEREST

The author confirms that he has no conflict of interest to declare for this publication.

ACKNOWLEDGEMENTS

Professor Shin'ichi Nakatsuji from University of Hyogo, Japan is acknowledged for his support and fruitful collaboration.

This work was supported by a grant of the Romanian National Authority for Scientific Research and Innovation, CNCS/CCCDI –UEFISCDI, project number PN-III-P4-ID-PCE-2016-0508, within PNCDI III.

REFERENCES

[1] Hussain, A.; Pina, A.S.; Roque, A.C.A. Bio-recognition and detection using liquid crystals. *Biosens. Bioelectron.,* **2009**, *25*(1), 1-8.
[http://dx.doi.org/10.1016/j.bios.2009.04.038] [PMID: 19477113]

[2] Astruc, D. Why is ferrocene so exceptional? *Eur. J. Inorg. Chem.,* **2017**, *1*, 6-29.
[http://dx.doi.org/10.1002/ejic.201600983]

[3] Deschenaux, R.; Goodby, J.W. *Ferrocenes: Homogeneous Catalysis, Organic Synthesis, Materials Science*; Togni, A.; Hayashi, T., Eds.; VCH: Weinheim, Germany, **1995**.

[4] Giroud-Godquin, A.M.; Maitlis, P.M. Metallomesogens – metal – complexes in organized fluid phases. *Angew. Chem. Int. Ed. Engl.,* **1991**, *30*, 375-402.
[http://dx.doi.org/10.1002/anie.199103751]

[5] Espinet, P.M.; Esteruelas, A.; Oro, L.A.; Serrano, J.L.; Sola, E. Transition – metal liquid crystals – advanced materials within the reach of the coordination chemist. *Coord. Chem. Rev.,* **1992**, *117*, 215-274.
[http://dx.doi.org/10.1016/0010-8545(92)80025-M]

[6] Hudson, S.A.; Maitlis, P.M. Calamitic metallomesogens – metal – containing liquid – crystals with rodlike shapes. *Chem. Rev.,* **1993**, *93*, 861-885.
[http://dx.doi.org/10.1021/cr00019a002]

[7] Serrano, J.L. *Metallomesogens: Synthesis, Properties and Applications*; VCH: Weinheim, Germany, **1996**.

[8] Bruce, D.W. *Inorganic Materials*; Bruce, D.W.; O'Hare, D., Eds.; Wiley: Chichester, **1996**, p. 429.

[9] Giroud-Godquin, A.M. Coord. My 20 years of research in the chemistry of metal containing liquid crystals. *Chem. Rev.,* **1998**, *178*, 1485-1499.
[http://dx.doi.org/doi.org/10.1016/S0010-8545(98)00081-2]

[10] Bruce, D.W.; Donnio, B. *Structure and Bonding*; Mingos, D.M.P., Ed.; Springer: Berlin, Heidelberg, **1999**, Vol. 95, p. 193.

[11] Gimenez, R.; Lydon, D.P.; Serrano, J.L. Metallomesogens: a promise or a fact? *Curr. Opin. Solid State Mater. Sci.,* **2002**, *6*, 527-535.
[http://dx.doi.org/10.1016/S1359-0286(03)00009-3]

[12] Donnio, B.; Guillon, D.; Deschenaux, R.; Bruce, D.W. *Comprehensive Coordination Chemistry II: from Biology to Nanotechnology*; McCleverty, J.A.; Meyer, T.J., Eds.; Elsevier: Oxford, **2003**, Vol. 7, p. 357.
[http://dx.doi.org/10.1016/B0-08-043748-6/06150-8]

[13] Imrie, C.; Engelbrecht, P.; Loubser, C.; McCleland, C.W. Monosubstituted thermotropic ferrocenomesogens: an overview 1976-1999. *Appl. Organomet. Chem.,* **2001**, *15*, 1-15.
[http://dx.doi.org/10.1002/1099-0739(200101)15:1<1::AID-AOC109>3.0.CO;2-3]

[14] Donnio, B.; Guillon, D.; Bruce, D.W.; Deschenaux, R. *Comprehensive Organometallic Chemistry III: from Fundamentals to Applications*; Crabtree, R.H.; Mingos, D.M.P., Eds.; Elsevier: Oxford, **2006**, Vol. 12, p. 195.

[15] Deschenaux, R.; Donnio, B.; Guillon, D. Liquid-crystalline fullerodendrimers. *New J. Chem.,* **2007**, *31*, 1064-1073.
[http://dx.doi.org/10.1039/b617671m]

[16] Kadkin, O.N.; Galyametdinov, Yu.G. Ferrocene-containing liquid crystals. *Russ. Chem. Rev.,* **2012**, *81*(8), 675-699.
[http://dx.doi.org/10.1070/RC2012v081n08ABEH004270]

[17] Donnio, B. Liquid-crystalline metallodendrimers. *Inorg. Chim. Acta,* **2014**, *409*, 53-67.
[http://dx.doi.org/10.1016/j.ica.2013.07.045]

[18] Gray, G.W. *Liquid Crystals and Plastic Crystals*; Gray, G.W.; Winsor, P.A., Eds.; Ellis Harwood Ltd.: Chichester, England, **1974**, Vol. 1.

[19] Gray, G.W.; Mosley, A. Mesomorphic transition-temperatures for homologous series of 4-N-alkyl-4'-cyanotolanes and other related compounds. *Mol. Cryst. Liq. Cryst. (Phila. Pa.),* **1976**, *37*, 213-231.
[http://dx.doi.org/10.1080/15421407608084357]

[20] Demus, D. 100 Years liquid-crystal chemistry. *Mol. Cryst. Liq. Cryst. (Phila. Pa.),* **1988**, *165*, 45-84.
[http://dx.doi.org/doi.org/10.1080/00268948808082196]

[21] Woodward, R.B.; Rosenblum, M.; Whiting, M.C. A new aromatic system. *J. Am. Chem. Soc.,* **1952**, *74*, 3458-3450.
[http://dx.doi.org/10.1021/ja01133a543]

[22] Scutaru, D.; Tataru, L.; Mazilu, I.; Vata, M.; Lixandru, T.; Simionescu, Cr. Contributions to the synthesis of some ferrocene-containing antibiotics. *Appl. Organomet. Chem.,* **1993**, *7*, 225-231.
[http://dx.doi.org/10.1002/aoc.590070402]

[23] Islam, M.T.; Best, S.P.; Bourke, J.D.; Tantau, L.J.; Tran, C.Q.; Wang, F.; Chantler, C.T. Accurate X-ray absorption spectra of dilute systems: absolute measurements and structural analysis of ferrocene and decamethylferrocene. *J. Phys. Chem. C,* **2016**, *120*, 9399-9418.
[http://dx.doi.org/10.1021/acs.jpcc.6b00452]

[24] Loubser, C.; Imrie, C. Thermal properties of monosubstituted ferrocene derivatives: A series of new ferrocenomesogens. *J. Chem. Soc., Perkin Trans. 2,* **1997**, *3*, 399-409.
[http://dx.doi.org/10.1039/a600739b]

[25] Imrie, C.; Loubser, C.; Engelbrecht, P.; McCleland, C.W.; Zheng, Y. The synthesis and liquid crystal

behaviour of monosubstituted ferrocenomesogens. *J. Organomet. Chem.,* **2003**, *665*, 1-2, 48-64.
[http://dx.doi.org/10.1016/S0022-328X(02)02044-2]

[26] Loubser, C.; Imrie, C.; van Rooyen, P.H. The synthesis and crystal-structure of a liquid-crystalline ester of 4-carboxyphenylferrocene. *Adv. Mater.,* **1993**, *5*, 45-47.
[http://dx.doi.org/10.1002/adma.19930050109]

[27] Tamaoki, N. Cholesteric liquid crystals for color information technology. *Adv. Mater.,* **2001**, *13*(15), 1135-1147.
[http://dx.doi.org/10.1002/1521-4095(200108)13:15<1135::AID-ADMA1135>3.0.CO;2-S]

[28] Tamaoki, N.; Parfenov, A.V.; Masaki, A.; Matsuda, H. Rewritable full-color recording on a thin film of a cholesteric low-molecular-weight compound. *Adv. Mater.,* **1997**, *9*, 1102-1104.
[http://dx.doi.org/10.1002/adma.19970091408]

[29] Tamaoki, N.; Kruk, G.; Matsuda, H. Optical and thermal properties of cholesteric solid from dicholesteryl esters of diacetylenedicarboxylic acid. *J. Mater. Chem.,* **1999**, *9*, 2381-2384.
[http://dx.doi.org/10.1039/a904276h]

[30] Kruk, G.; Tamaoki, N.; Matsuda, H.; Kida, Y. Optical and liquid crystalline properties of dicholesteryl derivatives having different configurations of the hydrocarbon linkages. *Liq. Cryst.,* **1999**, *26*(11), 1687-1693.
[http://dx.doi.org/10.1080/02678292.1999.11509452]

[31] Carlescu, I.; Scutaru, A-M.; Apreutesei, D.; Alupei, V.; Scutaru, D. The liquid crystalline properties of some ferrocene containing Schiff bases. *Appl. Organomet. Chem.,* **2007**, *21*, 661-669.
[http://dx.doi.org/10.1002/aoc.1242]

[32] Carlescu, I.; Hurduc, N.; Scutaru, D.; Catanescu, O.; Chien, L-C. Monosubstituted ferrocene-containing liquid crystals. *Mol. Cryst. Liq. Cryst. (Phila. Pa.),* **2005**, *439*, 107-123.
[http://dx.doi.org/10.1080/15421400590956298]

[33] Onofrei, R-M.; Carlescu, I.; Epure, L.; Scutaru, D. Synthesis and liquid crystalline properties of some esters of 4-ferrocenyl-4'-hydroxyazobenzene. *Acta Chim. Slov.,* **2013**, *60*(3), 604-616.
[PMID: 24169715]

[34] Kadkin, O.N.; Kim, E.H.; Rha, Y.J.; Kim, S.Y.; Tae, J.; Choi, M-G. Novel tetrahedratic smectic C and nematic mesophases in unsymmetrically 1,1'-bis-substituted ferrocenomesogens. *Chemistry,* **2009**, *15*(40), 10343-10347.
[http://dx.doi.org/10.1002/chem.200901349] [PMID: 19739219]

[35] Carlescu, I.; Scutaru, A-M.; Apreutesei, D.; Alupei, V.; Scutaru, D. The liquid crystalline behaviour of ferrocene derivatives containing azo and imine linking groups. *Liq. Cryst.,* **2007**, *34*, 775-785.
[http://dx.doi.org/10.1080/02678290701343190]

[36] Onofrei, R-M.; Carlescu, I.; Lisa, G.; Silion, M.; Hurduc, N.; Scutaru, D. Synthesis and liquid crystalline behavior of some monosubstituted ferrocene containing Schiff bases. *Revista de Chimie,* **2012**, *63*, 139-145.

[37] Carlescu, I.; Lisa, G.; Scutaru, D. Thermal stability of some ferrocene containing Schiff bases. *J. Therm. Anal. Calorim.,* **2008**, *91*, 535-540.
[http://dx.doi.org/10.1007/s10973-007-8450-8]

[38] Reinitzer, F. Beiträge zur kenntniss des cholesterins. *Wiener Monatsh. Chem.,* **1888**, *9*, 421-441.
[http://dx.doi.org/10.1007/BF01516710]

[39] Akiyama, H.; Mallia, V.A.; Tamaoki, N. Synthesis, liquid-crystalline properties, and photo-optical studies of photoresponsive oligomeric mesogens as dopants in a chiral glassy liquid crystal. *Adv. Funct. Mater.,* **2006**, *16*, 477-484.
[http://dx.doi.org/10.1002/adfm.200500176]

[40] Imrie, C.T.; Henderson, P.A. Liquid crystal dimers and higher oligomers: between monomers and polymers. *Chem. Soc. Rev.,* **2007**, *36*(12), 2096-2124.

[http://dx.doi.org/10.1039/b714102e] [PMID: 17982523]

[41] Yelamaggad, C.V.; Shanker, G.; Hiremath, U.S.; Prasad, S.K. Cholesterol-based nonosymmetric liquid crystal dimmers: an overview. *J. Mater. Chem.*, **2008**, *18*, 2927-2949.
[http://dx.doi.org/10.1039/b804579h]

[42] Ercole, F.; Whittaker, M.R.; Quinn, J.F.; Davis, T.P. Cholesterol modified self-assemblies and their application to nanomedicine. *Biomacromolecules*, **2015**, *16*(7), 1886-1914.
[http://dx.doi.org/10.1021/acs.biomac.5b00550] [PMID: 26098044]

[43] Thiemann, T.; Vill, V. Homologous series of liquid crystalline steroidal lipids. *J. Phys. Chem. Ref. Data*, **1997**, *26*(2), 291-333.
[http://dx.doi.org/10.1063/1.556007]

[44] Lisa, G.; Apreutesei Wilson, D.; Scutaru, D.; Tudorachi, N.; Hurduc, N. Investigation of thermal degradation of some ferrocene liquid crystals. *Thermochim. Acta*, **2010**, *507-08*, 49-59.
[http://dx.doi.org/10.1016/j.tca.2010.04.030]

[45] Apreutesei, D.; Lisa, G.; Hurduc, N.; Scutaru, D. Investigations on the thermal behavior of some ferrocene derivatives with liquid crystalline properties. *Sci Study Res.*, **2005**, *VI*, 165-172.

[46] Apreutesei, D.; Lisa, G.; Akutsu, H.; Hurduc, N.; Nakatsuji, S.; Scutaru, D. Thermotropic properties of ferrocene derivatives bearing a cholesteryl unit: structure-properties correlations. *Appl. Organomet. Chem.*, **2005**, *19*, 1022-1037.
[http://dx.doi.org/10.1002/aoc.953]

[47] Apreutesei, D.; Lisa, G.; Scutaru, D.; Hurduc, N. Investigations on thermal stability of some ferrocene liquid crystals bearing azo, ferrocenyl and cholesteryl units. *J. Optoelectron. Adv. Mater.*, **2006**, *8*, 738-741.

[48] Apreutesei, D.; Lisa, G.; Hurduc, N.; Scutaru, D. Thermal behavior of some cholesteric esters. *J. Therm. Anal. Calorim.*, **2006**, *83*, 335-340.
[http://dx.doi.org/10.1007/s10973-005-6522-1]

[49] Apreutesei, D.; Mehl, G.H.; Scutaru, D. Ferrocene-containing liquid crystals bearing a cholesteryl unit. *Liq. Cryst.*, **2007**, *34*, 819-831.
[http://dx.doi.org/10.1080/02678290701478715]

[50] Lisa, G.; Apreutesei Wilson, D.; Curteanu, S.; Lisa, C.; Piuleac, C-G.; Bulacovschi, V. Ferrocene derivatives thermostability prediction using neural networks and genetic algorithms. *Thermochim. Acta*, **2011**, *521*, 26-36.
[http://dx.doi.org/10.1016/j.tca.2011.03.037]

CHAPTER 3

Progress in the Chemistry of Phosphorothioates

Mihaela Gulea[*]

Laboratoire d'Innovation Thérapeutique (LIT, UMR 7200), Université de Strasbourg, CNRS, 74 Route du Rhin, 67401 Illkirch, France

Abstract: The present review focuses on the synthesis and applications of phosphorothioates (phosphorothioic esters) with the general structure $R_2P(O)SR'$, which are molecules belonging to the large family of organophosphorus compounds. They found important applications as biological active molecules, as well as synthetic intermediates to access valuable phosphorus and/or sulfur-containing structures. The reactivity of both heteroatoms, phosphorus and sulfur, is discussed by showing how they are involved as electrophilic, nucleophilic, or radical species in different transformations. The general methods for the preparation of phosphorothioates are described according to the various strategies involving P-S bond formation from one sulfur-reagent and one phosphorus-reagent, which include recent metal-catalyzed cross-coupling reactions. Particular attention is paid to synthetic applications of phosphorothioates occurring *via* the P-S bond cleavage, including for example anionic rearrangements and cascades, formation of sulfur-heterocycles, homolytic processes for olefins phosphorylation, and metal-catalyzed alkynes thiophosphorylation. The literature of the last two decades is covered, highlighting selected contributions in the chemistry of this class of compounds.

Keywords: Anionic Rearrangement, Mercaptophosphonate, Phosphorus, Phosphorothioates, Phosphorothioic esters, P-S Bond Cleavage, P-S Bond Formation, P-S Coupling, Sulfur, Sulfur Heterocycles, Sigmatropic Rearrangement, Thiophosphates.

INTRODUCTION

Compounds that contain phosphorus and sulfur atoms bonded to each other represent an important class of compounds due to their biological properties and fields of applications, in particular agrochemistry (as pesticides) [1], or in medicinal chemistry (as enzymes inhibitors or antivirals) [2]. As the literature covering this topic is very broad, we decided to discuss herein only one among the three categories of PS-compounds illustrated in Fig. (**1**) (phosphorothioates **A**, phosphorothionates **B**, and phosphorodithioates **C**), that of phosphorothioates **A**.

---

[*] Corresponding author Mihaela Gulea: Laboratoire d'Innovation Thérapeutique (LIT, UMR 7200), Université de Strasbourg, CNRS, 74 Route du Rhin, 67401 Illkirch, France; Tel: 33 + (0)3688 54148; E-mail: gulea@unistra.fr

Atta-ur-Rahman (Ed.)
All rights reserved-© 2018 Bentham Science Publishers

As far as concerns the chemistry of specific P-S compounds such as the tetraphosphorus decasulfide (P_4S_{10}) or the Lawesson's reagent (LR), it was the topic of recent reviews [3].

$$\underset{\underset{R}{R}}{\overset{\overset{O}{\|}}{P}}-SR' \qquad \underset{\underset{R}{R}}{\overset{\overset{S}{\|}}{P}}-OR' \qquad \underset{\underset{R}{R}}{\overset{\overset{S}{\|}}{P}}-SR'$$
$$\quad\textbf{A} \qquad\qquad\qquad \textbf{B} \qquad\qquad\qquad \textbf{C}$$

R = Alkyl, Aryl, OAlkyl, OAryl

Fig. (1). General structures of phosphorothioates **A**, phosphorothionates **B**, and phosphorodithioates **C**.

One widely known example is the incorporation of the P(O)-S moiety as an essential structural element in modified oligonucleotides for therapeutic applications and studies of processes involving RNA [4]. Additionally, phosphorothioates **A** are useful synthetic tools in various chemical transformations, most of them involving the cleavage of the P-S bond. Some enantiopure phosphorothioic acids have been developed for their chiral recognition ability [5].

In the first section of this chapter, general approaches to access phosphorothioic esters will be described, according to the various methods for the P-S or C-S bond formation (Scheme **1**: **Syntheses**). The second section will deal with the rich chemistry of phosphorothioates due to the presence of the both P and S heteroatoms directly linked and their synthetic applications (Scheme **1**: **Reactions**). Contributions of the literature of the last two decades have been selected and highlighted, demonstrating the interest and the advances in the chemistry of these compounds.

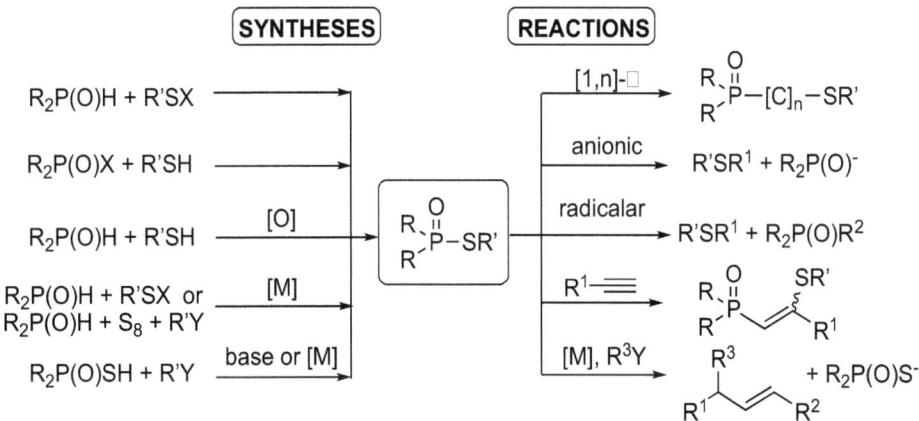

Scheme 1. Overview of selected syntheses and reactions of phosphorothioates covered in this chapter.

SYNTHESES OF PHOSPHOROTHIOIC ESTERS

The different strategies used to prepare phosphorothioic esters either consist of a form a P-S bond by using appropriate phosphorous and sulfur partners as electrophiles, nucleophiles, or radicals, or to introduce the P(O)S-function directly, *via* a C-S bond formation. The overall collection of methods enables access to a broad variety of structures of both S-alkyl and S-aryl phosphorothioate derivatives.

P-S Bond Formation from P-nucleophile and S-electrophile

The main phosphorus reagents used in these methods are P(IV) species of type $R_2P(O)H$, with rare cases of P(III) species involved in Arbuzov-type reactions. On the other hand, different sources of sulfur have been used, including elemental sulfur, disulfides, thiocyanates, as well as S-oxidized species.

A simple and general method has been developed for the synthesis of S-alkyl and S-benzyl phosphorothioates *via* a one-pot reaction of alkyl or benzyl halides (or tosylates) with a mixture of diethyl phosphite, ammonium acetate, and sulfur powder, in the presence of acidic alumina, under solvent-free conditions, using microwave irradiation (Scheme **2**) [6, 7]. Less reactive alkyl halides such as 1-hexyl chloride also reacted efficiently.

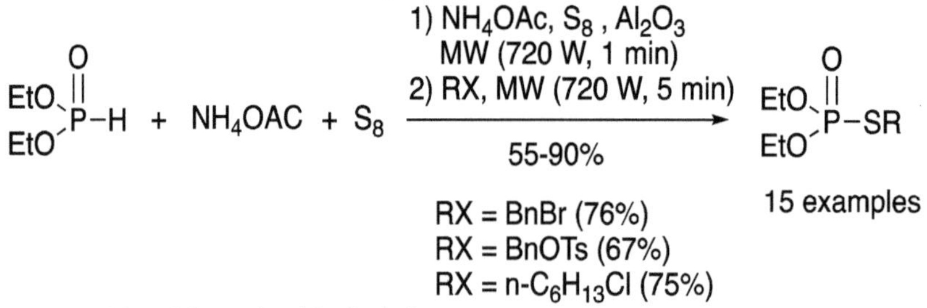

$$(EtO)_2P(O)H + NH_4OAc + S_8 \xrightarrow[\text{2) RX, MW (720 W, 5 min)}]{\text{1) } NH_4OAc, S_8, Al_2O_3 \text{ MW (720 W, 1 min)}} (EtO)_2P(O)\text{-SR}$$

55-90%

RX = BnBr (76%)
RX = BnOTs (67%)
RX = n-C_6H_{13}Cl (75%)

15 examples

Scheme 2. One-pot P-S bond formation / S-alkylation.

Renard, Mioskowski, and co-workers reported a mild synthetic method involving a thiocyanate as the sulfur-source and H-phosphine oxides, phosphinates, or phosphonates, as the phosphorus-source (Scheme **3**) [8]. Thiocyanates [9] proved their compatibility with a wider variety of functional groups than other previously described sulfur species. The reaction takes place under catalysis by hindered non-nucleophilic bases, such as phosphazene P4-tBu. The authors applied the method to achieve the synthesis of **Ph-VX** as a substrate for antibody-catalyzed hydrolysis. This phenylphosphorothioate represents a less toxic analogue of methylphosphorothioate **VX**, which has been described as the most powerful acetylcholinesterase inhibitor. As the thiocyanate partner already bears a basic hindered amine moiety, the use of an additional base was not necessary.

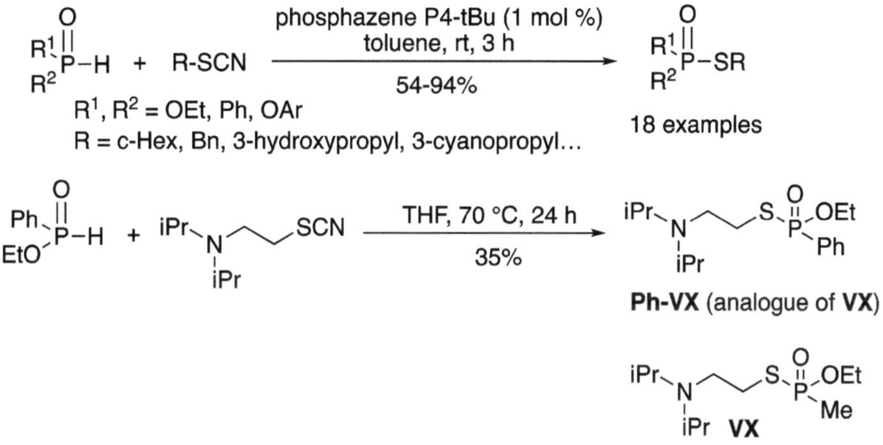

Scheme 3. Thiocyanates as sulfur-source to access phosphorothioates.

S-glycosyl anomeric thiocyanates have been transformed in the corresponding phosphorothioates by Michaelis-Arbuzov type reaction with triethylphosphite, dimethyl phenylphosphonite, or methyl diphenylphosphinite (Scheme 4) [10].

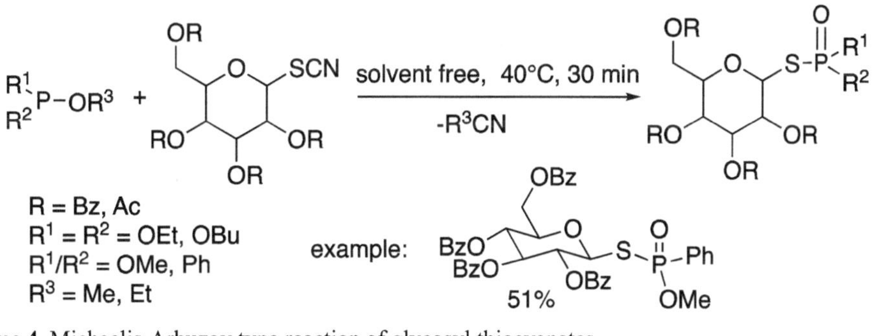

Scheme 4. Michaelis-Arbuzov type reaction of glycosyl thiocyanates.

Nguyen et al. described the phosphorylation of disulfides by using N-heterocyclic carbenes (NHCs) as efficient organocatalysts to promote the reaction under mild conditions (Scheme 5) [11]. The protocol is simple, metal- and oxidant-free, and needs very short reaction times. Imidazolylidene catalysts (5 mol%) gave the best results for this P-S coupling reaction. In the mechanism proposed by the authors, the NHC acts as a Brønsted base catalyst to activate the phosphorus center *via* a proton-transfer to form an ion-pair intermediate. Then, the nucleophilic phosphorus reacts with the disulfide to form the phosphorothioate product. Another procedure to prepare S-phenyl phosphonothioate from phosphites and disulfides employed zinc dust, in ethanol, at room temperature [12].

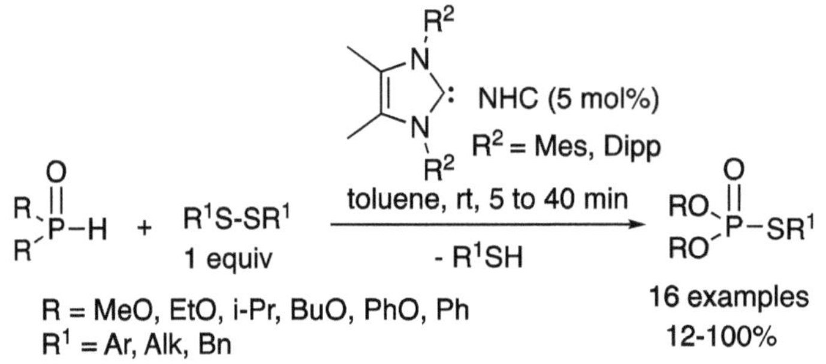

Scheme 5. NHC-catalyzed reaction of phosphites with disulfides.

Li *et al.* have developed a synthesis for S-methyl and S-butyl phosphorothioates starting from various phosphites and dimethyl or dibutyl sulfoxide, activated by 2,4,6-trichloro [1, 3, 5]-triazine (cyanuric chloride, TCT) [13]. The procedure was also applied to the introduction of SCD_3 group into bioactive phosphorothioates. The high purity of the obtained deuterated derivatives allows their use in mechanistic biological studies (Scheme 6).

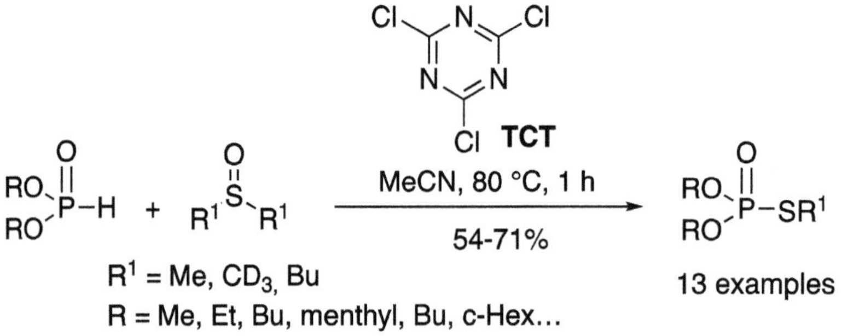

Scheme 6. Reaction of phosphites with TCT-activated sulfoxides.

A metal- and oxidant-free S–P(O) bond construction has been developed by directly reacting P(O)H compounds with sulfinic acids [14]. The method is compatible with a broad range of substituents on various substrates including aliphatic, aromatic, and heterocyclic moieties, as well as optically active P-chiral partners, which gave a stereospecific coupling (Scheme 7). The authors proposed a mechanism involving the nucleophilic attack of the P(III) form of the phosphite R_2P-OH on the sulfinyl cation, which is generated *in situ* by dehydration of sulfinic acid. Finally, the formed intermediate undergoes reduction by triphenyl phosphine to afford the desired phosphorothioate. The authors suggested that maybe other electrophilic species could be involved.

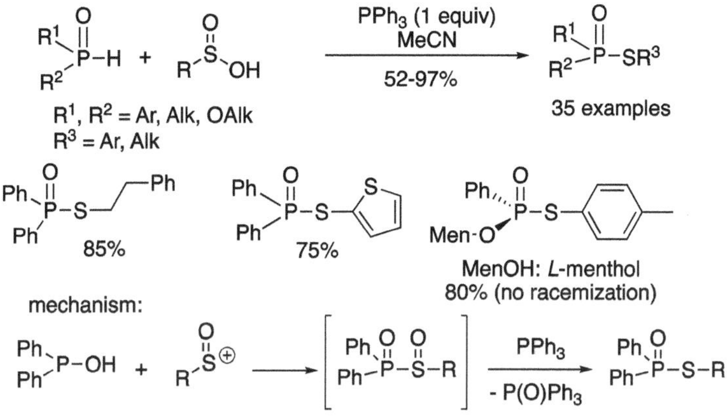

Scheme 7. Reaction of phosphites with sulfinic acids.

P-S Bond Formation from P-electrophile and S-nucleophile

A priori a routine method, the strategy employing a phosphorus-electrophile and a sulfur-nucleophile has been less employed than the previous one, probably because the sensitivity of P(O)X electrophilic reagents. In the main cases the thiolate is generated from the thiol by treatment with a base, then phosphoryl chloride is added to give the expected phosphorothioate (Scheme **8**).

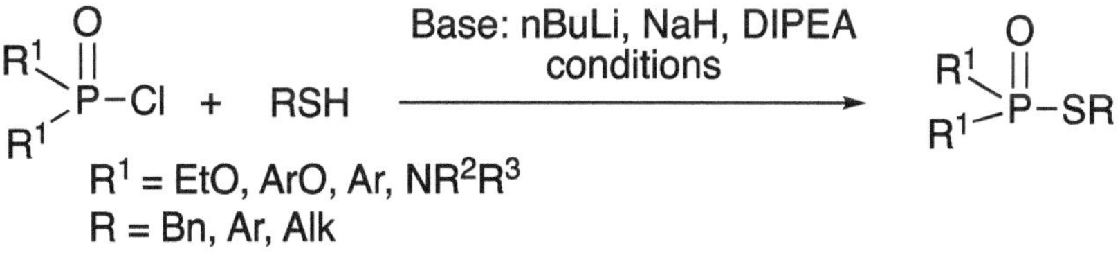

Scheme 8. P-S formation from P(O)Cl reagent and thiolates.

Katritzky *et al.* used instead chlorophosphates the more stable P-benzotriazole (P-Bzt) derivatives, giving access also to more sensitive biphenol derivatives (Scheme **9**) [15].

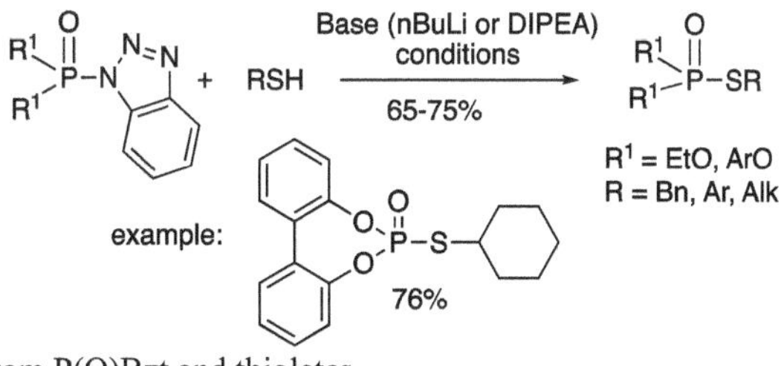

Scheme 9. P-S formation from P(O)Bzt and thiolates.

Odorless and stable sodium arylsulfinates have been used as precursors of arylsulfenyl radicals, which reacted with H-phosphine oxides or dialkyl phosphites to afford phosphorothioates in moderate to good yields (Scheme **10**) [16]. The yield was poor when sodium methanesulfinate was used instead of a sodium arylsulfinate. The reaction occurred under acidic and reductive conditions, in the presence of sulfuric acid and water, the reducing agent being the phosphorus partner. The mechanism proposed by the authors is also given in Scheme **10**. The resulting arylsulfenyl radical reacts with a P-electrophile intermediate leading to the phosphorothioate. The authors showed that the reaction works also by using diaryl chlorophosphines, without acid.

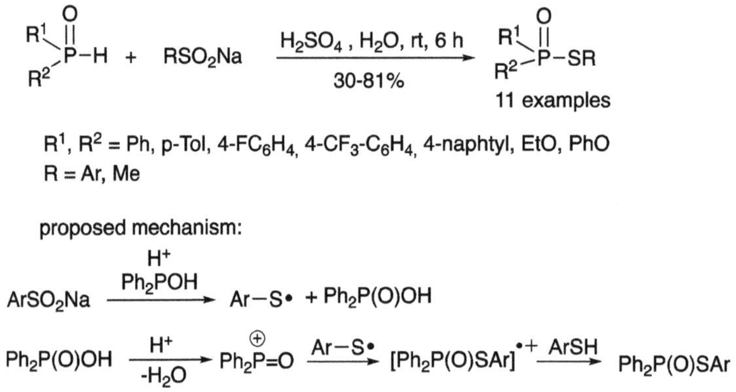

R^1, R^2 = Ph, p-Tol, 4-FC$_6$H$_4$, 4-CF$_3$-C$_6$H$_4$, 4-naphtyl, EtO, PhO
R = Ar, Me

proposed mechanism:

Scheme 10. P-S formation from sodium sulfinates and phosphites.

Metal-free Oxidative P-S Bond Formation

Some methods to access phosponothioates are based on the P-S formation from phosphites and thiols, both of them nucleophiles, in the presence of an oxidant.

N-chlorosuccinimide (NCS) [17] and 1,3-dichloro-5,5-dimethylhydantoin (DCDMH) [18] have been used by two different research groups to promote the P-S coupling between thiols and dialkylphosphites (Scheme **11**). The authors proposed the formation of a sulfenyl chloride (R^{1}SCl) as the reaction intermediate, which reacted as the electrophile with the nucleophilic phosphorus partner. Both aromatic and aliphatic thiols were tolerated in the reaction conditions.

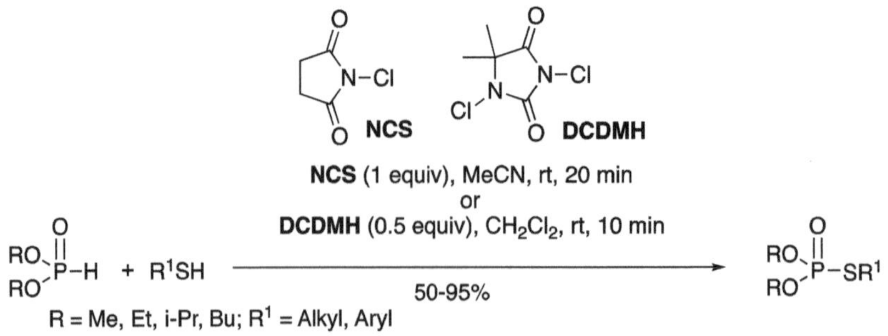

R = Me, Et, i-Pr, Bu; R^1 = Alkyl, Aryl

Scheme 11. NCS or DCDMH-promoted reaction between thiols and phosphites.

Pan, Wu, and coworkers described the use of di-tert-butyl peroxide (DTBP) [19] and tert-butyl peroxybenzoate (TBPB) [20] to promote the P-S bond formation between thiols and phosphites (Scheme **12**). In the case of the method using DTBP it was demonstrated that disulfides could be also used instead the corresponding thiols and only aromatic thiols were used as partners, however in a large variety of structures. The authors suggested that the thiyl radical is probably the reaction intermediate (see mechanism in Scheme **12**). In the case of TBPB-promoter the reaction has been done mainly between H-phosphine oxides or H-phosphinates and aliphatic thiols, leading to the expected phosphorothioates in good yields.

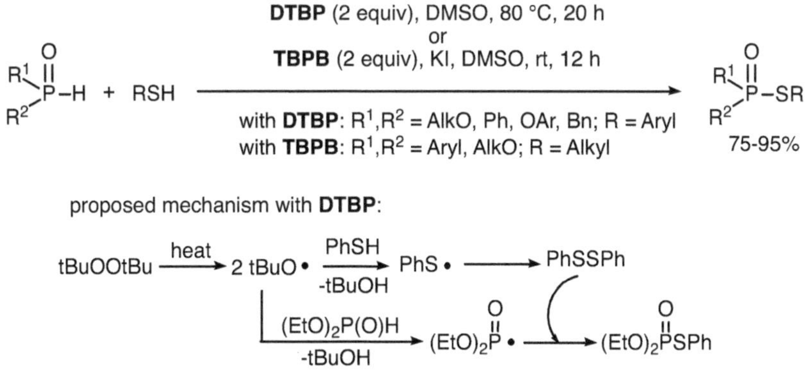

Scheme 12. Peroxide-promoted reaction between thiols and phosphites.

A iodine-catalyzed cross-coupling reaction between phosphonates and diaryl disulfides with H_2O_2 as oxidant was also reported (Scheme **13**) [21]. Various H-phosphonates were employed for coupling with diaryl disulfides bearing an electron-donating or an electron-withdrawing substituent on para- or meta-position.

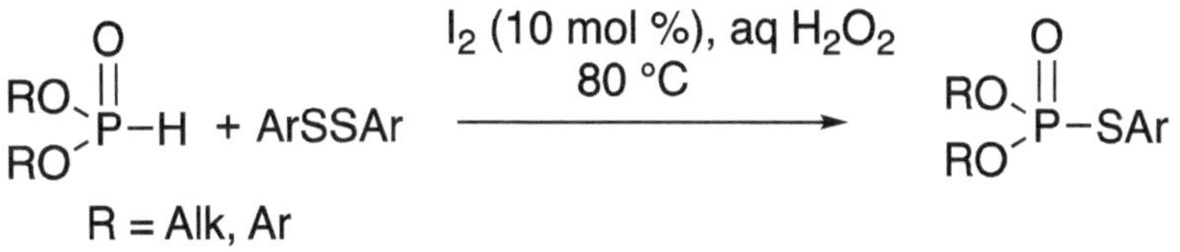

Scheme 13. I_2/H_2O_2 mediated reaction between disulfides and phosphites.

An oxidative cross-coupling of thiols with P(O)H compounds was achieved *via* organic dye-sensitized photocatalysis [22]. The commercially available and inexpensive Rose Bengal was used as the photocatalyst, and air as a convenient green oxidant (Scheme **14**). The method was efficient for the synthesis of a range of phosphorothioates in moderate to excellent yields and showed good functional-group tolerance.

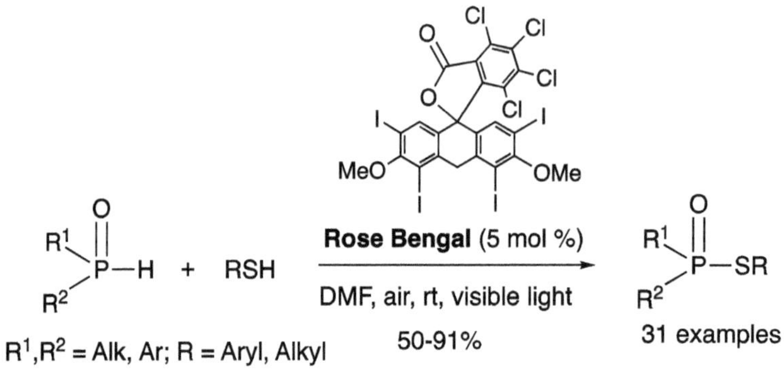

Scheme 14. Visible-light-mediated oxidative coupling between thiols and phosphites.

A cheap method for the synthesis of S-aryl phosphorothioates consisted in the cross-dehydrogenative coupling of P(O)H compounds and aryl thiols, in the presence of Na_2CO_3 as the base and air as the oxidant. The reaction was performed under mild conditions and led to the expected products in moderate to good yields. The proposed mechanism involved the sodium phosphite salt as the nucleophile and the disulfide resulted by the oxidation of the thiol in air, as the electrophile (Scheme **15**) [23].

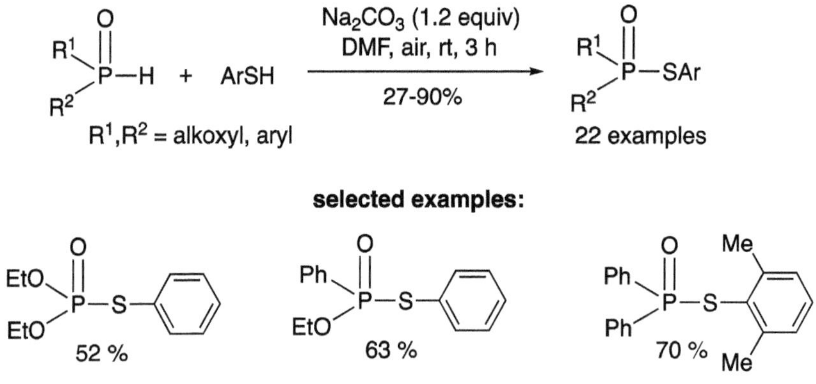

Scheme 15. Na_2CO_3/air mediated reaction between thiols and phosphites.

A catalytic version of the P-S cross-dehydrogenative coupling was described using Cs_2CO_3 as the base-catalyst and molecular oxygen as the oxidant (Scheme **16**) [24]. The method provided good functional-group tolerance, a broad substrate scope, and could be applied on gram scale and in the preparation of P-S containing bioactive molecules such as demeton, echothiopate, and an AZT-derivative.

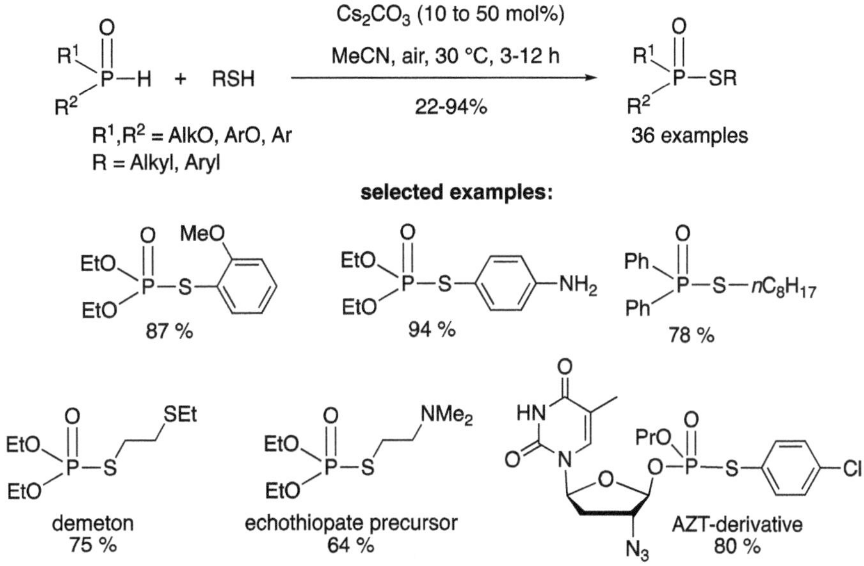

Scheme 16. Cs$_2$CO$_3$/air mediated reaction between thiols and phosphites.

Metal-catalyzed Processes Involving Phosphorus and Sulfur Sources

Transition metals as copper or palladium have been used to form a P-S bond from phosphites and various sulfur-partners. Specifically, efforts have been made to replace thiols by others sulfur reagents that are air-stable and odorless. Another synthetic strategy consists in the one-pot formation of P-S and S-C bonds by using three reaction partners, including the phosphorous and the sulfur sources, and a coupling partner, in the presence of a metal catalyst. In these cases, the metal is mainly necessary for the C-S cross-coupling step.

Metal-catalyzed P-S Formation

Tang, Zhao, and co-workers reported a practical method for the synthesis of S-aryl phosphorothioates under copper-catalysis [25]. P-S bond was formed by coupling dialkyl phosphites with diaryl disulfides, in the presence of catalytic amounts of copper iodide and diethylamine, in DMSO at 30 °C, in air atmosphere (Scheme **17**). A nucleotide P-S derivative was synthesized by this procedure in 92% yield. Moreover, the method was extended to selenium and tellurium counterparts.

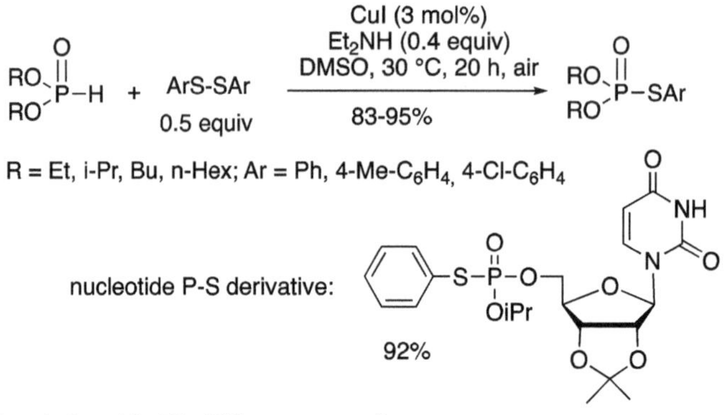

Scheme 17. Cu-catalyzed phosphite/disulfide cross-coupling.

Kaboudin *et al.* published a copper-catalyzed coupling between phosphites and thiophenols [26]. The reaction took place under mild conditions (room temperature), using copper iodide and triethylamine, in DMF, 5 hours, in air (Scheme **18**). The use of different substituents on the aromatic ring had not influence on the yield of S-aryl phosphorothioates.

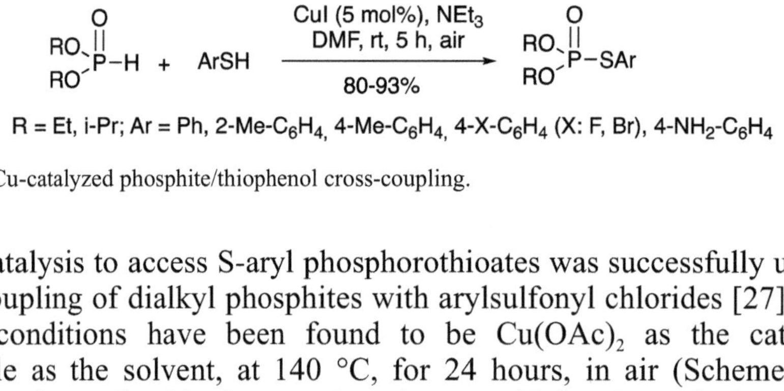

Scheme 18. Cu-catalyzed phosphite/thiophenol cross-coupling.

Copper-catalysis to access S-aryl phosphorothioates was successfully used in the case of coupling of dialkyl phosphites with arylsulfonyl chlorides [27]. The best reaction conditions have been found to be $Cu(OAc)_2$ as the catalyst and acetonitrile as the solvent, at 140 °C, for 24 hours, in air (Scheme **19**). The reaction works well with electron donating or withdrawing substituents on the aromatic ring. 2-Naphtyl and 2-thienyl derivatives have been also obtained in good yields. The authors proposed a mechanism based on single-electron transfer (SET) involving a $ArS-Cu^{III}(OAc)P(O)$ species.

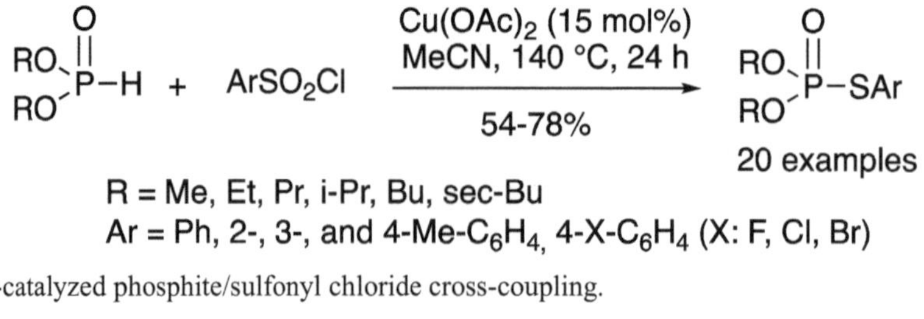

Scheme 19. Cu-catalyzed phosphite/sulfonyl chloride cross-coupling.

An improved procedure for the Cu(II)-catalyzed coupling of sulfuryl chloride with P(O)–H was recently described (Scheme 20) [28]. L-proline was used as the ligand of choice, which allowed a smaller catalyst loading and also expanded the substrate scope, in particular the sulfur-partner. Moreover, the Cu^{II}(proline)$_2$ catalyst complex can be prepared in gram-scale, is easy to handle, and could be recovered and reused up to ten runs.

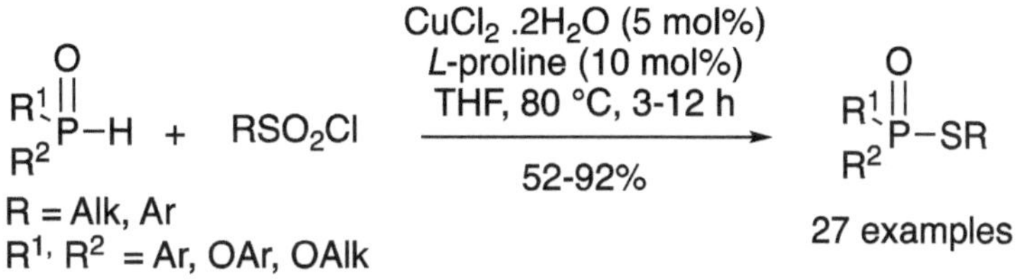

$$R^1_{}\!\!\!\underset{R^2}{\overset{O}{\|}}\!\!P\text{-}H + RSO_2Cl \xrightarrow[52\text{-}92\%]{\substack{CuCl_2 \cdot 2H_2O\ (5\ mol\%) \\ L\text{-proline}\ (10\ mol\%) \\ THF,\ 80\ °C,\ 3\text{-}12\ h}} R^1_{}\!\!\!\underset{R^2}{\overset{O}{\|}}\!\!P\text{-}SR$$

R = Alk, Ar
R^1, R^2 = Ar, OAr, OAlk

27 examples

Scheme 20. Cu(II)/proline-catalyzed phosphite/sulfonyl chloride cross-coupling.

Sulfonyl hydrazides have also been used as a thiol surrogate in the copper-catalyzed P-S cross-coupling with and various phosphorous P-H partners (phosphites, phosphinites, phosphine oxides, phosphine boranes) [29]. The optimized reaction conditions were the following: copper iodide as the catalyst, in 1,4-dioxane, at 80 °C, in air (Scheme 21). The presence of a base was not necessary, whereas the use of a Boc-protected hydrazide function inhibited the reaction. In this case not only variously substituted S-aryl phosphorothioates have been obtained, but also S-alkyl and S-benzyl derivatives were accessible in good yields. In the case of the use of a P-stereogenic enantiopure substrate, the P-configuration was not preserved during the transformation.

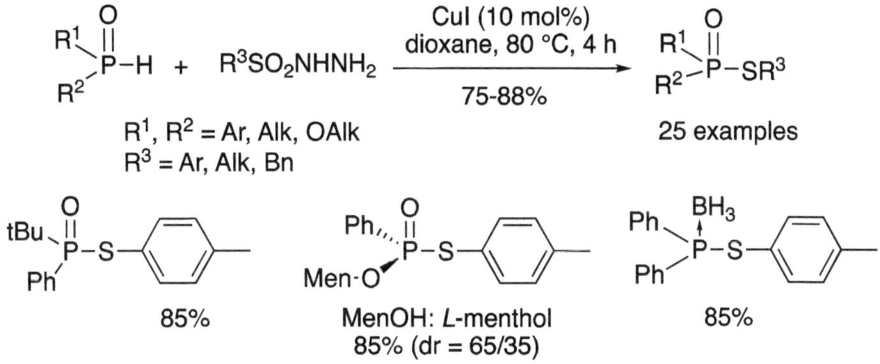

$$R^1_{}\!\!\!\underset{R^2}{\overset{O}{\|}}\!\!P\text{-}H + R^3SO_2NHNH_2 \xrightarrow[75\text{-}88\%]{\substack{CuI\ (10\ mol\%) \\ dioxane,\ 80\ °C,\ 4\ h}} R^1_{}\!\!\!\underset{R^2}{\overset{O}{\|}}\!\!P\text{-}SR^3$$

R^1, R^2 = Ar, Alk, OAlk
R^3 = Ar, Alk, Bn

25 examples

tBu, Ph–P(=O)–S–C$_6$H$_4$–Me
85%

Ph, MenO–P(=O)–S–C$_6$H$_4$–Me
MenOH: L-menthol
85% (dr = 65/35)

Ph, Ph–P(→BH$_3$)–S–C$_6$H$_4$–Me
85%

Scheme 21. Cu-catalyzed phosphite/sulfonyl hydrazide cross-coupling.

A general access to a large variety of phosphorothioates consist in the Pd-catalyzed dehydrogenative phosphorylation of thiols (Scheme 22) [30]. Both aromatic and aliphatic thiols coupled with three kinds of P(O)H compounds: phosphites (including cyclic derivatives), phosphinates, and secondary phosphine oxides. Electron-rich and electron-deficient thiophenols with various functional groups worked efficiently. Notably, this catalytic process allowed access to P-chiral phosphorothioates, the reaction occurring with retention of configuration at phosphorus.

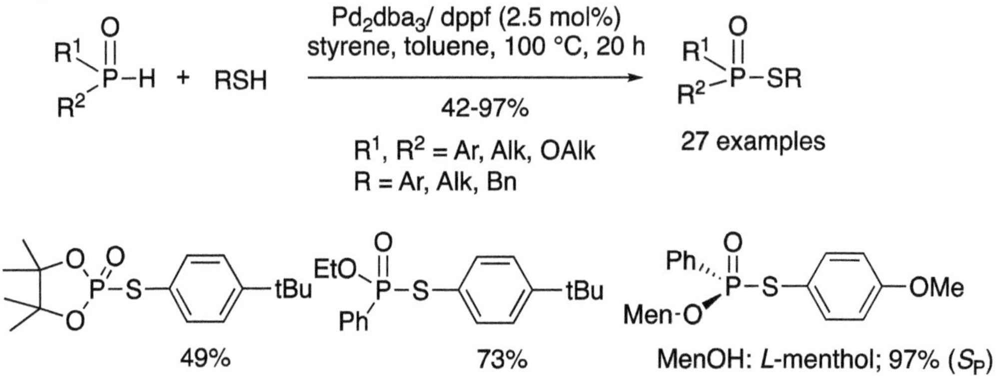

Scheme 22. Pd-catalyzed phosphite/thiol cross-coupling.

A publication mentioned the palladium-catalyzed formation of an S-aryl phosphorothioate from a phosphite and a sodium arylsulfinate, in the presence of silver acetate [31]. The product was in this case undesired, as the expected reaction was the desulfitative C-P coupling leading to the aryl phosphonate.

Metal-catalyzed Three-component Reactions

A highly efficient three-component copper-catalyzed reaction has been developed to access S-aryl phosphorothioates from phosphites, elemental sulfur, and aryl boronic acids (Scheme 23) [32]. A broad variety of functionalized boronic acids have been used successfully, including electron rich and deficient aromatics, as well as sterically hindered substrates such 2,6-dimethyl phenylboronic acid. The authors showed that diethyl phosphite and octasulfur react first in the presence of triethylamine to give the phosporothioic acid ammonium salt, which then reacts with the phenylboronic acid under Chan-Evans-Lam copper-catalyzed conditions to give S-phenyl phosphorothioate.

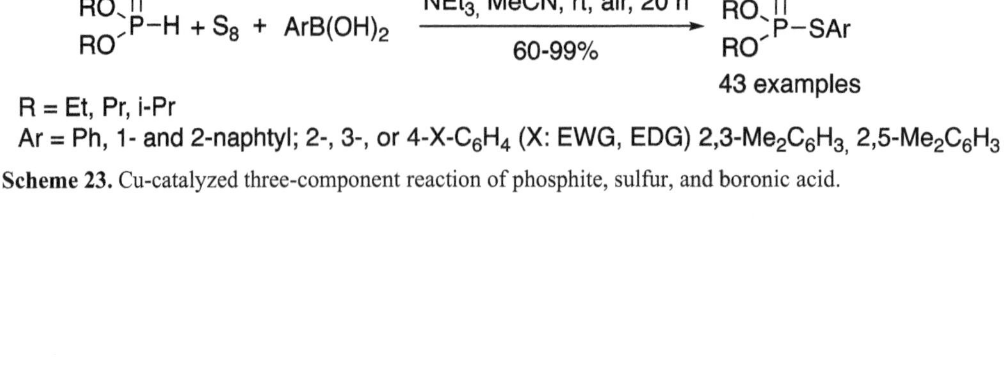

R = Et, Pr, i-Pr
Ar = Ph, 1- and 2-naphtyl; 2-, 3-, or 4-X-C_6H_4 (X: EWG, EDG) 2,3-$Me_2C_6H_3$, 2,5-$Me_2C_6H_3$

Scheme 23. Cu-catalyzed three-component reaction of phosphite, sulfur, and boronic acid.

Another three-component copper-catalyzed ligand-free reaction has been developed to access S-aryl phosphorothioates from phosphites, sulfur powder, and diaryliodonium or aryl diazonium salts (Scheme **24**) [33]. The aryl coupling partners can be readily prepared from simple corresponding arene precursors, aryl iodides or aromatic amines. The method covers a broad scope of substrates. When diphenylphosphine oxide was used instead dialkyl phosphite, a yield of only 22% was obtained.

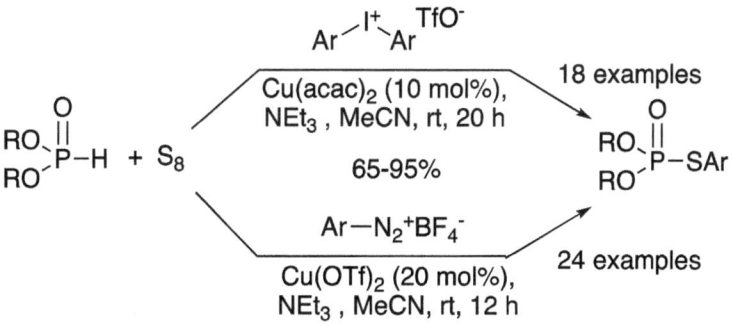

R = Me, Et, i-Pr, Bn
Ar: Ph, 2-, 3-, or 4-X-C_6H_4 (X: EWG, EDG) 2,3-$Me_2C_6H_3$, 2,5-$Me_2C_6H_3$

Scheme 24. Cu-catalyzed three-component reaction of phosphite, sulfur, and diaryliodonium or diazonium salt.

A Fe_3O_4-functionalized Cu-BTC (H_3BTC = 1,3,5-benzenetricarboxylic acid) was prepared and used as a heterogeneous catalyst in the synthesis of S-aryl phosphorothioates (Scheme **25**) [34]. The catalyst could be magnetically recovered and reused without significant loss of its activity after six cycles. The reaction involves three components: aniline, phosphite, and elemental sulfur. Aryl diazonium salt is generated *in situ* from aniline in the presence of tBuONO and $MeSO_3H$, then reacted with the phosphite and the sulfur powder under copper-catalysis. A large variety of substrates have been successfully used, including electron-rich and -deficient aryl amines, and various phosphites. Moreover, gram-scale synthesis was performed with high yield (82%), which provides the possibility to apply this method in practical scale-up synthesis.

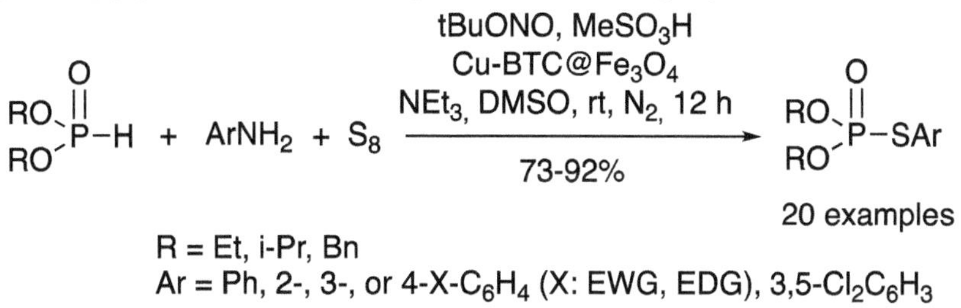

$$\underset{RO}{\overset{RO}{>}}\overset{O}{\underset{\|}{P}}-H + ArNH_2 + S_8 \xrightarrow[\text{73-92\%}]{\substack{\text{tBuONO, MeSO}_3\text{H} \\ \text{Cu-BTC@Fe}_3\text{O}_4 \\ \text{NEt}_3\text{, DMSO, rt, N}_2\text{, 12 h}}} \underset{RO}{\overset{RO}{>}}\overset{O}{\underset{\|}{P}}-SAr$$

20 examples

R = Et, i-Pr, Bn
Ar = Ph, 2-, 3-, or 4-X-C_6H_4 (X: EWG, EDG), 3,5-$Cl_2C_6H_3$

Scheme 25. Three-component reaction catalyzed by a Cu-composite of phosphite, aniline, and sulfur.

From Phosphorothioic Acids or their Salts

By S-alkylation

Wu and coworkers reported a transition-metal-free synthetic method to access phosphorothioate esters from allylic alcohols or their ethers and phosphorothioic acids (Scheme 26) [35]. The reaction was performed under photochemical conditions (UV light). A wide range of acyclic and cyclic substituted allylic alcohols and methyl ethers have been used with good results. The reaction scope was then expanded to involve alcohols protected with groups such as benzyl, *tert*-butyldimethylsilyl (TBS), and benzoyl ethers (Bzl). To determine the reaction stereospecificity, one experiment was carried out using an enantioenriched (53% ee) allylic ether. A nearly racemic (< 10% ee) phosphorothioate ester was obtained. The authors explained this result by postulating intermediates such as allylic carbocations and/or allylic radicals generated from the photochemical cleavage of the C-S bond of the product.

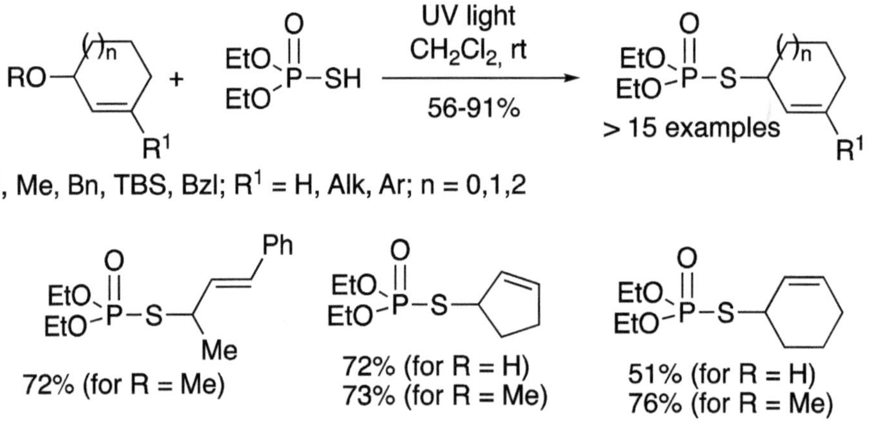

Scheme 26. Photochemical-promoted reaction of alcohols/ethers with phosphorothioic acid.

The same research group reported the direct substitution of various alcohols with *O,O*-diethyl phosphorothioic acid, catalyzed by Ga(OTf)$_3$ as an inexpensive, hydrolytically stable Lewis acid (Scheme 27) [36]. A wide range of primary and secondary alcohols has been successfully used, however only in benzylic or allylic series. The method was applicable for tertiary alcohols bearing alkyl substituents. When the reaction was done on the enantiopure (*S*)-α-methylbenzyl alcohol the expected phosphorothioate was isolated in nearly racemic form (8% ee).

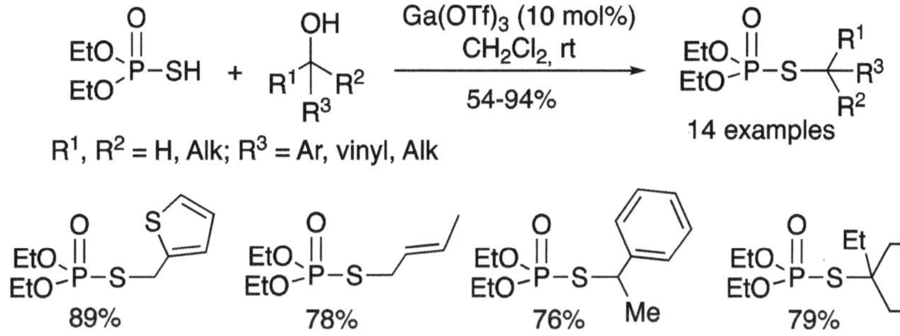

Scheme 27. Ga(OTf)₃-catalyzed reaction of phosphorothioic acid with alcohols.

Porter and coworkers used the thioiminium salt prepared from N,N-dimethylthioformamide and Meerwein's salt as the coupling agent for the direct conversion of primary and secondary alcohols to the corresponding phosphorothioic esters by reaction with sodium phosphorothioate (Scheme **28**) [37]. Alternatively, the reaction can be performed with phosphorothioic acid in the presence of imidazole, under vigorous stirring. The mechanism proposed by the authors consists in the nucleophilic addition of the alcohol to the thioiminium salt with loss of ethanethiolate, then nucleophilic displacement of dimethylformamide (DMF) by the phosphorothioate salt to afford the phosphorothioic ester. Both of primary and secondary alcohols were converted to the corresponding phosphorothioates in good yields. (*E*)-Cinnamyl alcohol could be converted to the desired product, without allylic rearrangement. The stereochemical course of the reaction was studied by using (*S*)-α-methylbenzyl alcohol. The corresponding phosphorothioate was obtained with inversion of configuration.

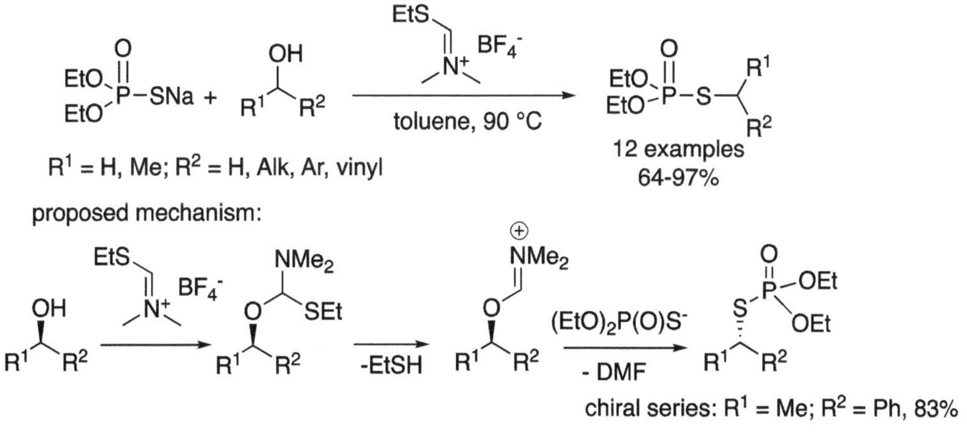

Scheme 28. Ga(OTf)₃-catalyzed Conversion of alcohols to phosphorothioates using a thioiminium coupling agent.

By Metal-mediated C-S Bond Formation

Some publications described metal-catalyzed reactions consisting in the direct C-S coupling between a phosphorothioate salt and an aryl or vinyl partner.

S-vinyl phosphorothioates have been prepared in good yields, under mild reaction conditions, from potassium phosphorothioate and vinyl(phenyl) iodonium salts, under copper-catalysis (Scheme **29**) [38]. The process occurred with retention of the double bond geometry.

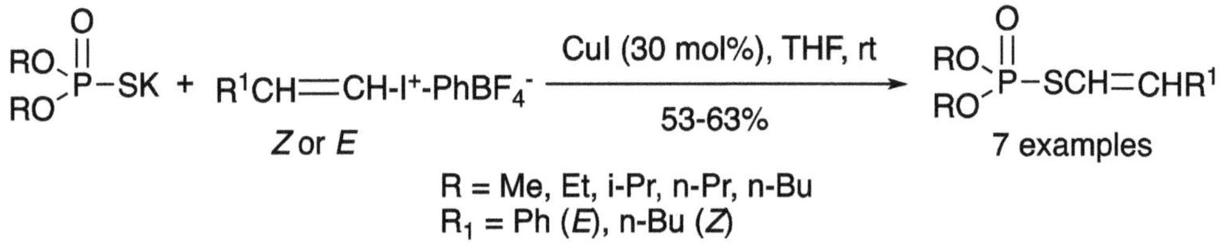

$$\text{(RO)}_2\text{P(O)-SK} + R^1\text{CH}=\text{CH-I}^+\text{-PhBF}_4^- \xrightarrow[53-63\%]{\text{CuI (30 mol\%), THF, rt}} \text{(RO)}_2\text{P(O)-SCH}=\text{CHR}^1$$

Z or E

R = Me, Et, i-Pr, n-Pr, n-Bu
R$_1$ = Ph (E), n-Bu (Z)

7 examples

Scheme 29. Cu-catalyzed C-S coupling of phosphorothioate salts and hypervalent iodine.

In one publication the phosphorothioate disulfide species was involved in C-S Ni- or Pd-catalyzed reactions with aryl halides to afford S-aryl phosphorothioates [39]. For the mechanistic study, potassium phosphorothioate was reacted with the nickel(II)-complex obtained from bromobenzene in the presence of triphenylphosphine ligand, affording S-phenyl phosphorothioate in good yield (Scheme **30**).

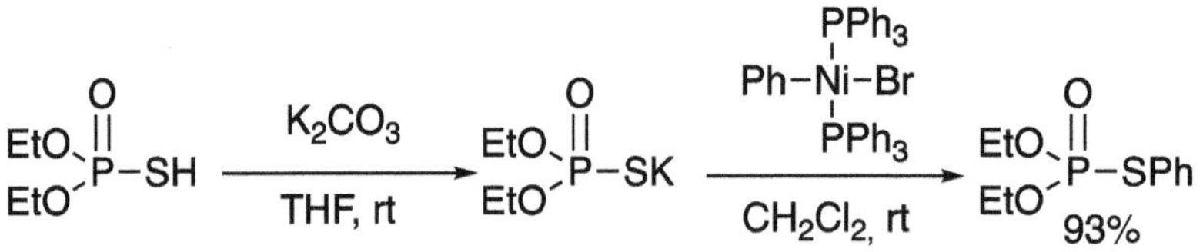

$$\text{(EtO)}_2\text{P(O)-SH} \xrightarrow[\text{THF, rt}]{\text{K}_2\text{CO}_3} \text{(EtO)}_2\text{P(O)-SK} \xrightarrow[\text{CH}_2\text{Cl}_2,\text{ rt}]{\text{Ph-Ni(PPh}_3)_2\text{-Br}} \text{(EtO)}_2\text{P(O)-SPh} \quad 93\%$$

Scheme 30. Ni-mediated C-S coupling of phosphorothioate salts.

REACTIONS OF PHOSPHOROTHIOIC ESTERS

Specific reactivities of phosphorothioates are based on the ionic, radicalar, or metal-mediated cleavage of the P-S bond. In processes involving an anionic P-S cleavage that include the rearrangement reactions, the nucleophilic attack of a lithium carbanion or of an alkoxide takes place on the phosphorus atom, with releasing of a thiolate. In the case of a homolytic P-S cleavage, the radical species attacks the sulfur atom, with releasing of a P-centered radical. The metal-mediated P-S cleavage was less studied and probably involves an S-M-P(O) intermediate (M = Pd, Pt). Phosphorothioate was also used as a living group in allylic substitutions with carbon-nucleophiles (Grignard reagents) and fluorine nucleophiles, and the differences and benefits in comparison with other leaving groups (*i.e.* chloride, carbonate) were demonstrated.

Rearrangement Reactions

The different types of anionic rearrangements described below consist in the migration of the electrophilic phosphorus from the sulfur-atom to the carbanion generated in the 2-, 3-, or 4- position of the sulfur by treatment with a lithium-base (*via* deprotonation or halogen/lithium exchange), with release of the thiolate function that can be trapped by protonation or alkylation to afford the corresponding P,S-difunctionalized products (Scheme **31**) [40]. Several of the obtained compounds have shown interesting metal-chelating properties due to both P and S functionalities, and have been used in the synthesis of metal complexes such as platinum complexes [41 - 43] Fig. (**2**) or as enzymes inhibitors (metallo-enzymes) that are based on Zn-chelation [44].

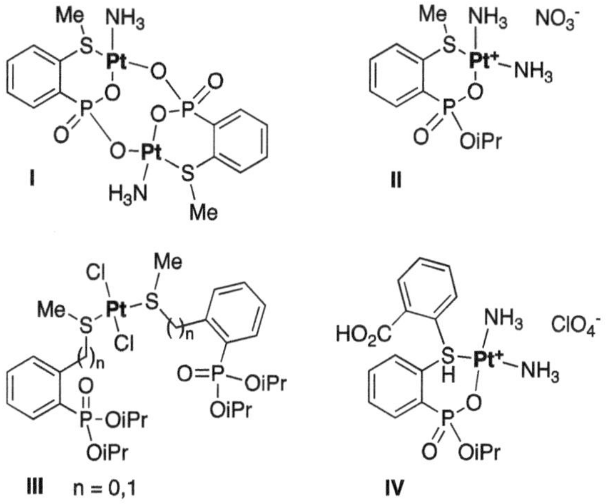

Fig. (2). Examples of Pt-complexes with difunctionalized P,S-compounds obtained *via* anionic P-S to P-C rearrangement.

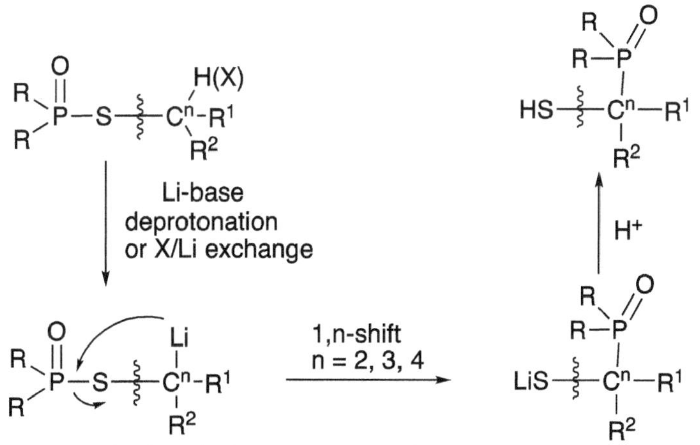

Scheme 31. General reaction for the anionic rearrangement of phosphorothioate into mercaptophosphonate.

Anionic 1,2-Rearrangement

While the rearrangement of the hydroxymethylene bis(phosphonate) into the corresponding phosphonomethyl phosphate occurring in basic media is the favored process in the oxygen series, the reverse rearrangement, S-phosphonomethyl phosphorothioate into the α-thiolato bis(phosphonate), has been described in the sulfur series (Scheme 32) [45]. The 1,2-shift occurred at -35 °C from the carbanion generated by deprotonation of the S-phosphonomethyl phosphorothioate by BuLi. The thiolate can readily loose sulfur at higher temperature leading to the methylene bis(phosphonate) carbanion, however the corresponding thiol and various sulfides have been prepared by rapid protonation or addition of alkylating agents at -40 °C.

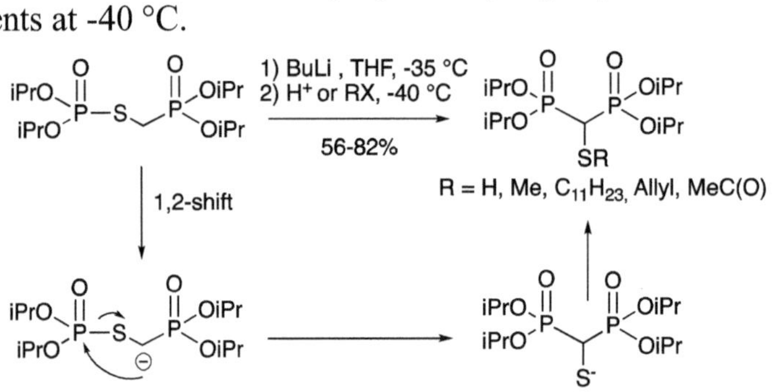

Scheme 32. Phosphorothioate-mercaptophosphonate rearrangement.

After several years, the rearrangement of S-alkyl O,O-dialkyl phosphorothioates to α-mercaptophosphonates and its mechanism have been studied [46]. The S-benzyl phosphorothioate was metallated alpha to the S-atom at low temperature by trityllithium and LiTMP (lithium 2,2,6,6-tetramethylpiperidide) to give a stabilized carbanion, which rearranged to α-mercaptophosphonate in 45% yield (Scheme 33). The mechanistic study using the chiral enantiopure substrate monodeuterated in α-position to the S-atom, showed that the 1,2-migration proceeds with retention of configuration.

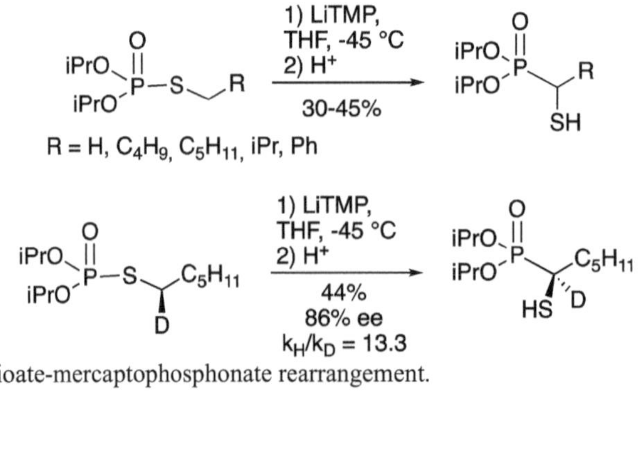

Scheme 33. Phosphorothioate-mercaptophosphonate rearrangement.

Another method to generate the carbanion in α-position to the S-atom, consisted in the thiophilic attack of silyl thioketones with lithium diethylphosphite [47]. Then, the reaction proceeded in the phosphorothioate to mercaptophosphonate rearrangement followed by the migration of the silyl group from the carbon to the sulfur atom (thia-Brook rearrangement), leading to the S-silylated sulfanylphosphonate carbanion (Scheme **34**).

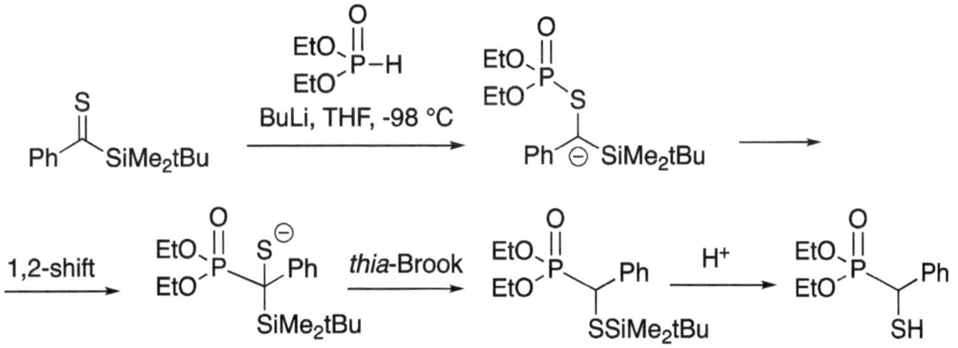

Scheme 34. Phosphorothioate-mercaptophosphonate/thia-Brook rearrangements.

Anionic 1,3-Rearrangement

The P-S to P-C 1,3-rearrangement of aryl or heteroaryl (such as pyridyl and thienyl) phosphorothioates into ortho-sulfanyl (hetero)aryl phosphonates has been developed by Masson and co-workers [48]. For the success of this anionic rearrangement which takes place after ortho-lithiation, it was necessary to use bulky isopropoxy groups on the phosphorus atom, as this prevents direct nucleophilic attack of the base at the P-atom (Scheme **35**).

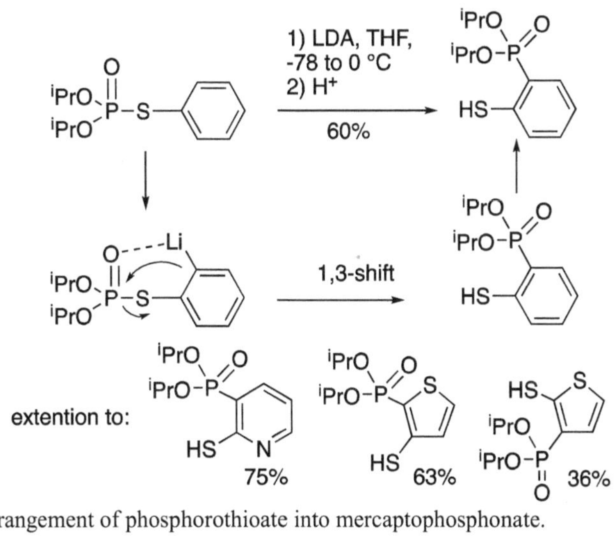

Scheme 35. 1,3-Rearrangement of phosphorothioate into mercaptophosphonate.

The same type of rearrangement has also been studied in phosphoramide series (Scheme **36**). By treatment of the substrate with sec-BuLi and subsequent methylation of the thiolate resulted after the P-S to P-C shift, the corresponding 2-methylsulfanyl phenylphosphoramide was obtained [49]. In an asymmetric version, starting from the chiral substrate derived from (*S*)-(+)-2-(anilinomethyl)-pyrrolidine, the ortho-lithiation by LDA, followed by 1,3-rearrangement and protonation, led to the expected thiophenol derivative in low yield (20%) [50].

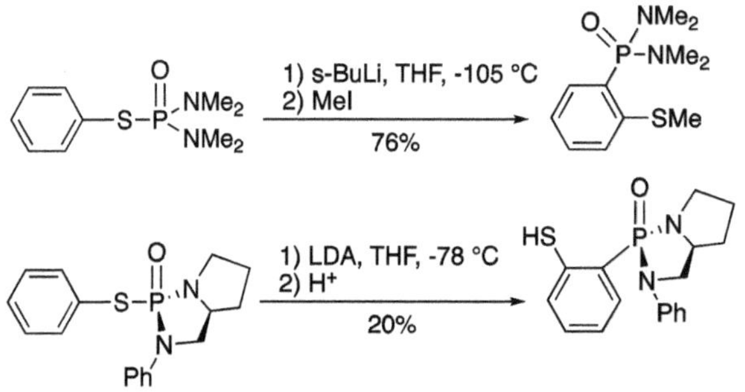

Scheme 36. 1,3-Rearrangement in phosphoramide series.

The rearrangement has been extended by Bonini *et al.* [51] in ferrocenyl phosphorothioate series and the resulting thiolate was trapped by protonation or methylation to give a racemic mixture of products (Scheme **37**). When the thiolate was tosylated with a chiral aminoalcohol derivative, or was used as the nucleophile for the ring-opening of a chiral epoxide, a mixture of separable diastereomers was obtained. An asymmetric version using a chiral phosphoramide enabled access to enantiomerically pure planar ferrocenyl derivatives. In this case the rearrangement required the use of tBuLi in refluxing ether.

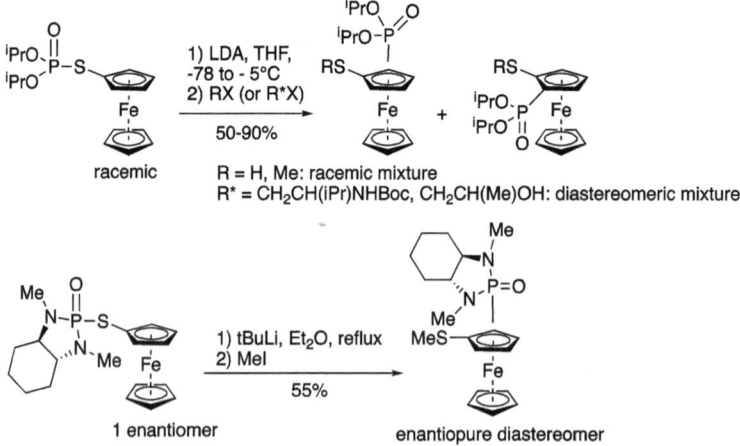

Scheme 37. 1,3-Rearrangement in ferrocenyl series.

The migration of the phosphinyl group from the sulfur to the carbon atom has also been extended to P-stereogenic phosphinothioate, leading stereoselectively to the corresponding ortho-mercaptophenyl phenyl-tert-butyl phosphine oxide, with retention of configuration at phosphorus (Scheme 38) [52].

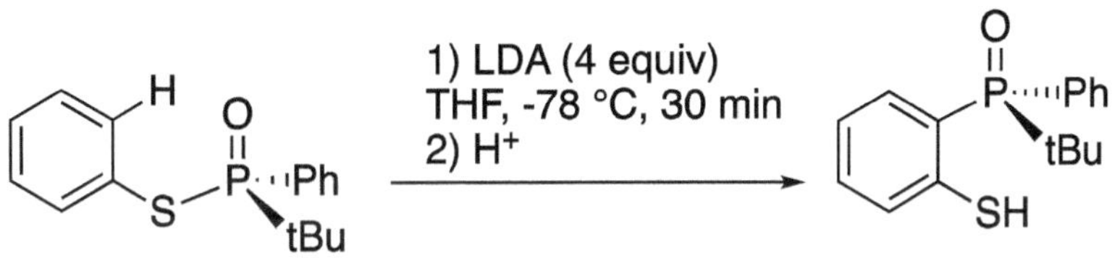

Scheme 38. Phenyl phosphinothioate to 2-mercaptophenyl phosphine oxide rearrangement.

Anionic 1,4-Rearrangement

The 1,4-sigmatropic rearrangement was used to transform *S*-(2-iodobenzyl) *O,O*-dialkyl phosphorothioate into the corresponding 2-(sulfanylmethyl)phenylphosphonate [43]. The iodide-lithium exchange by reacting the iodoaryl precursor with tert-butyllithium is followed by the 1,4-migration of the phosphoryl group from the sulfur atom to the carbanion of the aromatic ring (Scheme 39). At the end of the reaction, the acidic hydrolysis of the sulfanyl led to the benzylic thiol derivative in 70% yield. This example of 1,4 P-S to P-C shift appears to be the only one of this type.

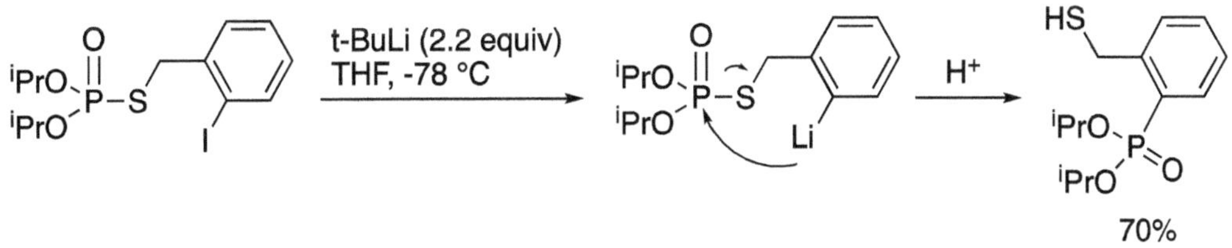

Scheme 39. 1,4-Rearrangement of phosphorothioate into mercaptophosphonate

Thiophosphate to Phosphoroamidate Rearrangement

The rearrangement of *S*-(2-aminoethyl) phosphorothioates to *N*-(2-mercaptoethyl) phosphoramidates has been studied by Gothelf and co-workers (Scheme 40) [53]. The transformation took place at room temperature, under basic conditions and the resulting thiols were rapidly oxidized into the corresponding disulfides. Alternatively, in a *one-pot* procedure starting from 2-bromoethylamine hydrobromide and potassium phosphorothioate, the thiol resulting from the P-S to P-N rearrangement has been subsequently S-benzylated.

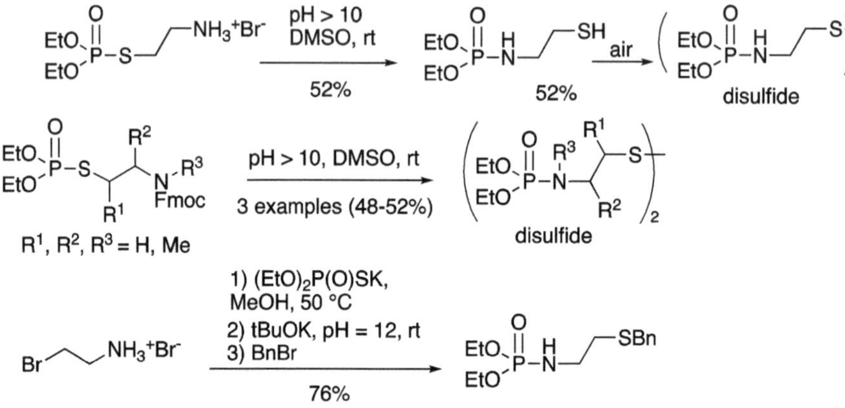

Scheme 40. Phosphorothioate to phosphoroamidate rearrangement.

Anionic P-S Bond Cleavage with C-S Bond Formation

Wu and Robertson developed a methodology to access sulfides (thioethers) by using the phosphorothioic acid as a safer and odorless surrogate for H_2S. First they described the synthesis of allylic sulfides starting from the corresponding phosphorothioic esters and alcohols (Scheme **41**). The reaction occurred *via* a double substitution process: the attack of the alkoxide to the P-atom of the corresponding acyclic or cyclic allylic phosphorothioate with release of the thiolate and subsequent displacement by this latter of the phosphate group. Mainly allylic and benzylic alcohols have been used, and various functional groups were tolerated. It was demonstrated that the reaction is stereospecific leading to enantioenriched sulfides from the corresponding chiral secondary alcohols [54].

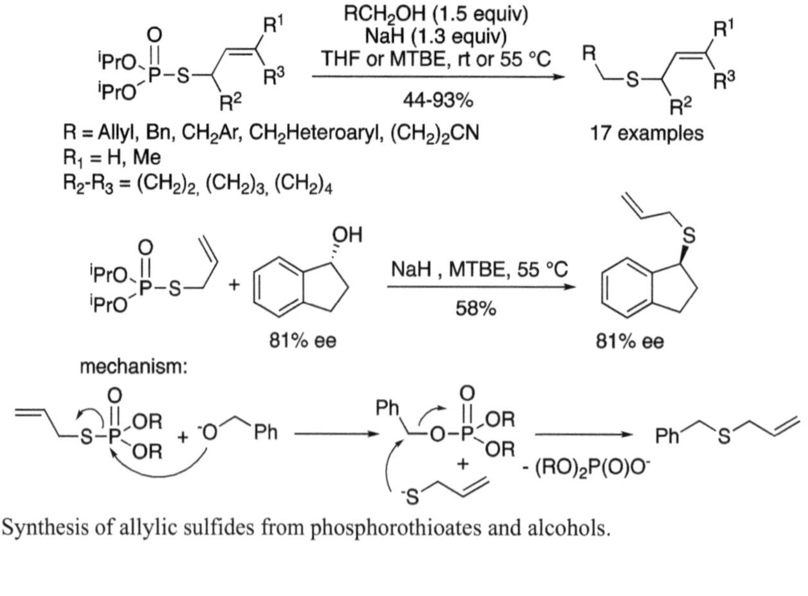

Scheme 41. Synthesis of allylic sulfides from phosphorothioates and alcohols.

An intramolecular version has been then reported by the same authors starting from γ-thiophosphoryl ketones and enabled the synthesis of enantioenriched 2-substituted tetrahydrothiophenes (thiolanes) (Scheme **42**) [55]. The transformation consists in the asymmetric CBS reduction of the ketone into the enantioenriched alcohol of which alkoxide attacks in intramolecular on the phosphorus center releasing the thiolate. Then, the intramolecular substitution (S_N2-type) of the phosphate by the thiolate led to the five-membered sulfur-heterocycle. The method was extended to the synthesis of a chiral thietane (four-membered thiacycle).

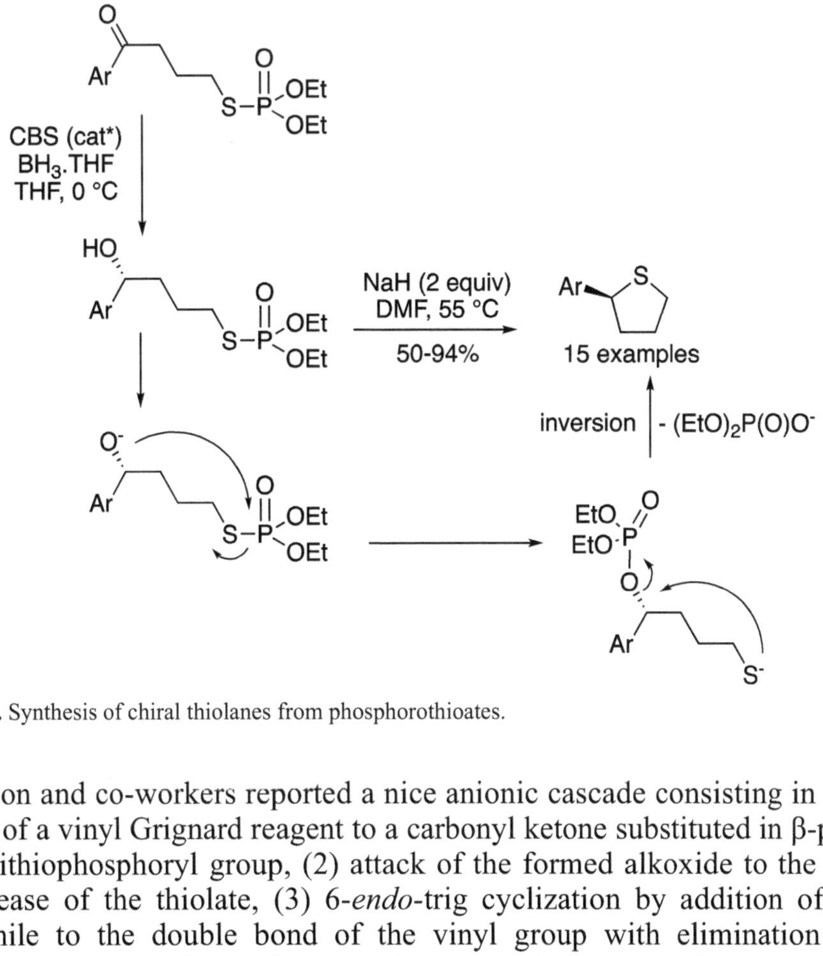

Scheme 42. Synthesis of chiral thiolanes from phosphorothioates.

Njardarson and co-workers reported a nice anionic cascade consisting in the: (1) addition of a vinyl Grignard reagent to a carbonyl ketone substituted in β-position with a dithiophosphoryl group, (2) attack of the formed alkoxide to the P-atom with release of the thiolate, (3) 6-*endo*-trig cyclization by addition of the S-nucleophile to the double bond of the vinyl group with elimination of the thiophosphate group [56]. The overall process led to the formation of the thiopyran ring (Scheme **43**). Depending on the substrates and on the reaction conditions the 4-*exo* cyclization can also occur, leading to the corresponding 4-membered thietanes. By using a carboxylic ester instead the ketone and a vinyl cerium reagent, 4-vinyl thiopyran products have been obtained.

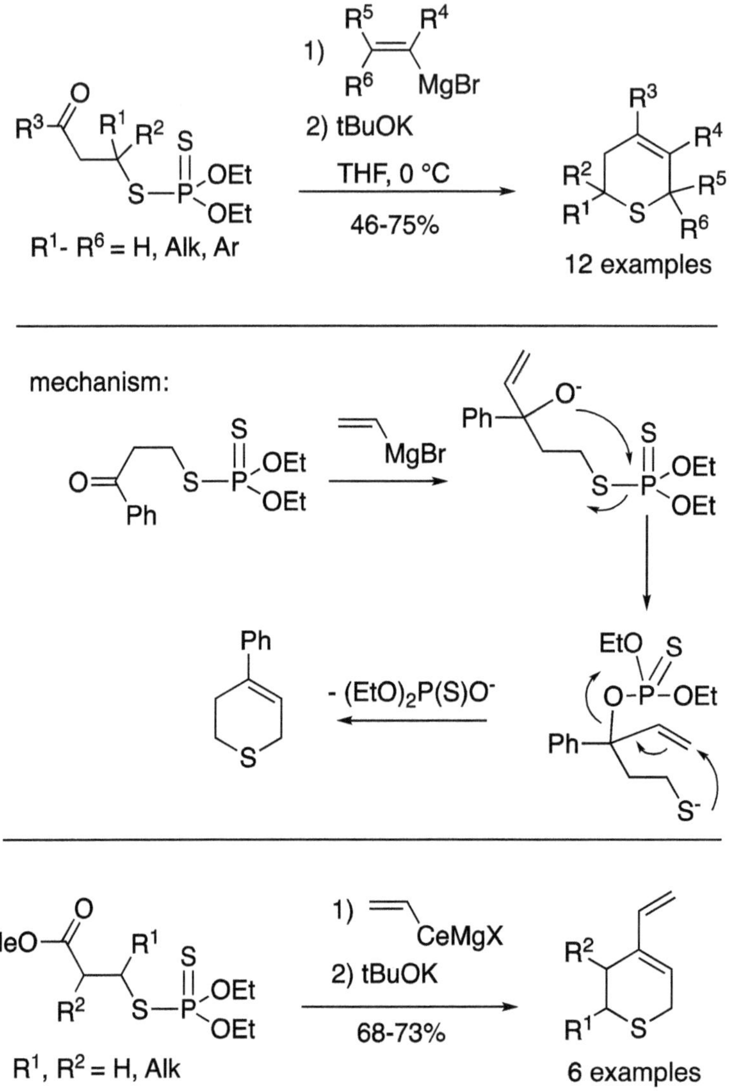

Scheme 43. Synthesis thiopyrans from phosphorodithioates *via* an anionic cascade.

Homolytic P-S Cleavage

Fensterbank, Lacôte, Malacria and co-workers employed phosphorothioate bromobenzene substrates to generate a phosphorus-centered radical *via* a homolytic substitution at the sulfur atom (Scheme **44**) [57, 58]. The P-radical was trapped *in-situ* by an olefin to give the desired phosphorylated adduct and the dihydrobenzothiophene as the by-product. Phosphonyl, phosphinoyl and diaminophosphonyl radicals have been successfully produced and reacted with both rich and poor double bonds. A tin-free version was then proposed starting from a phosphorothioate linked to a terminal alkyne. In this case the reaction is initiated by a thiyl radical (from thiophenol) in the presence of the olefin, and sulfanylated thiolane is obtained together with the P-adduct. If the substrate is placed in the same reactions conditions without the olefin, a phosphorylated sulfur-heterocycle is obtained as the product of the radical thiophosphinoylation of triple bond.

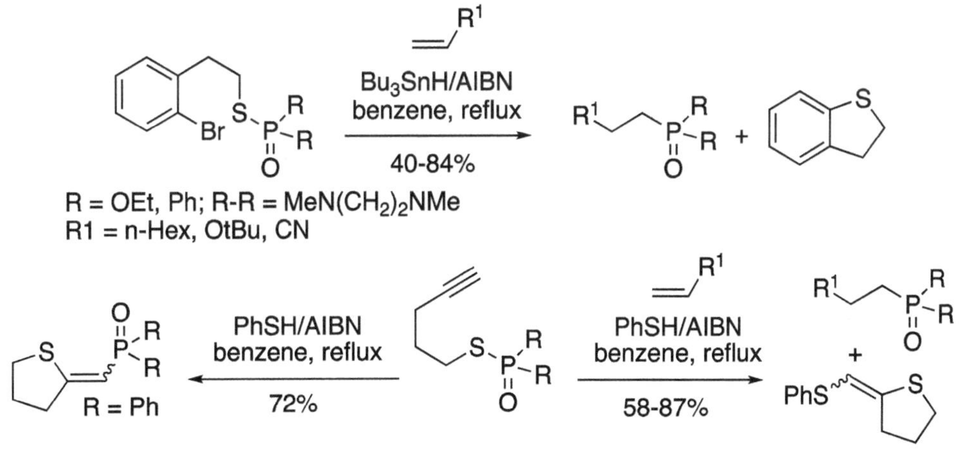

Scheme 44. Domino radical process *via* homolytic P-S cleavage.

Metal-catalyzed P-S Cleavage; Thiophosphorylation of Triple Bonds

Han and Tanaka reported the Pd-catalyzed thiophosphorylation of terminal alkynes [59]. The 1-phosphonylated 2-sulfanylated alkenes were obtained with good yields (Scheme **45**). Functionalized aliphatic alkynes led to excellent results in terms of stereoselectivity (major Z) and regioselectivity, whereas arylacetylene gave a mixture of Z/E isomers. Internal alkenes were unreactive. The authors proposed a mechanism involving Pd-insertion in the P-S bond with formation of [S-Pd-P] species.

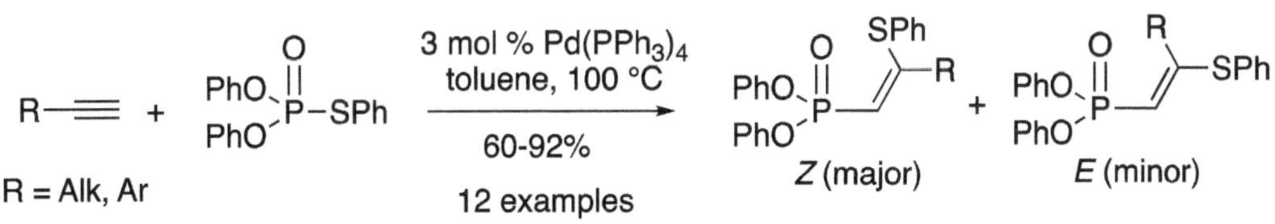

Scheme 45. Pd-catalyzed thiophosphorylation of terminal alkynes.

Allylic Substitution of Phosphorothioate Group

Wu and coworkers developed metal-free or metal-catalyzed allylic substitution reactions, which employ a phosphorothioate as the leaving group.

The first example consisted in the transition-metal-free allylic substitution reaction between Grignard reagents as carbon-nucleophiles and allylic phosphorothioate esters (Scheme **46**) [35]. The reaction worked with various aryl Grignard reagents with both electron-withdrawing and donating groups, with vinyl derivatives, as well as with both primary and secondary alkyl Grignard reagents. As in principle the reaction can lead to two regioisomers, it was demonstrated by the authors that the use of aromatic and alkenyl nucleophiles led to the α-regioisomer, whereas the use of secondary aliphatic ones resulted in the favored opposite γ-regioselectivity. These regioselectivities have been partially rationalized by a radical-based mechanism in the case of aliphatic Grignard reagents and an ionic mechanism in the case of aromatic counterparts.

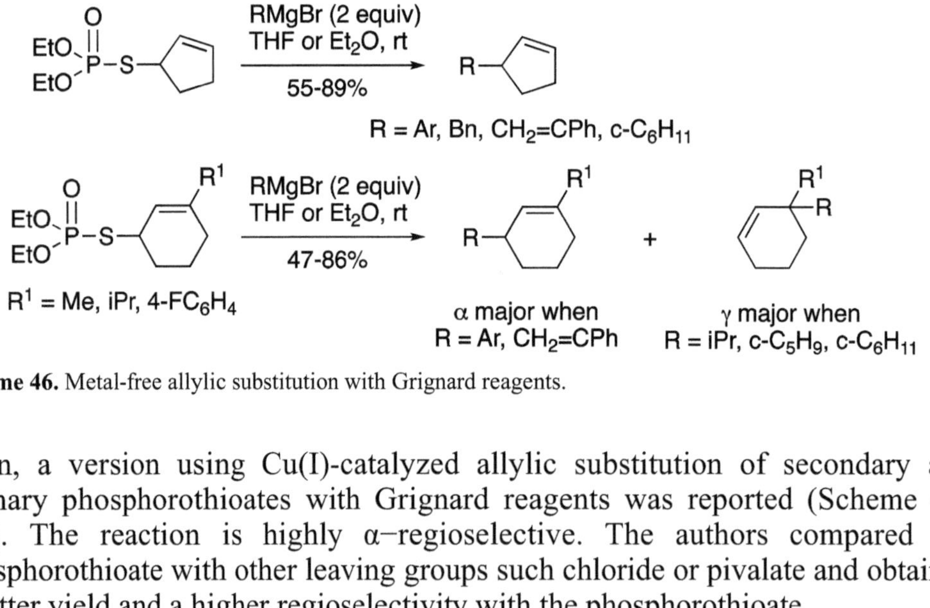

Scheme 46. Metal-free allylic substitution with Grignard reagents.

Then, a version using Cu(I)-catalyzed allylic substitution of secondary and primary phosphorothioates with Grignard reagents was reported (Scheme **47**) [60]. The reaction is highly α–regioselective. The authors compared the phosphorothioate with other leaving groups such chloride or pivalate and obtained a better yield and a higher regioselectivity with the phosphorothioate.

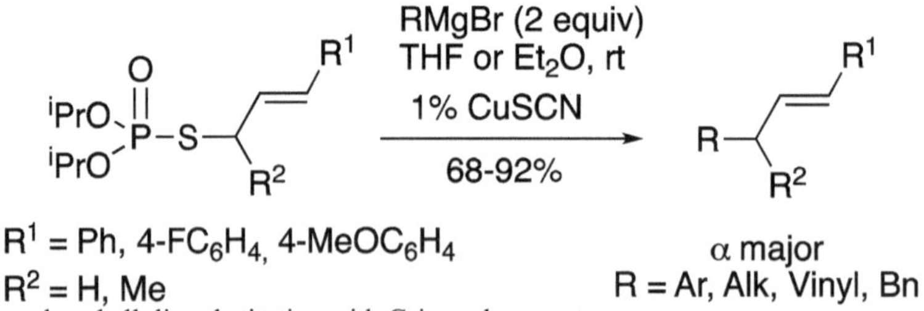

Scheme 47. Cu-catalyzed allylic substitution with Grignard reagents.

The same research group reported the regioselective, palladium-catalyzed allylic fluorination of phosphorothioates with silver fluoride (Scheme **48**) [61]. The method is efficient compared to other ones used to introduce fluorine in allylic position, by reducing the undesired diene formation, which is obtained in this case as a minor product. When R^1 = Me, the ratio allyl fluoride/diene in the crude mixture varied between 4:1 and 5:1. The mechanistic studies show that the process proceeds through a palladium π-allyl intermediate and is stereospecific, providing fluorinated products with retention of stereochemical configuration.

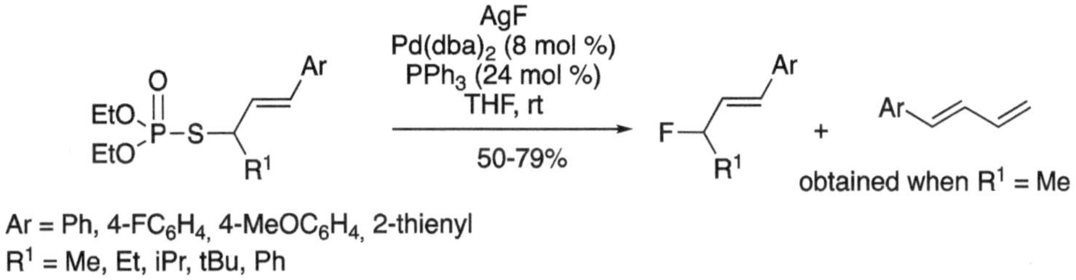

Ar = Ph, 4-FC$_6$H$_4$, 4-MeOC$_6$H$_4$, 2-thienyl
R^1 = Me, Et, iPr, tBu, Ph

Scheme 48. Pd-catalyzed allylic fluorination.

CONCLUSION

The excellent contributions reported in this chapter clearly demonstrate the remarkable efforts and progress made in the past 20 years in the chemistry of phosphorothioates. Due to the development of elegant and efficient access to these species, as well as of diverse innovative and useful synthetic transformations, it is obvious that a larger scientific community will be attracted by this versatile functional group.

CONSENT FOR PUBLICATION

Not applicable.

CONFLICT OF INTEREST

The author confirms that he has no conflict of interest to declare for this publication.

ACKNOWLEDGEMENTS

M.G. thanks the "Centre National de la Recherche Scientifique (CNRS)" and the "Ministère de l'Enseignement Supérieur, de la Recherche et de l'Innovation (M.E.S.R.I)" for recurring financial support.

REFERENCES

[1] (a). Storm, J.E. Organophosphorus Pesticides.*Patty's Toxicology*; J. Wiley and Sons, **2012**, 95, pp. 1077-1234.
[http://dx.doi.org/10.1002/0471435139.tox095.pub2]
(b). Hodgson, E. *Pesticide Biotransformation and Disposition, Academic Press*; Elsevier, **2012**.

[2] (a). Lu, H.; Berkman, C.E. Stereoselective inhibition of glutamate carboxypeptidase by chiral phosphonothioic acids. *Bioorg. Med. Chem.*, **2001**, *9*, 395-402.
[http://dx.doi.org/10.1016/S0968-0896(00)00254-6]
(b). Fraietta, J.A.; Mueller, Y.M.; Do, D.H.; Holmes, V.M.; Howett, M.K.; Lewis, M.G.; Boesteanu, A.C.; Alkan, S.S.; Katsikis, P.D. *Antimicrob. Agents Chemother.*, **2010**, *54*, 4064-4073.
[http://dx.doi.org/10.1128/AAC.00367-10]
(c). Xie, R.; Zhao, Q.; Zhang, T.; Fang, J.; Mei, X.; Ning, J.; Tang, Y. Design, synthesis and biological evaluation of organophosphorous-homodimers as dual binding site acetylcholinesterase inhibitors. *Bioorg. Med. Chem.*, **2013**, *21*, 278-282.
[http://dx.doi.org/10.1016/j.bmc.2012.10.030]
(d). Ye, J.D.; Barth, C.D.; Anjaneyulu, P.S.R.; Tuschl, T.; Piccirilli, J.A. Reactions of phosphate and phosphorothiolate diesters with nucleophiles: comparison of transition state structures. *Org. Biomol. Chem.*, **2007**, *5*, 2491-2497.
[http://dx.doi.org/10.1039/b707205h]

[3] (a). Ozturk, T.; Ertas, E.; Mert, O. A Berzelius Reagent, Phosphorus Decasulfide (P4S10), in Organic Syntheses. *Chem. Rev.*, **2010**, *110*, 3419-3478.
[http://dx.doi.org/10.1021/cr900243d]
(b). Ozturk, T.; Ertas, E.; Mert, O. Use of Lawesson's Reagent in Organic Syntheses. *Chem. Rev.*, **2007**, *107*, 5210-5278.
[http://dx.doi.org/10.1021/cr040650b]

[4] (a). Li, N.S.; Frederiksen, J.K.; Piccirilli, J.A. Synthesis, Properties, and Applications of Oligonucleotides Containing an RNA Dinucleotide Phosphorothiolate Linkage. *Acc. Chem. Res.*, **2011**, *44*, 1257-1269.
[http://dx.doi.org/10.1021/ar200131t]
(b). Kumar, T.S.; Yang, T.; Mishra, S.; Cronin, C.; Chakraborty, S.; Shen, J.B.; Liang, B.T.; Jacobson, K.A. 5′-Phosphate and 5′-Phosphonate Ester Derivatives of (N)-Methanocarba Adenosine with *in Vivo* Cardioprotective Activity. *J. Med. Chem.*, **2013**, *56*, 902-914.
[http://dx.doi.org/10.1021/jm301372c]
(c). Wu, S.Y.; Yang, X.; Gharpure, K.M.; Hatakeyama, H.; Egli, M.; McGuire, M.H.; Nagaraja, A.S.; Miyake, T.M.; Rupaimoole, R.; Pecot, C.V.; Taylor, M.; Pradeep, S.; Sierant, M.; Rodriguez-Aguayo, C.; Choi, H.J.; Previs, R.A.; Armaiz-Pena, G.N.; Huang, L.; Martinez, C.; Hassell, T.; Ivan, C.; Sehgal, V.; Singhania, R.; Han, H.D.; Su, C.; Kim, J.H.; Dalton, H.J.; Kovvali, C.; Keyomarsi, K.; McMillan, N.A.J.; Overwijk, W.W.; Liu, J.; Lee, J.S.; Baggerly, K.A.; Lopez-Berestein, G.; Ram, P.T.; Nawrot, B.; Sood, A.K. 2′-OMe-phosphorodithioate-modified siRNAs show increased loading into the RISC complex and enhanced anti-tumour activity. *Nat. Commun.*, **2014**, *5*, 3459.
[http://dx.doi.org/10.1038/ncomms4459]
(d). Stec, W.J.; Wilk, A. Stereocontrolled Synthesis of Oligo(nucleoside phosphorothioate)s. *Angew. Chem. Int. Ed. Engl.*, **1994**, *33*, 709-722.
[http://dx.doi.org/10.1002/anie.199407091]
(e). Eckstein, F. Phosphorothioates, Essential Components of Therapeutic Oligonucleotides. *Nucleic Acid Ther.*, **2014**, *24*, 374-387.
[http://dx.doi.org/10.1089/nat.2014.0506]
(f). Guga, P.; Koziołkiewicz, M. Phosphorothioate Nucleotides and Oligonucleotides – Recent Progress in Synthesis and Application. *Chem. Biodivers.*, **2011**, *8*, 1642-1681.
[http://dx.doi.org/10.1002/cbdv.201100130]

[5] (a). Omelanczuk, J.; Mikolajczyk, M. Chiral t-butylphenylphosphinothioic acid: A useful chiral solvating agent for direct determination of enantiomeric purity of alcohols, thiols, amines, diols, aminoalcohols and related compounds. *Tetrahedron Asymmetry*, **1996**, *7*, 2687-2694.
[http://dx.doi.org/10.1016/0957-4166(96)00345-X]
(b). Kobayashi, Y.; Morisawa, F.; Saigo, K. Enantiopure O-Substituted Phenylphosphonothioic Acids: Chiral Recognition Ability during Salt Crystallization and Chiral Recognition Mechanism. *J. Org. Chem.*, **2006**, *71*, 606-615.
[http://dx.doi.org/10.1021/jo052020v]

[6] Kaboudin, B.; Farjadian, F. Synthesis of phosphorothioates using thiophosphate salts. *Beilstein J. Org. Chem.*, **2006**, *2*, 1-5.
[http://dx.doi.org/10.1186/1860-5397-2-4]

[7] Kaboudin, B. A simple and new method for the synthesis of thiophosphates. *Tetrahedron Lett.*, **2002**, *43*, 8713-8714.
[http://dx.doi.org/10.1016/S0040-4039(02)02136-6]

[8] Renard, P-Y.; Schwebel, H.; Vayron, P.; Josien, L.; Valleix, A.; Mioskowski, C. Easy Access to Phosphonothioates. *Chemistry*, **2002**, *8*, 2910-2916.
[http://dx.doi.org/10.1002/1521-3765(20020703)8:13<2910::AID-CHEM2910>3.0.CO;2-R]

[9] Castanheiro, T.; Suffert, J.; Donnard, M.; Gulea, M. Recent advances in the chemistry of organic thiocyanates. *Chem. Soc. Rev.*, **2016**, *45*, 494-505.
[http://dx.doi.org/10.1039/C5CS00532A]

[10] Piekutowska, M.; Pakulski, Z. The Michaelis–Arbuzov rearrangement of anomeric thiocyanates: synthesis and application of S-glycosyl thiophosphates, thiophosphonates and thiophosphinates as glycosyl donors. *Tetrahedron Lett.*, **2007**, *48*, 8482-8486.
[http://dx.doi.org/10.1016/j.tetlet.2007.09.162]

[11] Crocker, R.D.; Hussein, M.A.; Ho, J.; Nguyen, T.V. NHC-Catalyzed Metathesis and Phosphorylation Reactions of Disulfides: Development and Mechanistic Insights. *Chemistry*, **2017**, *23*, 6259-6263.
[http://dx.doi.org/10.1002/chem.201700744]

[12] Mitra, S.; Mukherjee, S.; Sen, S.K.; Hajra, A. Environmentally benign synthesis and antimicrobial study of novel chalcogenophosphates. *Bioorg. Med. Chem. Lett.*, **2014**, *24*, 2198-2201.
[http://dx.doi.org/10.1016/j.bmcl.2014.03.008]

[13] Liu, T.; Cui, X.; Yu, Z.; Li, C. A New Method of Introducing SCH3 and SCD3 Groups to Phosphorothioates. *Phosphorus Sulfur Silicon Relat. Elem.*, **2012**, *187*, 606-611.
[http://dx.doi.org/10.1080/10426507.2011.634467]

[14] Moon, Y.; Moon, Y.; Choi, H.; Hong, S. Metal- and oxidant-free S–P(O) bond construction via direct coupling of P(O)H with sulfinic acids. *Green Chem.*, **2017**, *19*, 1005-1013.
[http://dx.doi.org/10.1039/C6GC03285K]

[15] Panmand, D.S.; Tiwari, A.D.; Panda, S.S.; Monbaliu, J-C.M.; Beagle, L.K.; Asiri, A.M.; Stevens, C.V.; Steel, P.J.; Hall, C.D.; Katritzky, A.R. New benzotriazole-based reagents for the phosphonylation of various N-, O-, and S-nucleophiles. *Tetrahedron Lett.*, **2014**, *55*, 5898-5901.
[http://dx.doi.org/10.1016/j.tetlet.2014.07.057]

[16] Lin, Y.; Lu, G.; Wang, G.; Yi, W. Acid/Phosphide-Induced Radical Route to Alkyl and Alkenyl Sulfides and Phosphonothioates from Sodium Arylsulfinates in Water. *J. Org. Chem.*, **2017**, *82*, 382-389.
[http://dx.doi.org/10.1021/acs.joc.6b02459]

[17] Liu, Y-C.; Lee, C-F. N-Chlorosuccinimide-promoted synthesis of thiophosphates from thiols and phosphonates under mild conditions. *Green Chem.*, **2014**, *16*, 357-364.
[http://dx.doi.org/10.1039/C3GC41839A]

[18] Bi, X.; Li, J.; Meng, F.; Wang, H.; Xiao, J. DCDMH-promoted synthesis of thiophosphates by coupling of H-phosphonates with thiols. *Tetrahedron*, **2016**, *72*, 706-711.
[http://dx.doi.org/10.1016/j.tet.2015.12.020]

[19] Wang, J.; Huang, X.; Ni, Z.; Wang, S.; Pan, Y.; Wu, J. Peroxide promoted metal-free thiolation of phosphites by thiophenols/disulfides. *Tetrahedron*, **2015**, *71*, 7853-7859.
[http://dx.doi.org/10.1016/j.tet.2015.08.025]

[20] Wang, J.; Huang, X.; Ni, Z.; Wang, S.; Wu, J.; Pan, Y. TBPB-promoted metal-free synthesis of thiophosphinate/phosphonothioate by direct P–S bond coupling. *Green Chem.*, **2015**, *17*, 314-319.
[http://dx.doi.org/10.1039/C4GC00944D]

[21]   Wu, X-M.; Hong, Y-X. Iodine-catalyzed Sulfenylation of H-phosphonates with Diaryl Disulfides under Metal, Base and Solvent-free Conditions. *Lett. Org. Chem.*, **2017**, *14*, 49-55.
[http://dx.doi.org/10.2174/1570178614666161214153431]

[22]   Sun, J-G.; Yang, H.; Li, P.; Zhang, B. Metal-Free Visible-Light-Mediated Oxidative Cross-Coupling of Thiols with P(O)H Compounds Using Air as the Oxidant. *Org. Lett.*, **2016**, *18*, 5114-5117.
[http://dx.doi.org/10.1021/acs.orglett.6b02563]

[23]   He, W.; Hou, X.; Li, X.; Song, L.; Yu, Q.; Wang, Z. Synthesis of P(O)-S organophosphorus compounds by dehydrogenative coupling reaction of P(O)H compounds with aryl thiols in the presence of base and air. *Tetrahedron*, **2017**, *73*, 3133-3138.
[http://dx.doi.org/10.1016/j.tet.2017.04.035]

[24]   Song, S.; Zhang, Y.; Yeerlan, A.; Zhu, B.; Liu, J.; Jiao, N. Cs_2CO_3-Catalyzed Aerobic Oxidative Cross-Dehydrogenative Coupling of Thiols with Phosphonates and Arenes. *Angew. Chem. Int. Ed.*, **2017**, *56*, 2487-2491.
[http://dx.doi.org/10.1002/anie.201612190]

[25]   Gao, Y.X.; Tang, G.; Cao, Y.; Zhao, Y.F. A Novel and General Method for the Formation of S-Aryl, Se-Aryl, and Te-Aryl Phosphorochalcogenoates. *Synthesis*, **2009**, *2009*, 1081-1086.
[http://dx.doi.org/10.1055/s-0028-1088012]

[26]   Kaboudin, B.; Abedi, Y.; Kato, J-Y.; Yokomatsu, T. Copper(I) Iodide Catalyzed Synthesis of Thiophosphates by Coupling of H-Phosphonates with Benzenethiols. *Synthesis*, **2013**, *45*, 2323-2327.
[http://dx.doi.org/10.1055/s-0033-1339186]

[27]   Bai, J.; Cui, X.; Wang, H.; Wu, Y. Copper-catalyzed reductive coupling of aryl sulfonyl chlorides with H-phosphonates leading to S-aryl phosphorothioates. *Chem. Commun. (Camb.)*, **2014**, *50*, 8860-8863.
[http://dx.doi.org/10.1039/C4CC02693D]

[28]   Zhang, X.; Wang, D.; An, D.; Han, B.; Song, X.; Li, L.; Zhang, G.; Wang, L. Cu(II)/Proline-Catalyzed Reductive Coupling of Sulfuryl Chloride and P(O)–H for P–S–C Bond Formation. *J. Org. Chem.*, **2018**, *83*, 1532-1537.
[http://dx.doi.org/10.1021/acs.joc.7b02608]

[29]   Kumaraswamy, G.; Raju, R. Copper(I)-Induced Sulfenylation of H-Phosphonates, H-Phosphonites and Phosphine Oxides with Aryl/alkylsulfonylhydrazides as a Thiol Surrogate. *Adv. Synth. Catal.*, **2014**, *356*, 2591-2598.
[http://dx.doi.org/10.1002/adsc.201400116]

[30]   Zhu, Y.; Chen, T.; Li, S.; Shimada, S.; Han, L.B. Efficient Pd-Catalyzed Dehydrogenative Coupling of P(O)H with RSH: A Precise Construction of P(O)–S Bonds. *J. Am. Chem. Soc.*, **2016**, *138*, 5825-5828.
[http://dx.doi.org/10.1021/jacs.6b03112]

[31]   Li, J.; Bi, X.; Wang, H.; Xiao, J. Palladium-catalyzed desulfitative C–P coupling of arylsulfinate metal salts and H-phosphonates. *RSC Advances*, **2014**, *4*, 19214-19217.
[http://dx.doi.org/10.1039/C4RA01270D]

[32]   Xu, J.; Zhang, L.; Li, X.; Gao, Y.; Tang, G.; Zhao, Y. Phosphorothiolation of Aryl Boronic Acids Using P(O)H Compounds and Elemental Sulfur. *Org. Lett.*, **2016**, *18*, 1266-1269.
[http://dx.doi.org/10.1021/acs.orglett.6b00118]

[33]   Zhang, L.; Zhang, P.; Li, X.; Xu, J.; Tang, G.; Zhao, Y. Synthesis of S-Aryl Phosphorothioates by Copper-Catalyzed Phosphorothiolation of Diaryliodonium and Arenediazonium Salts. *J. Org. Chem.*, **2016**, *81*, 5566-5573.
[http://dx.doi.org/10.1021/acs.joc.6b00925]

[34]   Wang, L.; Yang, S.; Chen, L.; Yuan, S.; Chen, Q.; He, M-Y.; Zhang, Z-H. Magnetically recyclable Cu-BTC@Fe_3O_4 composite-catalyzed C(aryl)–S–P bond formation using aniline, P(O)H compounds and sulfur powder. *Catal. Sci. Technol.*, **2017**, *7*, 2356-2361.
[http://dx.doi.org/10.1039/C7CY00467B]

[35]   Han, X.; Zhang, Y.; Wu, J. Mild Two-Step Process for the Transition-Metal-Free Synthesis of Carbon–Carbon Bonds from Allylic Alcohols/Ethers and Grignard Reagents. *J. Am. Chem. Soc.,* **2010**, *132*, 4104-4106.
[http://dx.doi.org/10.1021/ja100747n]

[36]   Han, X.; Wu, J. Ga(OTf)$_3$-Catalyzed Direct Substitution of Alcohols with Sulfur Nucleophiles. *Org. Lett.,* **2010**, *12*, 5780-5782.
[http://dx.doi.org/10.1021/ol102565b]

[37]   Grounds, H.; Ermanis, K.; Newgas, S.A.; Porter, M.J. Conversion of Alcohols to Phosphorothiolates Using a Thioiminium Salt as Coupling Agent. *J. Org. Chem.,* **2017**, *82*, 12735-12739.
[http://dx.doi.org/10.1021/acs.joc.7b01657]

[38]   Yan, J.; Chen, Z-C. Hypervalent Iodine in Synthesis. XXXVII: A Convenient Synthesis of S-Vinyl O,O-dialkyl Phosphorothioates. *Synth. Commun.,* **1999**, *29*, 3605-3612.
[http://dx.doi.org/10.1080/00397919908085995]

[39]   Okamoto, K.; Housekeepe, J.B.; Luscombe, C.K. Room-temperature carbon–sulfur bond formation from Ni(II) σ-aryl complex via cleavage of the S–S bond of disulfide moieties. *Appl. Organomet. Chem.,* **2013**, *27*, 639-643.
[http://dx.doi.org/10.1002/aoc.2975]

[40]   Gulea, M.; Masson, S. Recent Advances in the Chemistry of Difunctionalized Organo-Phosphorus and -Sulfur Compounds.*New Aspects in Phosphorus Chemistry III. Topics in Current Chemistry*; Majoral, J.P., Ed.; Springer: Berlin, Heidelberg, **2004**, Vol. 35, pp. 161-198.
[http://dx.doi.org/10.1002/chin.200437268]

[41]   Mauger, C.; Vazeux, M.; Albinati, A.; Kozelka, J. Reaction between the diaqua form of cisplatin and 2-methylsulfanylphenylphosphonic acid yields a dinuclear phosphonato-bridged complex via NH$_3$ elimination. *Inorg. Chem. Commun.,* **2000**, *3*, 704-707.
[http://dx.doi.org/10.1016/S1387-7003(00)00170-2]

[42]   Hamel, M.; Lecinq, M.; Gulea, M.; Kozelka, J. Ortho-(methylsulfanyl)phenylphosphonates and derivatives: Synthesis and applications as mono- or bidentate ligands for the preparation of platinum complexes. *J. Organomet. Chem.,* **2013**, *745-746*, 206-213.
[http://dx.doi.org/10.1016/j.jorganchem.2013.07.069]

[43]   Hamel, M.; Rizzato, S.; Lecinq, M.; Sene, A.; Vazeux, M.; Gulea, M. Albinati, A.; Kozelka, J. Study of Intramolecular Competition between Carboxylate and Phosphonate for Pt(II) with the Aid of a Novel Tridentate Carboxylato-Thioether-Phosphonato Ligand. *Chem. Eur. J.,* **2007**, *13*, 5441-5449.
[http://dx.doi.org/10.1002/chem.200601573]

[44]   Lassaux, P.; Hamel, M.; Gulea, M.; Delbrueck, H.; Mercuri, P.S.; Horsfall, L.; Dehareng, D.; Kupper, M.; Frère, J-M.; Hoffmann, K.; Galleni, M.; Bebrone, C. Mercaptophosphonate Compounds as Broad-Spectrum Inhibitors of the Metallo-β-lactamases. *J. Med. Chem.,* **2010**, *53*, 4862-4876.
[http://dx.doi.org/10.1021/jm100213c]

[45]   Masson, S.; Saquet, M.; Marchand, P. First Synthesis of a (Mercaptomethylene) Diphosphonate *via* a Phosphorothiolate -Mercaptophosphonate Rearrangement. *Tetrahedron,* **1998**, *54*, 1523-1528.
[http://dx.doi.org/10.1016/S0040-4020(97)10376-3]

[46]   Philippitsch, V.; Hammerschmidt, F. Rearrangement of lithiated S-alkyl O,O-dialkyl thiophosphates: Scope and stereochemistry of the thiophosphate–mercaptophosphonate rearrangement. *Org. Biomol. Chem.,* **2011**, *9*, 5220-5227.
[http://dx.doi.org/10.1039/c1ob05246b]

[47]   Takeda, K.; Sumi, K.; Hagisawa, S. Reaction of silyl thioketones with lithium diethylphosphite: first observation of Thia–Brook rearrangement. *J. Organomet. Chem.,* **2001**, 611-449.

[48]   (a). Masson, S.; Saint-Clair, J-F.; Saquet, M. Two Methods for the Synthesis of (2-Mercaptophenyl)phosphonic Acid. *Synthesis,* **1993**, *45*, 485-486.

[http://dx.doi.org/10.1055/s-1993-25889]
(b). Masson, S.; Saint-Clair, J-F.; Saquet, M. Synthesis of New Mercapto-Phosphono Substituted Heterocycles via a Thiophosphate -Mercaptophosphonate rearrangement. *Tetrahedron Lett.,* **1994**, *35*, 3083-3084.
[http://dx.doi.org/10.1016/S0040-4039(00)76834-1]
(c). Mauger, C.; Vazeux, M.; Masson, S. Synthesis of chiral ortho-thio-substituted phenyl phosphonodiamidates via a P–S to P–C rearrangement. *Tetrahedron Lett.,* **2004**, *45*, 3855-3859.
[http://dx.doi.org/10.1016/j.tetlet.2004.03.132]

[49] Watanabe, M.; Date, M.; Kawanishi, K.; Akiyoshi, R.; Furukawa, S. Synthesis of chiral ortho-thi--substituted phenyl phosphonodiamidates via a P–S to P–C rearrangement. *J. Heterocycl. Chem.,* **1991**, *28*, 173-176.
[http://dx.doi.org/10.1002/jhet.5570280130]

[50] Legrand, O.; Brunel, J.M.; Buono, G. Scope and Limitations of the Aromatic Anionic [1,3] P–O to P–C Rearrangement in the Synthesis of Chiral o-Hydroxyaryl Diazaphosphonamides. *Tetrahedron,* **2000**, *56*, 595-603.
[http://dx.doi.org/10.1016/S0040-4020(99)00966-7]

[51] Bonini, B.F.; Femoni, C.; Fochi, M.; Gulea, M.; Masson, S.; Ricci, A. Asymmetric Versions of P-S to P-C [1,3]-Sigmatropic Rearrangement in Ferrocene Series. *Tetrahedron Asymmetry,* **2005**, *16*, 3003-3010.
[http://dx.doi.org/10.1016/j.tetasy.2005.07.018]

[52] Au-Yeung, T-L.; Chan, K-Y.; Haynes, R.K.; William, I.D.; Yeung, L.L. Completely stereoselective P-C bond formation via base-induced [1,3]- and [1,2]-intramolecular rearrangements of aryl phosphinates, phosphinoamidates and related compounds: generation of P-chiral β-hydroxy, β-mercapto- and α-amino tertiary phosphine oxides and phosphine sulfides. *Tetrahedron Lett.,* **2001**, *42*, 457-460.
[http://dx.doi.org/10.1016/S0040-4039(00)01952-3]

[53] Chen, M.; Maetzke, A.; Jensen, S.J.K.; Gothelf, K.V. Rearrangement of S-(2-Aminoethyl) Thiophosphates to N-(2-Mercaptoethyl)-phosphoramidates. *Eur. J. Org. Chem.,* **2007**, 5826-5833.
[http://dx.doi.org/10.1002/ejoc.200700646]

[54] Robertson, F.J.; Wu, J. Convenient Synthesis of Allylic Thioethers from Phosphorothioate Esters and Alcohols. *Org. Lett.,* **2010**, *12*, 2668-2671.
[http://dx.doi.org/10.1021/ol1009202]

[55] Robertson, F.J.; Wu, J. Phosphorothioic Acids and Related Compounds as Surrogates for H_2S-Synthesis of Chiral Tetrahydrothiophenes. *J. Am. Chem. Soc.,* **2012**, *134*, 2775-2780.
[http://dx.doi.org/10.1021/ja210758n]

[56] Li, F.; Calabrese, D.; Brichacek, M.; Lin, I.; Njardarson, J.T. Efficient Synthesis of Thiopyrans Using a Sulfur-Enabled Anionic Cascade. *Angew. Chem. Int. Ed.,* **2012**, *51*, 1938-1941.
[http://dx.doi.org/10.1002/anie.201108261]

[57] Carta, P.; Puljic, N.; Robert, C.; Dhimane, A-L.; Fensterbank, L.; Lacôte, E.; Malacria, M. Generation of Phosphorus-Centered Radicals via Homolytic Substitution at Sulfur. *Org. Lett.,* **2007**, *9*, 1061-1063.
[http://dx.doi.org/10.1021/ol0631096]

[58] Carta, P.; Puljic, N.; Robert, C.; Dhimane, A-L.; Fensterbank, L.; Lacôte, E.; Malacria, M. Intramolecular homolytic substitution at the sulfur atom: an alternative way to generate phosphorus- and sulfur-centered radicals. *Tetrahedron,* **2008**, *64*, 11865-11875.
[http://dx.doi.org/10.1016/j.tet.2008.08.108]

[59] Han, L-B.; Tanaka, M. Novel palladium-catalyzed thiophosphorylation of alkynes with phosphorothioate: An efficient route to (Z)-1-(diphenoxyphosphinyl)-2-(phenylthio)alkenes. *Chem. Lett.,* **1999**, 863-864.
[http://dx.doi.org/10.1246/cl.1999.863]

[60] Lauer, A.M.; Mahmud, F.; Wu, J. Cu(I)-Catalyzed, α-Selective, Allylic Alkylation Reactions between Phosphorothioate Esters and Organomagnesium Reagents. *J. Am. Chem. Soc.,* **2011**, *133*, 9119-9123.
[http://dx.doi.org/10.1021/ja202954b]

[61] Lauer, A.M.; Wu, J. Palladium-Catalyzed Allylic Fluorination of Cinnamyl Phosphorothioate Esters. *Org. Lett.,* **2012**, *14*, 5138-5141.
[http://dx.doi.org/10.1021/ol302263m]

CHAPTER 4

Kinetic Resolution Using Diastereoselective Acylating Agents as a Synthetic Approach to Enantiopure Amines

Galina L. Levit*, Dmitry A. Gruzdev and Victor P. Krasnov

Postovsky Institute of Organic Synthesis, Russian Academy of Sciences (Ural Branch), Ekaterinburg, Russia

Abstract: The kinetic resolution (KR) of racemates is one of the most important modern approaches to prepare optically pure substances. This method is based on the difference in the reaction rates of enantiomers under the action of chiral resolving agents or catalysts. KR as a result of the acylation reaction (acylative KR) has been widely used to obtain enantiopure amines that are valuable precursors and structural fragments of biologically active compounds and pharmaceuticals. The nonenzymatic acylative KR of racemic amines is usually carried out in the presence of chiral acyl-transfer catalysts or by the action of enantioselective or diastereoselective low-molecular weight acylating agents.

This review surveys the research work on KR of racemic amines under the action of chiral diastereoselective acylating agents. This approach has recently received a very significant development, which is due to the availability of various reagents of this kind, as well as the ease of process implementation. Application of derivatives of chiral acids, including natural amino acids, 2-arylpropionic acids, 2-oxy-carboxylic acids, *etc.* as acylating agents in the KR processes makes it possible to obtain amines of various classes and related compounds with a very high enantiomeric purity, which is often inaccessible in other nonenzymatic methods. Diastereoisomerically enriched (pure) amides formed as a result of acylative KR of racemic amines may themselves be of interest for further synthetic transformations.

Keywords: Acylation, Acyl Chlorides, Amino Acids, 2-Arylpropionic Acids, Chiral Amines, Enantiomers, 2-Hydroxy Carboxylic Acids, Kinetic Resolution, Optical Purity, Resolving Agents, Stereoselectivity.

---

* **Corresponding author Galina L. Levit:** Postovsky Institute of Organic Synthesis, Russian Academy of Sciences (Ural Branch), Ekaterinburg, Russia; Tel: +73433623057; Fax: +73433693058; E-mail: ca512@ios.uran.ru

Atta-ur-Rahman (Ed.)
All rights reserved-© 2018 Bentham Science Publishers

INTRODUCTION

The development of methods for the preparation of enantiopure compounds is one of the priorities of modern organic chemistry. The importance of this problem is due, first and foremost, to the exclusive role played by optically pure compounds in the process of design and application of novel pharmaceutical agents because the enantiomers of chiral biologically active compounds exhibit different, sometimes opposite, activity. It is also extremely important that the development of original methods for the synthesis of stereoisomers of chiral compounds promotes the development of new methods of fine organic synthesis and a deeper understanding of the features of the mechanisms of chemical and biochemical reactions.

Enantiopure substances can be obtained by chemical transformation of chiral natural compounds, asymmetric synthesis or by resolution of racemates, being the last, the most employed at the industry [1 - 4]. This can be explained first of all by the simplicity of the optical resolution procedures that are convenient, robust and reliable and suitable for large-scale applications. Despite the fact that great advances have been made in asymmetric synthesis, it applied in only one out of 10 cases in industry because of a complexity of designing the catalytic processes and uncertainty of outcome [2]. One more circumstance, resolution of racemates can still be a straightforward alternative, especially when both enantiomers are required. Currently, enzymatic methods are becoming more efficient, but they still require the right matching of enzymes and strictly controlled conditions for the process implementation. Although chromatographic techniques appear to be most convenient, they are also quite expensive and are thus not feasible to be scaled-up [5].

KR as a result of acylation reaction is one of the most important modern approaches for preparation of the optically pure amines and their derivatives from racemates. Acylative KR is usually carried out in the presence of chiral catalysts (enzymes or synthetic catalysts of acyl transfer) or under the action of chiral resolving acylating agents. Resolving agents can be divided into two main groups: enantioselective agents (a chiral center remains in the leaving group of the agent) and diastereoselective agents (a chiral center of the acyl fragment is transferred to the reaction product). This Chapter is devoted to the KR of racemic amines using diastereoselective acylating agents. In the last two decades, a vast experience has been accumulated in the investigation of the acylative KRs, certain regularities of their reactions have been revealed, original techniques and methods of realizing on an industrial scale have been found.

Basic Principles of Kinetic Resolution

KR of racemic compounds is the chemical process in which one enantiomer forms a product more rapidly than the other under the action of a chiral non-racemic agent (reagent, catalyst, solvent, *etc.*) (Scheme 1) [6 - 11]. The principles of the KR method, its varieties, and methods for calculating the basic parameters of the processes are described in detail in the classical works of Henri Kagan [6, 8]. In this section, we present a summary of the KR method necessary for better understanding the information provided.

$$S_R \xrightarrow{k_R} P_R \quad k_R > k_S$$

or

$$S_S \xrightarrow{k_S} P_S \quad k_S > k_R$$

Scheme 1. Schematic representation of KR: S_R and S_S are the (*R*)- and (*S*)-enantiomers of a substrate; P_R and P_S are the products formed from (*R*)- and (*S*)-enantiomers of a substrate, respectively; k_R and k_S are the rate constants for (*R*)- and (*S*)-enantiomers, respectively.

The KR efficiency is determined by the ratio of the rate constants of two independent reactions of fast- and slow-reacting enantiomers that is called the selectivity factor (or Kagan's factor) $s = k_{fast} / k_{slow}$ [6, 10]. In addition, the enantiomeric excess (*ee*) and/or diastereoisomeric excess (*de*) of the reaction products and the starting materials are often used to evaluate the efficiency of the KR process. The *ee* value is calculated from the relative content of enantiomers in the mixture: $ee = ([R]-[S])/([R]+[S])$ (in the case if the (*R*)-enantiomer predominates), where [*R*] and [*S*] are the content of (*R*)-enantiomer and (*S*)-enantiomer in the mixture, respectively. The *de* value is calculated in a similar way: $de = ([R,S]-[S,S])/([R,S]+[S,S])$ (in case of predominance of (*R,S*)-diastereoisomer), where [*R,S*] and [*S,S*] are the contents of the (*R,S*)- and (*S,S*)-diastereoisomers, respectively.

The selectivity factor *s* can be calculated by the formula [6, 11]:

$$s = \frac{\ln[(1-C)(1-ee_S)]}{\ln[(1-C)(1+ee_S)]}, \qquad (1)$$

where *C* is the conversion of the starting racemate, ee_S is the enantiomeric excess of unreacted substrate; or if the reaction product is chiral, the selectivity factor is calculated by the formula:

$$s = \frac{\ln[1-C(1+ee_P)]}{\ln[1-C(1-ee_P)]}, \qquad (2)$$

where ee_P is the enantiomeric excess of the reaction product.

Expressions (1) and (2) are applicable to the reactions that meet the following requirements: *i*) the reaction should be of the pseudo-first order with respect to substrate (and any other order for chiral reagent or catalyst); *ii*) the reaction mechanism should not change with time (reaction products should not catalyze or inhibit the process) [8].

It should be noted that although the reaction order is not always known and can differ from first order for various reasons, these ratios are used to calculate the *s* value for a comparative evaluation of similar reactions.

Conversion (*C*) at an arbitrary time can be calculated by the formula:

$$C = \frac{ee_S}{ee_S + ee_P}. \qquad (3)$$

In an ideal situation, only one of the stereoisomers enters the reaction, for example, $\mathbf{S}_R$ ($k_R \gg k_S$) (Scheme **1**). Then at *C* = 50%, a mixture containing 50% of $\mathbf{P}_R$ and 50% of $\mathbf{S}_S$ will be obtained. Such a mixture can be separated and (*S*)-enantiomer of a substrate can be obtained; in some cases, $\mathbf{S}_R$ can be also isolated after additional transformations of the product $\mathbf{P}_R$.

If racemic substrate and chiral reagent are taken in equimolar amounts, both enantiomers of the racemate will turn into products over time and separation will not occur. Therefore, it is important to stop the reaction before the racemate is completely converted, which is achieved by selecting a substrate–reagent molar ratio or by reducing the reaction time to the optimum one [6, 11].

In general, the KR method has a number of limitations: the theoretical yield of each of the enantiomers cannot exceed 50%; in most cases, only one stereoisomer is needed, and the other ("isomeric ballast") has little or no use; in many cases at a conversion close to 50%, ee_P or ee_S are quite low [12]. The possibility of isolation and racemization of unreacted enantiomer aimed at its re-use in the process is fundamentally important for increasing the KR efficiency.

It is generally accepted that the KR process can be useful from a preparative point of view in cases where *s* >10. If the *s* value exceeds 50, with a conversion close to 50%, it is possible to isolate both the reaction product and the unreacted substrate in optically pure form [9, 13]. It should be noted that the determination of *s* value more than 50 is not entirely accurate due to errors while taking the logarithm and inadequate accuracy in determining *ee* (or *de*) and conversion C [14, 15].

In order to enhance the process efficiency and to increase the yield of the target products, some modifications of the KR method were proposed and are being actively investigated, for example, dynamic KR and parallel KR.

Dynamic kinetic resolution (DKR) is a variant of the KR method that combines the resolution step of KR with *in situ* equilibration or racemization of the chirally labile substrate [1, 16 - 18]. If the rate of racemization of substrate **S** (k_{rac}) is significantly higher than the rate of formation of products $\mathbf{P}_R$ and $\mathbf{P}_S$, and stereoselectivity is high (for example, $k_R \gg k_S$), then this process, in principle, can lead to enantiomerically pure product $\mathbf{P}_R$ in up to 100% yield. Application of DKR for the preparation of chiral compounds, including amines and their derivatives, has been discussed in a number of reviews [19 - 23].

Parallel kinetic resolution (PKR) is the reaction of a racemate with a mixture of *quasi*-enantiomeric [24] resolving agents $\mathbf{Z}_1$ and $\mathbf{Z}_2$ (reagents or, more rarely, catalysts) that exhibit the same stereoselectivity towards opposite enantiomers $\mathbf{S}_R$ and $\mathbf{S}_S$ of a substrate (Scheme **2**) [25 - 29]. In this case, the ratio of enantiomers of a substrate is not changed throughout the process and is about 1:1, while the reaction products $\mathbf{P}_R$ and $\mathbf{Q}_S$ differing in structure have a high optical purity independent of the conversion of racemic substrate **S**. The resolving agents $\mathbf{Z}_1$ and $\mathbf{Z}_2$ have a very similar chemical structure and the opposite stereo configuration (*quasi*-enantiomers).

$$\mathbf{S}_R \xrightarrow[k_R]{\mathbf{Z}_1} \mathbf{P}_R$$

$$\mathbf{S}_S \xrightarrow[k_S]{\mathbf{Z}_2} \mathbf{Q}_S$$

$$k_R = k_S$$

Scheme 2. Schematic representation of parallel KR: $\mathbf{S}_R$ and $\mathbf{S}_S$ are the (*R*)- and (*S*)-enantiomers of a substrate; $\mathbf{Z}_1$ and $\mathbf{Z}_2$ are *quasi*-enantiomeric resolving agents; $\mathbf{P}_R$ and $\mathbf{Q}_S$ are the products formed from (*R*)- and (*S*)-enantiomers of a substrate, respectively; k_R and k_S are the rate constants for (*R*)- and (*S*)-enantiomers, respectively.

The main requirements for parallel reactions during PKR are the following: (*i*) parallel reactions should not compete with each other; *ii*) must have close reaction rates; *iii*) should proceed with complementary stereoselectivity; (*iv*) should lead to products that differ in structure [25]. In many cases, the differences in the structure of **P** and **Q** products make it easy to isolate them in high yields and *ee*. In general, PKR allows structurally different products $\mathbf{P}_R$ and $\mathbf{Q}_S$ to be obtained in up to 50% yields and with optical purity greater than in the case of traditional KR [27]. However, the disadvantage of this approach is the complexity of selecting and obtaining the *quasi*-enantiomeric resolving agents Z_1 and Z_2 in optically pure form.

Another version of the method, namely mutual KR is used when screening the efficient chiral resolving agents and studying the factors affecting stereochemical results of KR. This approach is based on the reaction between racemic substrate **S** and racemic resolving agent **Z** resulting in a mixture of four stereoisomers: $\mathbf{P}_{R,R}$, $\mathbf{P}_{R,S}$, $\mathbf{P}_{S,R}$ and $\mathbf{P}_{S,S}$ (Scheme 3). The ratio of reaction products $\mathbf{P}_{R^*,R^*}$ and $\mathbf{P}_{R^*,S^*}$ does not depend on the ratio of reagents **S** and **Z** and remains constant at any time, and the unreacted substrate remains racemic. In this case, the ratio of the resulting diastereoisomers $\mathbf{P}_{R^*,R^*}$ / $\mathbf{P}_{R^*,S^*}$ is equal to the selectivity factor s [8, 30, 31]. Therefore, mutual KR is a convenient and accurate method for determining the selectivity factor and allows screening the stereoselective reagents.

$$\mathbf{P}_{R^*,S^*} \begin{cases} \mathbf{P}_{R,S} \xleftarrow[k_{R\text{-}S}]{Z_S} S_R \xrightarrow[k_{R\text{-}R}]{Z_R} \mathbf{P}_{R,R} \\ \mathbf{P}_{S,R} \xleftarrow[k_{S\text{-}R}]{Z_R} S_S \xrightarrow[k_{S\text{-}S}]{Z_S} \mathbf{P}_{S,S} \end{cases} \mathbf{P}_{R^*,R^*}$$

$$\begin{array}{c} k_{R\text{-}R} = k_{S\text{-}S} \\ k_{R\text{-}S} = k_{S\text{-}R} \end{array} \qquad s = \frac{k_{R^*\text{-}R^*}}{k_{R^*\text{-}S^*}} = \frac{[\mathbf{P}_{R^*,R^*}]}{[\mathbf{P}_{R^*,S^*}]}$$

Scheme 3. Schematic representation of mutual KR: $\mathbf{S}_R$ and $\mathbf{S}_S$ are the (*R*)- and (*S*)-enantiomers of a substrate; $\mathbf{Z}_R$ and $\mathbf{Z}_S$ are the (*R*)- and (*S*)-enantiomers of resolving agent **Z**; $\mathbf{P}_{R,R}$ and $\mathbf{P}_{R,S}$ are the products formed from (*R*)-enantiomer of substrate; $\mathbf{P}_{S,R}$ and $\mathbf{P}_{S,S}$ are the products formed from (*S*)-enantiomer of substrate; respectively; $k_{R\text{-}R}$, $k_{R\text{-}S}$, $k_{S\text{-}R}$ and $k_{S\text{-}S}$ are the rate constants for the reaction between (*R*)- and (*S*)-enantiomers of substrate **S** and resolving agent **Z**.

In the case of the traditional KR (Scheme 1), the slow-reacting enantiomer of a substrate is accumulated as the process proceeds, and at a conversion close to 50%, the difference in the rates of formation of products $\mathbf{P}_R$ and $\mathbf{P}_S$ is substantially less than at the beginning of the reaction [8, 9, 26]. Dynamic, parallel, and mutual KRs are free of this drawback, because at any time the enantiomers of substrate are present in the reaction mixture in equal amounts. As a consequence, DKR and PKR are of considerable interest from the preparative point of view.

The KR method occupies a special place among the methods for obtaining enantiopure amines. The processes based on acylation, hydrolysis or alcoholysis of *N*-protected *N*-carboxyanhydrides [32 - 35], reduction of imines [36 - 39], DKR of azlactones [40 - 45], and enantioselective formation of *N*-oxides [46, 47] and others have been studied to date.

Acylative Kinetic Resolution of Racemic Amines

Chiral acylating agents used in acylative KR of racemic amines can be assigned to two main groups: 1) enantioselective reagents (the chiral center of the reagent remains in the leaving group; as a result, a mixture of enantioenriched amide and unreacted amine of the opposite configuration is formed); 2) diastereoselective reagents (the chiral center of the acyl moiety is transferred to the reaction product; the reaction results in a mixture of diastereoisomerically enriched amide and enantioenriched unreacted amine) [48].

The stereoselectivity of acylation depends significantly on many factors: the structures of racemic amine and acylating agents, as well as the process conditions (solvent, temperature, reagents' ratio, and reaction time). Either enantiomerically enriched unreacted amine or amide, the acylation product, can be the target of the synthesis, depending on which of the amine enantiomers is to be obtained. When preparing an enantioenriched amine from an amide, it is necessary that hydrolysis or other amide conversions are not accompanied by racemization.

DIASTEROSELECTIVE *N*-ACYLATION OF RACEMIC AMINES

In the process of acylative KR of racemic amines with chiral diastereoselective agents, the predominant formation of a diastereoisomeric amide from one enantiomer of amine occurs. The attractiveness of this approach is due, first of all, to the wide availability of chiral acids, such as derivatives of tartaric, camphoric, 2-arylpropionic acids, amino acids, and the simplicity of the process implementation. The additional advantages of acylative KR under the action of diastereoselective agents are that an increase in the optical purity of the amide can be achieved by conventional, non-stereospecific methods, and monitoring the optical purity of the reaction product can be easily carried out by HPLC or NMR spectroscopy.

Early Examples

Pioneer works on the diastereoselective acylation of amines began to appear in the late 1960s. Ugi and coworkers studied the acylation of various racemic amines and amino esters with mixed anhydrides formed *in situ* from *N*-protected amino acids and ethyl chloroformate [30]. Acylation was carried out in dichloromethane (0 °C) or diethyl ether (–60 °C) at an amine–mixed anhydride ratio of 10:1. The highest selectivity was observed during acylation of 1-phenylethylamine (**1**) with mixed anhydride (*R*)-**3** prepared from *N*-Cbz-(*R*)-phenylglycine [(*R*)-**2**] (Scheme **4**). In this case, the predominant formation of the (*R*,*R*)-diastereoisomeric amide **4** was observed. The selectivity factor s calculated based on the polarimetry data was 3.2 and 5.0 in dichloromethane and diethyl ether, respectively.

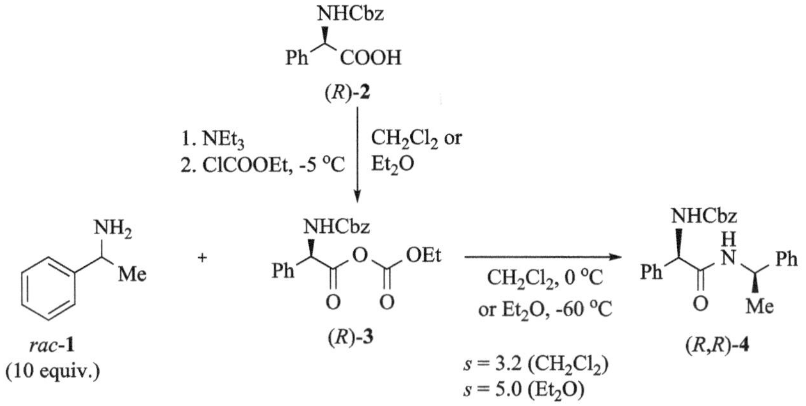

Scheme 4. KR of racemic 1-phenylethylamine (**1**) with mixed anhydride (*R*)-**3**.

Among first examples of KR of amines it is worth of mention the resolution of the synthetic racemic compound **5**, an analog of the product of oxidative aromatization of the delphinine alkaloid [49, 50] (Scheme 5). The reaction of racemic amine **5** with 1 equiv. of L-camphorsulfonyl chloride in pyridine led to sulfonamide **6** and enantio-enriched unreacted amine **5**. Recrystallization of unreacted compound **5** (as oxalate) yielded a sample identical to the product of natural delphinine degradation.

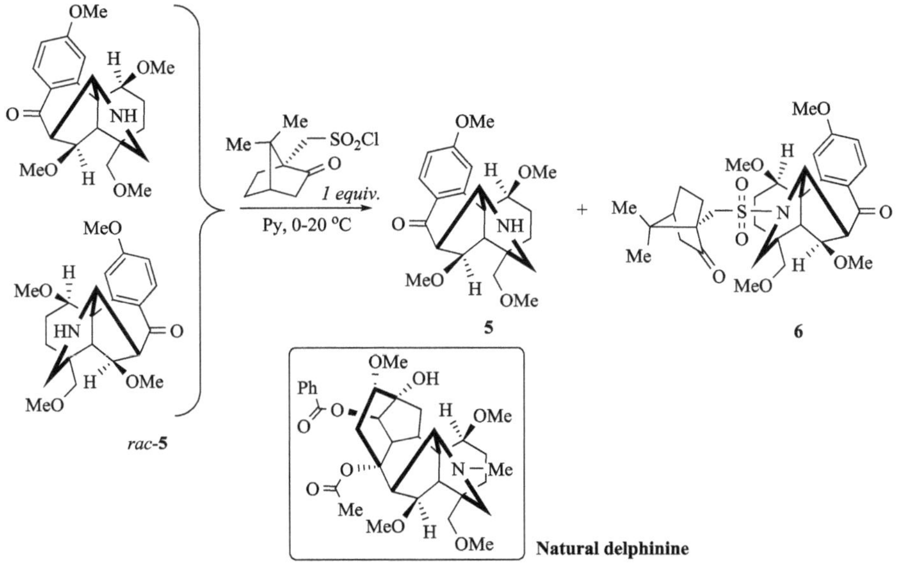

Scheme 5. KR of racemic compound **5** with L-camphorsulfonyl chloride.

In the 1960–1980s, it was reported about diastereoselective acylation of racemic amines with optically pure resolving agents, such as the derivatives of camphoric acid [51, 52], tartaric acid [53, 54], (*S*)-2-arylpropionic and (*S*)-2-phenylbutyric acids [55 - 59], as well as *N*-Cbz-(*S*)-phenylalanine [60] and oligopeptides [61]. The formation of isomerically enriched dipeptides, starting from racemic *N*- and *C*-protected amino acids, was studied [62]. However, the stereochemical outcomes of KR in most of the early examples were from moderate to low.

In 1981, Teramoto *et al.* used active esters of *N*-Cbz-amino acids (alanine and glycine) and optically pure *N*-hydroxy-(*R*,*R*)-tartrimides as acylating agents for diastereoselective peptide synthesis [63]. Interaction of *O*-(Cbz-L-alanyl)-*N*-hydroxy-(*R*,*R*)-tartrimide [(2*S*)-**7**] with ethyl DL-alaninate (2 equiv.) in THF in the presence of DCC at 0 °C resulted in dipeptide (*S*,*S*)-**9** in 93% yield and 100% *de* (Scheme 6). The reaction of tartrimide (2*R*)-**7** with DL-Ala-OEt afforded dipeptide (*R*,*R*)-**9** in 92% yield and 20.6% *de*. When DL-Ala-OEt was acylated with tartrimide (2*R*)-**8**, dipeptide (*R*,*R*)-**9** was formed in 92.5% yield and 100% *de* [63].

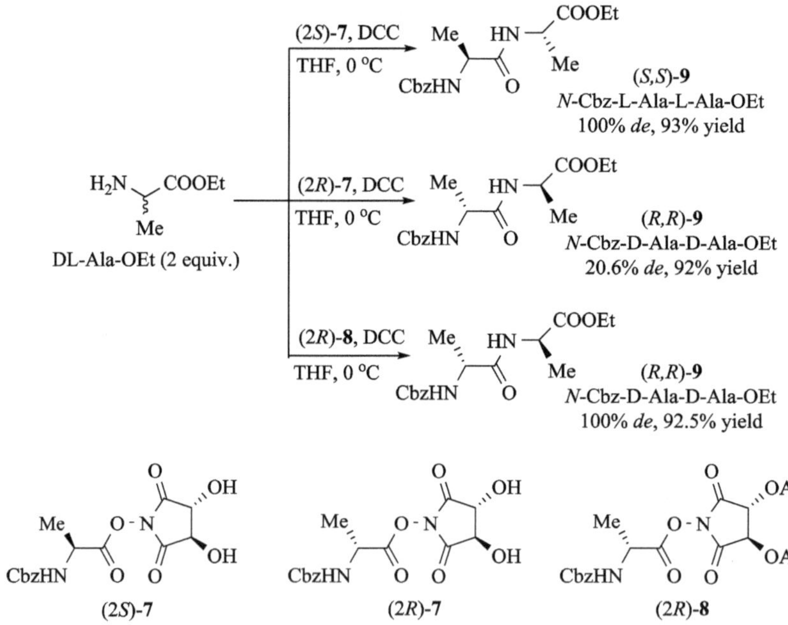

Scheme 6. Diastereoselective acylating agents based on active esters of *N*-Cbz-alanines in peptide synthesis.

Compounds **10-12** (Scheme 7) were obtained from *N*-Cbz-L-amino acids (alanine, valine and proline) and (+)-*N*-hydroxycamphorimide under the action of EDC×HCl in THF at 0 °C [64]. Acylation of racemic DL-amino esters (alanine, valine, leucine, phenylalanine) with compounds **10-12** (0.5 equiv.) was carried out in acetonitrile in the presence of TEA at room temperature for a long time (from 2 days to 3 months). For example, the reaction of compound **10** with ethyl DLleucinate in the presence of AcOH for 3 days resulted in Cbz-L-Ala-L-Leu-OEt in 91.2% yield and 100% *de*.

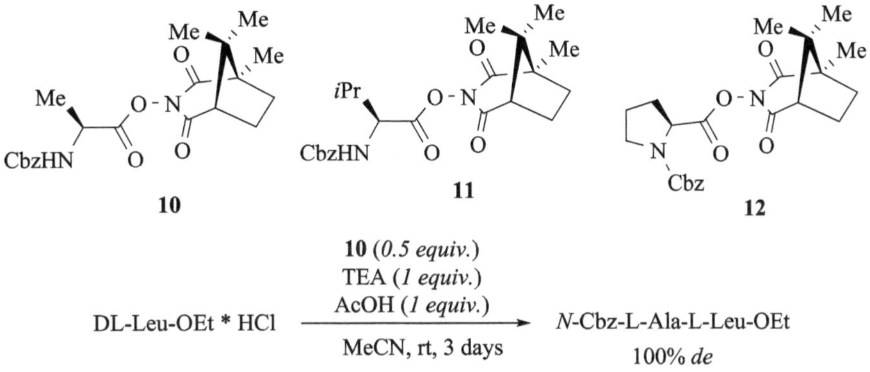

Scheme 7. Diastereoselective peptide synthesis using camphorimide esters of *N*-Cbz-amino acids.

The group of Prof. Kostyanovsky has developed an original approach for preparation of enantioenriched unsymmetrical *N*-substituted diaziridines [65 - 70]. Diastereoselective acylation of 1,3,3-trisubstituted diaziridines **13a-e** with *N*-tosyl-(*S*)-prolyl chloride (**14a**, 0.5 equiv.) led to *N*-acylhydrazones **15a-e** (as a result of the diaziridine–hydrazone rearrangement) and unreacted (1*R*,2*R*)-diaziridines **13a-e** (Scheme **8**). It has been found that the regiospecificity of acylation of diaziridines **13** at the N^1 nitrogen atom with the formation of *N*-acylhydrazones **15**, but not at the N^2 atom, is due to orbital and charge control [69]. The molecular structure of compound **15d** was determined by X-ray diffraction analysis [70]. The highest stereoselectivity was observed in the acylation of bicyclic diaziridines: 1-methyl-1,2-diazaspiro[5.2]octane (**13b**) and 5-methyl-1,6-diazabicyclo[3.1.0]hexane (**13e**); the *ee* of unreacted diaziridines **13b** and **13e** was 60 and 70%, respectively (according to polarimetry and ^{1}H NMR spectroscopy after derivatization with (*S*)-α-phenylethyl isocyanate) [69].

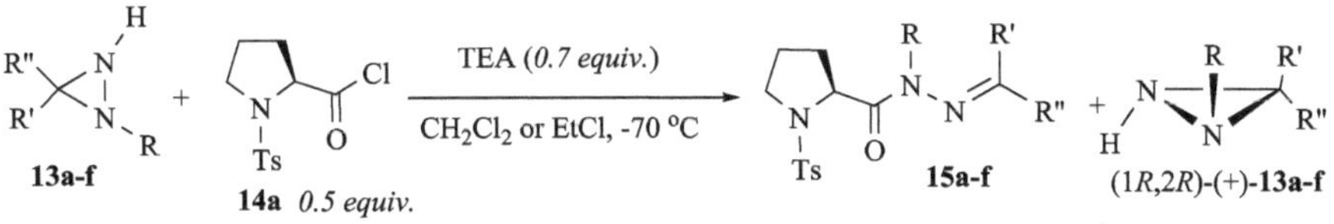

Diaziridine	R	R'	R"	ee, %	yield, %
13a	Me	Me	Me	44	34
13b	Me	(CH$_2$)$_5$		60	52
13c	Bn	Me	Me	11	74
13d	(CH$_2$)$_2$COMe	Me	Me	16	52
13e	(CH$_2$)$_3$		Me	75	50
13f	(CH$_2$)$_3$		CO$_2$Me	35	86

(ee, % and yield, % refer to unreacted (1*R*,2*R*)-(+)-**13**)

Scheme 8. KR of racemic diaziridines **13a-f** with *N*-tosyl-(*S*)-prolyl chloride.

In the case of KR of racemic diaziridine **13d** with *N*-(2,4-dinitrophenyl)-(*S*)-prolyl chloride (**14b**), the (–)-isomer **13d** (35.4% *ee*) predominated in the unreacted diaziridine (Scheme 9) [71], unlike the acylation with acyl chloride **14a**. Thus, by the example of KR of compound **13d** Kostyanovsky *et al*. demonstrated the possibility of preparation of enantioenriched (+)- and (–)-diaziridines as a result of acylation of racemate with reagents obtained on the basis of a single chiral precursor.

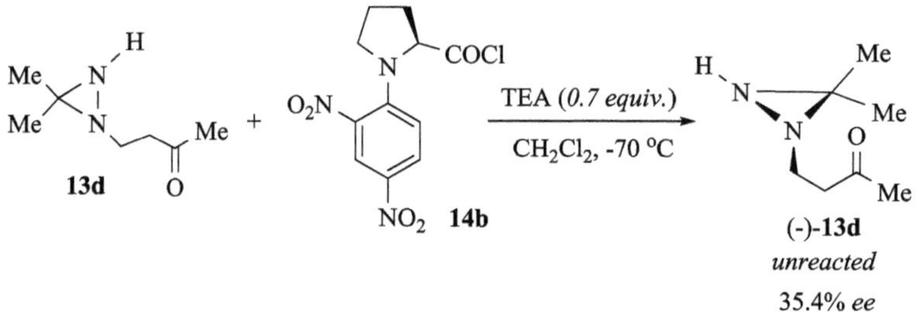

Scheme 9. KR of racemic diaziridine **13d** with *N*-(2,4-dinitrophenyl)-(*S*)-prolyl chloride.

Derivatives of 2-Arylpropionic Acids as Diastereoselective Acylating Agents

In 1999, our research group proposed to use (*S*)-naproxen acyl chloride (**16**) (Scheme **10**) as a resolving agent in KR of heterocyclic racemic amines, such as 3-methylbenzoxazines **17a,b** [72, 73]. The starting material in the synthesis of acyl chloride **16** is a nonsteroidal anti-inflammatory drug Naproxen ((+)-(*S*)-2-(6-methoxynaphth-2-yl)propanoic acid) [74] that is commercially available in enantiopure form. Naproxen has a characteristic UV absorption maximum at 230 nm with a high molar extinction coefficient, which enables the ease determination of its derivatives at UV detection. Naproxen and its derivatives are convenient resolving and derivatizing agents used to determine the optical purity of chiral compounds by HPLC [75 - 79] and other methods [41, 79 - 81]. However, before our studies, (*S*)-naproxen chloride has not been used for KR of racemic amines. Since that time, we have performed systematic investigations of KR of racemic *N*-heterocycles **17a-g** and their structural analogues **18a,b** and **19** (Scheme **10**) with (*S*)-naproxen chloride and other diastereoselective acylating agents.

Firstly, the acylation of amines **17a-c, e** was carried out with acyl chloride (*S*)-**16** (0.5 equiv.) in benzene at room temperature, which proceeded with high selectivity and resulted in the predominant formation of (*S,S*)-amides **20a-c,e** (*de* from 76 to 87% according to HPLC) (Scheme **11**) [72, 82]. Unreacted amines **17a-c,e** were enriched in (*R*)-enantiomers. In most cases, configuration of chiral centers in major (*S,S*)-amides **20a-c,e** was assigned using X-ray diffraction analysis based on the known configuration of the (*S*)-naproxen acyl fragment.

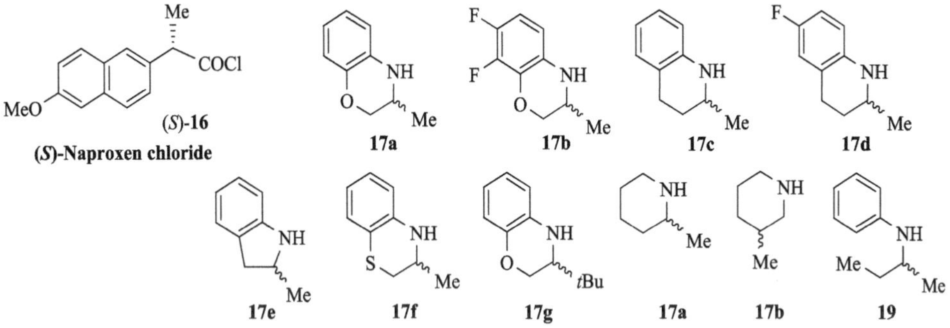

Scheme 10. Structures of (*S*)-naproxen chloride ((*S*)-**16**) and racemic amines **17-19**.

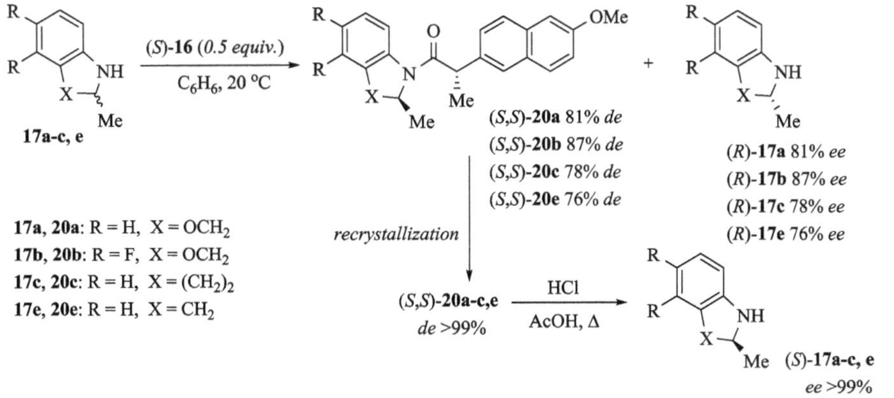

Scheme 11. KR of racemic amines **17a-c, e** with (*S*)-naproxen chloride ((*S*)-**16**) in benzene at +20 °C.

A detailed study of the effect of the reaction conditions (temperature, solvent, tertiary amine additives) on the diastereoisomeric composition of the acylation products has shown that at room temperature the solvent of choice in KR of amines **17a-c, e** is benzene [72, 82, 83]. The highest diastereoisomeric excess of amide (*S,S*)-**20a** (85.8% *de*) was observed during the reaction in benzene in the presence of *N*-methylmorpholine as the HCl acceptor [83]. A single recrystallization of (*S,S*)-amides from *n*-hexane (or *n*-hexane–ethyl acetate) resulted in diastereoisomerically pure amides (*S,S*)-**20a-c,e** (*de* ≥99%) with yields of 30-35% (relative to the starting racemic amines). Subsequent acidic hydrolysis afforded optically pure (*S*)-enantiomers of amines **17a-c,e** (*ee* ≥99%).

A comparative study of KR of racemic amines **17a,c** and their fluorine-substituted analogs **17b,d** under the action of acyl chloride (*S*)-**16** in dichloromethane for 6 h showed that the presence of fluorine atoms in the aromatic fragment of amine led to an increase in the acylation selectivity, *i.e.*, selectivity factor *s* (Scheme **12**) [84]. At the same time, the presence of fluorine atoms had practically no effect on the conversion (*C*) of starting racemate.

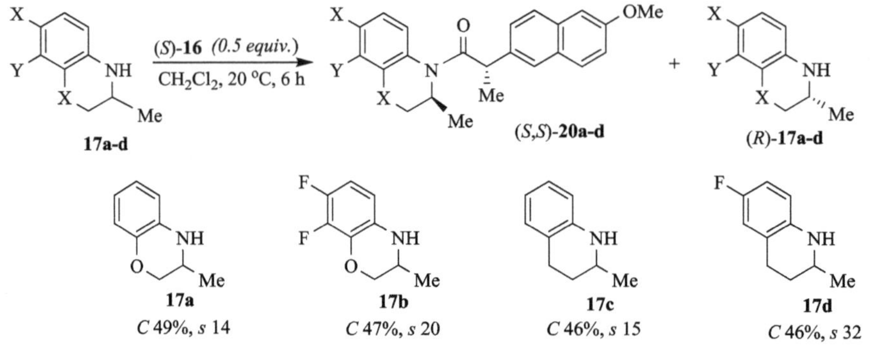

Scheme 12. KR of racemic amines **17a-d** with (*S*)-naproxen chloride ((*S*)-**16**) in dichloromethane at +20 °C.

(*S*)-7,8-Difluoro-3,4-dihydro-3-methyl-2*H*-[1,4]benzoxazine ((*S*)-**17b**) is of the greatest interest among the amines studied, since it is a key intermediate in the synthesis of antibacterial agent Levofloxacin. Using an approach based on the diastereoselective acylation of racemic amine with acyl chloride (*S*)-**16**, it became possible to develop an original method for the preparation of amine (*S*)-**17b** of *ee* >99.8% (Scheme **13**) [73, 85 - 87], which was suitable to develop a technological scheme for production of the active pharmaceutical substance of Levofloxacin.

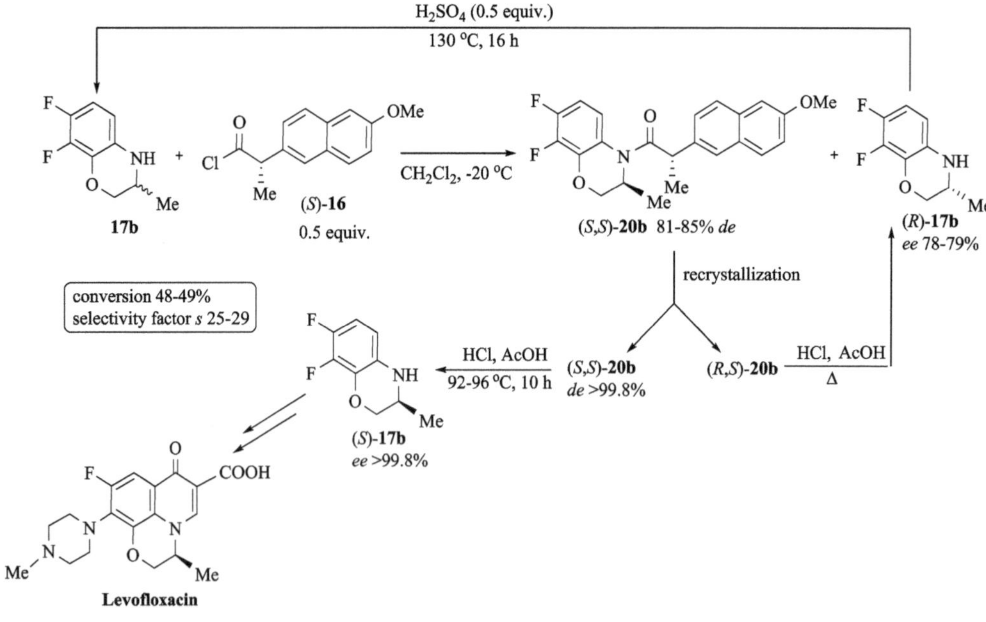

Scheme 13. Large-scale method for the preparation of (*S*)-7,8-difluoro-3,4-dihydro-3-methyl-2*H*-[1,4] benzoxazine ((*S*)-**17**) *via* KR of its racemate with (*S*)-naproxen chloride ((*S*)-**16**).

The diastereoselective acylation of racemic 7,8-difluoro-3,4-dihydro-3-methyl-2H-[1,4]benzoxazine (**17b**) with (*S*)-naproxen chloride (Scheme **13**), which underlies the method for preparation of amine (*S*)-**17b**, is the classic "ideal" case of acylative KR of racemic amine with a chiral resolving agent. The developed method is characterized by easy implementation, scalability (racemate loading up to 1 kg) and allows obtaining the target product of high optical purity and in high yield.

As a result of a large series of experiments, the following optimum conditions of the diastereoselective acylation of racemic amine **17b** with acyl chloride (*S*)-**16** were found: dichloromethane as a solvent, temperature –20 °C, starting amine concentration 0.3 M. The highest stereochemical results of KR ((*S*,*S*)-**20b**, 92.3% *de*, *s* 45) were observed when the acylation was carried out with small amounts of racemate (up to 1 g) in toluene at –20 °C; but a significant decrease in the diastereoisomeric purity of the resulting amide (*S*,*S*)-**20b** took place with increasing loading of the starting racemic amine. This can be explained by the fact that hydrochloride of racemic amine **17b** formed during the acylation is poorly soluble in toluene and partially precipitated, whereby the (*S*)-enantiomer of amine **17b** is removed from the reaction. Therefore, toluene was replaced with dichloromethane, in which the hydrochloride of racemic amine **17b** is soluble. The diastereoselective acylation step was carried out for 6 h, resulting in amide (*S*,*S*)-**20b** with 81–85% *de*. Subsequent recrystallization of amide (*S*,*S*)-**20b** resulted in the diastereoisomerically pure (*S*,*S*)-**20b** (*de* >99.8% according to HPLC and ^{1}H and ^{19}F NMR spectroscopy) in 73–80% yield (relative to the starting acyl chloride **16**).

Enantiopure amine (*S*)-**17b** (*ee* >99.8% according to chiral HPLC) was obtained in a high yield (up to 93%) as a result of acidic hydrolysis of diastereoisomerically pure amide (*S*,*S*)-**20b** (*de* >99.8%) under heating in a mixture of hydrochloric and acetic acids (Scheme **13**) [85, 86]. In this case, a loss of optical activity did not occur. The total yield of amine (*S*)-**17b** was 30-33% relative to the starting racemate. The high purity of the target amine (*S*)-**17b** made it possible to obtain this compound for the first time in crystalline form and to perform a comparative crystallographic study of its racemic and enantiopure forms by X-ray diffraction analysis [86].

To improve the efficiency of KR process, a route for racemization of the "isomeric ballast" containing unreacted (*R*)-enantiomer of amine **17b**, and the return of racemate to the synthesis scheme has been developed (Scheme **13**). It has been found that heating of the scalemic amine (*R*)-**17b** (78% *ee*) at 130 ° C in the presence of concentrated sulfuric acid (0.5 equiv.) for 16 h leads to the racemate **17b** in a yield of 89% (relative to the starting amine). Recycling of the "isomeric ballast" resulted in an increase in the total yield of enantiopure (S)-**17b** up to 55-60% (relative to racemate).

In 2005, the approach based on diastereoselective acylation with acyl chloride (S)-**16** was successfully applied for preparation of (S)-enantiomer of 3-benzyl-3, 4-dihydro-2H-[1,4]benzoxazine (**21**) (Scheme **14**) and assignment of the absolute configuration of (R)-enantiomer obtained as a result of enantioselective hydrogenation [88]. For this purpose, the acylation of racemic amine **21** with 0.5 equiv. of naproxen chloride ((S)-**16**) was carried out; then, the reaction product, amide (S,S)-**22** (de >99%) was isolated by column chromatography on silica gel; subsequent acidic hydrolysis of the diastereoisomerically pure amide (S,S)-**22** gave (S)-enantiomer of amine **21** in optically pure form (ee >99%). The configuration of chiral centers in amide (S,S)-**22** was determined by X-ray diffraction analysis.

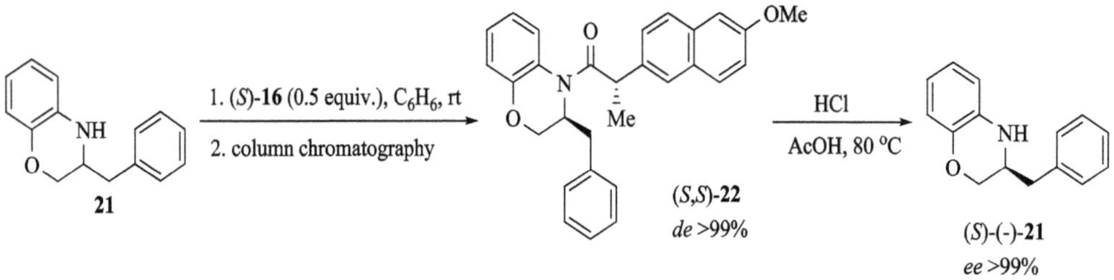

Scheme 14. Preparation of (S)-3-benzyl-3,4-dihydro-2H-[1,4]benzoxazine ((S)-**21**) via acylative KR of racemate with (S)-naproxen chloride.

In 2011–2013, a comparative study of the diastereoselective acylation of racemic heterocyclic amines **17a-c,e**, **18a**, and **19** with acyl chlorides **16**, **23a-f** from a series of 2-arylpropionic acids and their analogues was performed (Scheme **15**) [87, 89, 90]. To do this, an approach based on the reaction between racemic amines and racemic acylating agents (mutual KR) was used. In this case, the reaction leads to the formation of two pairs of diastereoisomeric amides: (S,S)–(R,R) and (R,S)–(S,R), a ratio of which (dr) corresponds to the selectivity factor $s = k_{fast}/k_{slow}$ [6, 31]. Moreover, a ratio of the concentrations of the diastereoisomeric reaction products is independent of either the starting reagent ratio or the reaction duration.

The acylation of aromatic heterocyclic amines **17a-c,e**, **18a** and **19** with racemic acyl chlorides **16** and **23a-f** was carried out at a 2:1 ratio in various solvents (toluene, dichloromethane, acetonitrile) at +20 and −20 °C. It has been found that in all cases of acylation of amines **17a-c,e** with acyl chlorides **16**, **23a-f**, the predominant formation of (S,S)–(R,R)-diastereoisomeric amides take place (Scheme **16**).

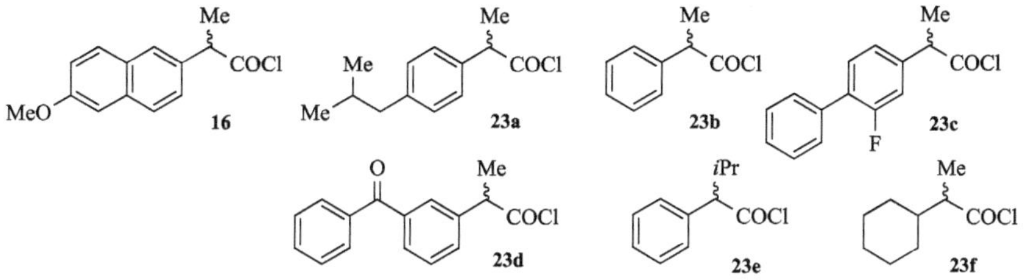

Scheme 15. Structures of racemic acyl chlorides 16 and 23a-f.

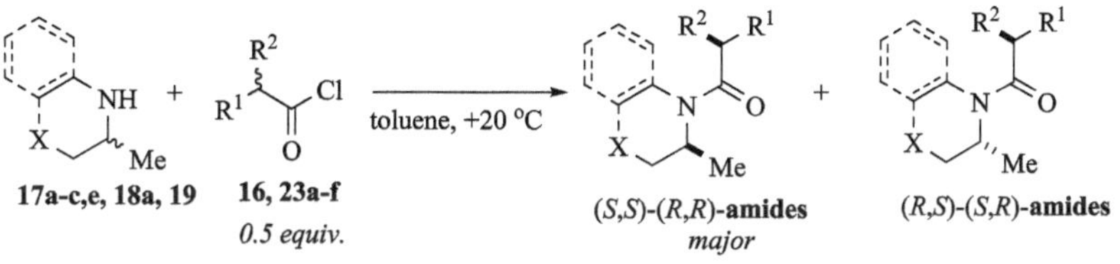

Racemic amine	Acyl chloride	dr of the amides formed (S,S)-(R,R)/(R,S)-(S,R)
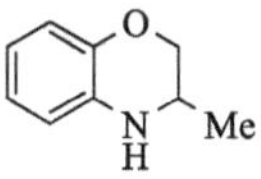 17a	16	97.0 : 3.0
	23a	98.2 : 1.8
	23b	97.0 : 3.0
	23c	94.4 : 5.6
	23d	93.6 : 6.4
	23e	95.0 : 5.0
	23f	80.6 : 19.4
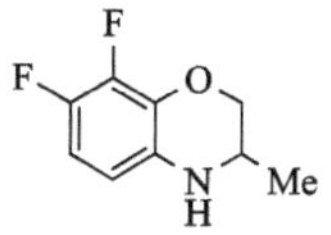 17b	16	97.8 : 2.2
	23a	97.9 : 2.1
	23b	96.9 : 3.1
	23c	96.0 : 4.0
	23d	95.7 : 4.3
	23e	95.5 : 4.5
	23f	75.5 : 24.5
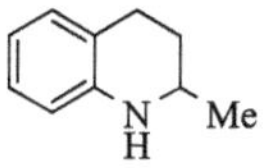 17c	16	96.8 : 3.2
	23a	97.0 : 3.0
	23b	95.9 : 4.1
	23c	95.2 : 4.8
	23d	94.0 : 6.0
	23e	95.0 : 5.0
	23f	78.3 : 21.7
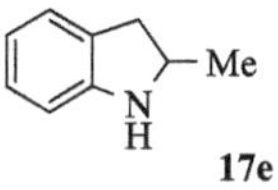 17e	16	93.5 : 6.5
	23a	94.7 : 5.3
	23b	93.5 : 6.5
	23c	88.5 : 11.5
	23d	88.4 : 11.6
	23e	91.2 : 8.8
	23f	67.2 : 32.8
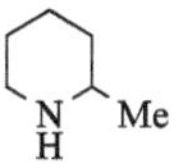 18a	16	60.0 : 40.0
	23a	65.0 : 35.0
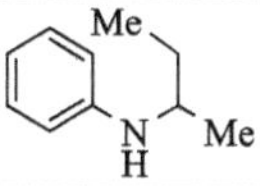 19	16	58.0 : 42.0
	23a	60.0 : 40.0

Scheme 16. Diastereoselective acylation of racemic amines 17a-c,e, 18a and 19 with racemic acyl chlorides 16, 23a-f.

The highest selectivity of acylation was observed in toluene, a decrease in selectivity occurred in more polar solvents; a decrease in the reaction temperature had an insignificant effect on the selectivity. Acyl chlorides **16, 23a-e** containing an aromatic substituent R^1 exhibit approximately equal selectivity under the same conditions. Ibuprofen chloride (2-(4-isobutylphenyl)propionyl chloride) (**23a**) showed the greatest diastereoselectivity compared to other acyl chlorides. It has also been found that replacement of the methyl substituent (**23b**, R^2 = Me) at the chiral center of acylating agent with isopropyl (**23e**, R^2 = iPr) results in a slight decrease in selectivity. Replacement of the phenyl substituent (**23b**, R^1 = Ph) by cyclohexyl (**23f**, R^1 = *cyclo*Hex) leads to a sharp decrease in the acylation selectivity, which indicates the significant role of aromatic interactions between the molecules of reagents.

The stereoselectivity of acylation of racemic *N*-(*sec*-butyl)aniline (**19**) and 2-methylpiperidine (**18a**), which can be considered as structural analogues of 2-methyl-1,2,3,4-tetrahydroquinoline (**17c**), with naproxen chloride (**16**) and ibuprofen chloride (**23a**) was significantly lower (the maximum value of *dr* was 65.0:35.0 in the acylation of amine **18a** with acyl chloride **23a**) than amines **17a-c,e** (Scheme 16). This fact suggested that for the realization of efficient KR in the acylation with 2-arylpropionyl chlorides, the amine molecule must contain a heterocyclic fragment which conformational mobility is additionally restricted by the fused aromatic system [90].

In 2005-2010, a systematic study of the mutual and parallel KR during diastereoselective acylation of chiral oxazolidin-2-ones and their analogs with active esters or acyl chlorides of 2-arylalkanoic acids was performed by the research group of J. Eames [91 - 99]. The deprotonated oxazolidinones were formed *in situ* under the action of a strong base (for example, butyllithium) in THF at –78 °C.

Initially it was found that the reaction between racemic oxazolidin-2-one **24a** and racemic 2-phenylpropionyl chloride (**23b**) resulted in isomerically enriched imide **25a** (Scheme 17) [91, 93]. The ratio of the resulting *syn-/anti*-isomers of compound **25a** depended on the nature of the base used for the deprotonation of oxazolidin-2-one **24a**. In the presence of lithium-containing bases, *anti*-**25a** predominated in the reaction product. If Na^+ or K^+ were used as counter-ions, the acylation resulted in predominant formation of *syn*-**25a**.

Acylation of racemic oxazolidin-2-ones **24a-e** with racemic pentafluorophenyl 2-phenylpropionate (**26**) proceeded with greater selectivity than in the case of acyl chloride **23b**, and the ratio of *syn-/anti*-isomers of the formed N-acyloxazolidin-2-ones **25a-e** reached 97:3 (Scheme 18) [92]. The selectivity of acylation of oxazolidin-2-ones **24a-c** derived from valine (**24a**: R = *i*Pr), phenylglycine (**24b**: R = Ph) and serine (**24c**: R = CO$_2$Et) was high (*syn-/anti-***25a-c** 95:5 and more). The reaction of ester **26** with norephedrine (**24d**: R = Me, R' = Ph) and phenylalanine (**24e**: R = CH$_2$Ph) derivatives proceeded with lower stereoselectivity. Diastereoselective acylation of compounds **24f-j**, structural analogs of 4-phenyloxazolidin-2-one **24b**, with ester **26** in all cases, except for triphenyl derivative **24j**, proceeded with high yields and diastereoselectivity (*syn-/anti-* 96:4 and more) (Scheme **18**) [97].

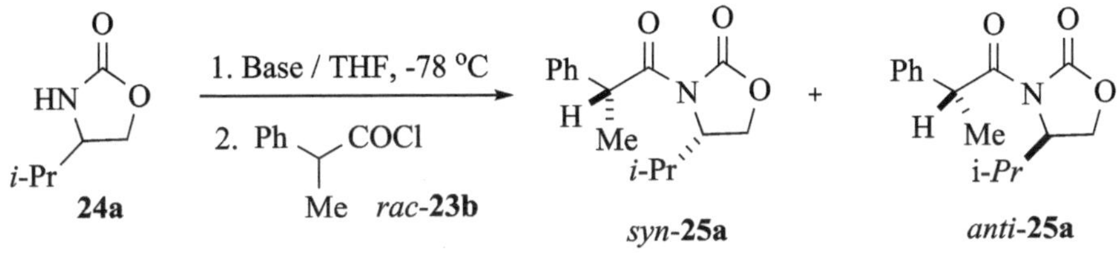

Base	*n*BuLi	LiHMDS	NaHMDS	KHMDS
anti / *syn*	70:30	64:36	34:66	38:62
Yield, %	60	70	70	58

Scheme 17. Mutual KR of racemic oxazolidin-2-one **24a** with racemic 2-phenylpropionyl chloride (**23b**).

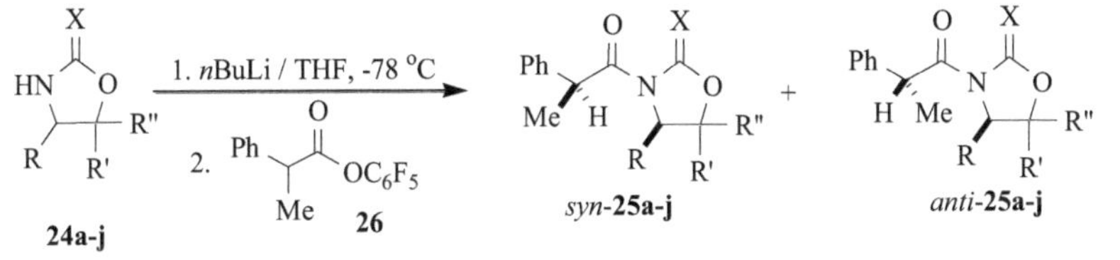

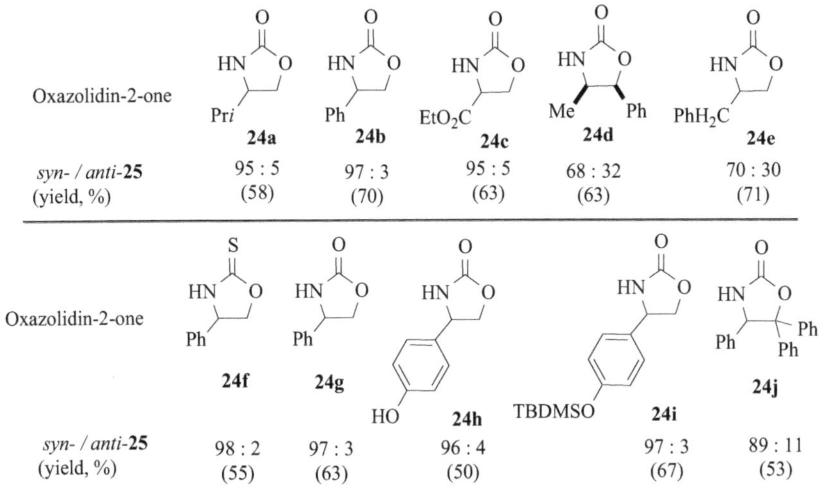

Oxazolidin-2-one	**24a** (P*ri*)	**24b** (Ph)	**24c** (EtO$_2$C)	**24d** (Me, Ph)	**24e** (PhH$_2$C)
syn- / *anti-***25** (yield, %)	95 : 5 (58)	97 : 3 (70)	95 : 5 (63)	68 : 32 (63)	70 : 30 (71)

Oxazolidin-2-one	**24f** (S, Ph)	**24g** (O, Ph)	**24h** (HO-C$_6$H$_4$)	**24i** (TBDMSO-C$_6$H$_4$)	**24j** (Ph, Ph)
syn- / *anti-***25** (yield, %)	98 : 2 (55)	97 : 3 (63)	96 : 4 (50)	97 : 3 (67)	89 : 11 (53)

Scheme 18. Diastereoselective acylation of oxazolidin-2-ones **24a-j** with active ester **26**.

A comparative study of the diastereoselectivity of acylation of oxazolidin-2-one **24j** with active esters of various 2-arylalkanoic acids **27a-e** (Scheme 19) has shown that acylation with ester **27a** with ethyl substituent at the chiral center proceeds with the highest selectivity [98]. A further increase in the volume of the alkyl substituent (**27b**: R = iPr) resulted in a decrease in stereoselectivity. Varying the substituents in the aromatic moiety (esters **27c-e**) did not significantly affect the stereochemical result of the reaction.

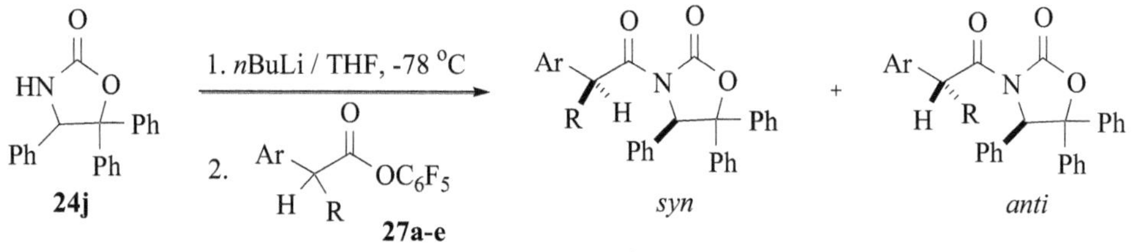

Activated ester	syn- / anti-	Yield, %
27a: R = Et, Ar = Ph	97 : 3	83
27b: R = iPr, Ar = Ph	78 : 22	18
27c: R = Me, Ar = 4-MeC$_6$H$_4$	93 : 7	60
27d: R = Me, Ar = 4-ClC$_6$H$_4$	92 : 8	58
27e: R = Me, Ar = 4-iBuC$_6$H$_4$	95 : 5	77

Scheme 19. Diastereoselective acylation of racemic oxazolidin-2-one **24j** with racemic active esters **27a-e**.

Diastereoselective acylation of 4-substituted oxazolidin-2-ones with pentafluorophenyl esters of optically pure acids, such as (R)-2-phenyl propionic acid ((R)-**26**), (S)-naproxen ((S)-**28a**) and (R)-ibuprofen ((R)-**28b**) was carried out (Scheme 20) [94, 96]. In all cases, selectivity of acylation was moderate: (*syn-/anti*-imides ratio about 80:20).

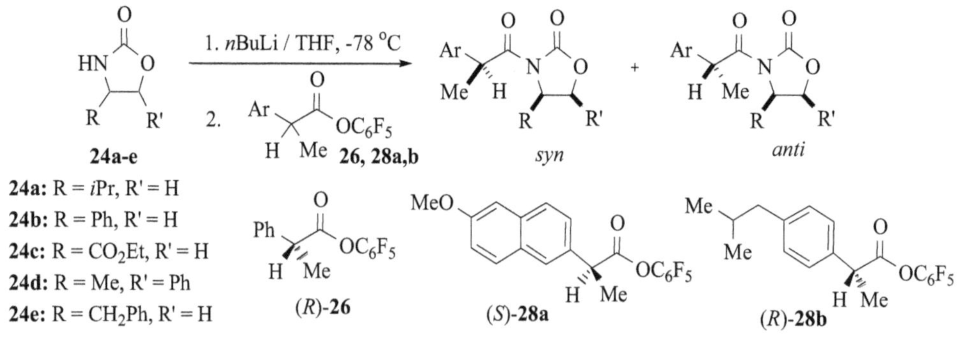

24a: R = iPr, R' = H
24b: R = Ph, R' = H
24c: R = CO$_2$Et, R' = H
24d: R = Me, R' = Ph
24e: R = CH$_2$Ph, R' = H

Scheme 20. Diastereoselective acylation of racemic oxazolidin-2-one **24a-e** with optically active esters **26**, **28a,b**.

Close stereoselectivity of acylation with esters (*R*)-**26**, (*S*)-**28a** and (*R*)-**28b** made it possible to use (*R*)-**26**–(*S*)-**28a** and (*S*)-**28a**–(*R*)-**28b** equimolar reagents' combinations for parallel KR of racemic oxazolidinones **24a-e** (Scheme **21**) [96].

Parallel KR of racemic *trans*-4,5-diphenyl-imidazolidin-2-thione (**31**) was carried out in a similar manner [99]. In preliminary experiments on the acylation of racemic **31** with active esters of optically pure and racemic acids: ibuprofen (**28b**) and 2-phenylbutyric acid (**28c**), it has been found that the reactions proceed with close selectivity (Scheme **22**). This made possible to carry out PKR using the *quasi*-racemic mixture of (*S*)-**28b** and (*R*)-**28c**; corresponding thioimides (2'*S*,4*R*,5*R*)-**32b** (82% *de*) and (2'*R*,4*S*,5*S*)-**32c** (90% *de*) were isolated in 28 and 24% yield, respectively. Acylation of racemate **31** with an equimolar mixture of active esters (*S*)-**28a**–(*R*)-**28c** resulted in the reaction products that were easily separated by column chromatography.

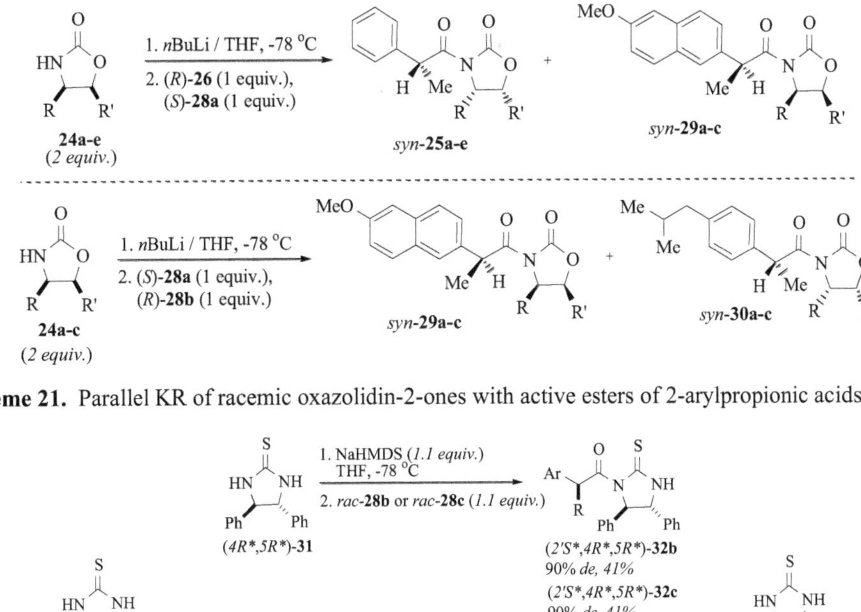

Scheme 21. Parallel KR of racemic oxazolidin-2-ones with active esters of 2-arylpropionic acids.

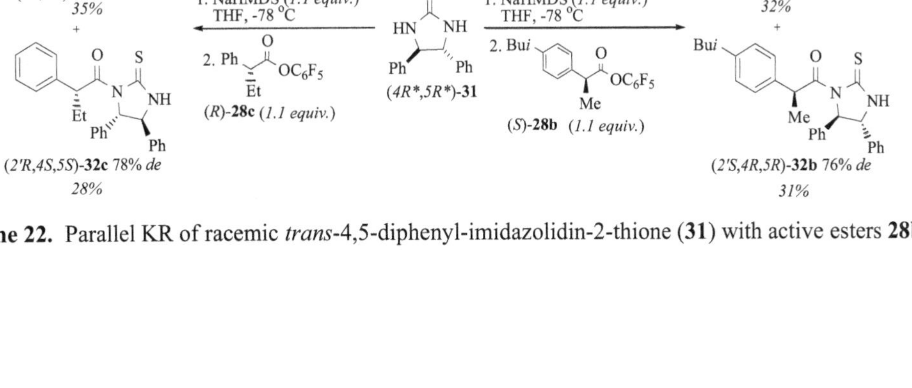

Scheme 22. Parallel KR of racemic *trans*-4,5-diphenyl-imidazolidin-2-thione (**31**) with active esters **28b** and **28c**.

Derivatives of 2-Hydroxy Carboxylic Acids as Diastereoselective Acylating Agents

While studying the diastereoselective acylation of chiral oxazolidinones with active esters of 2-methoxy and 2-phenoxy carboxylic acids, Eames and coworkers investigated the parallel KR of active pentafluorophenyl esters of 2-methoxy-2-phenylacetic, 2-phenoxy-2-phenylpropionic and Mosher's acids using a combination of *quasi*-enantiomeric oxazolidin-2-ones [96, 100]. So, in fact these processes can be considered as mutual KR. Thus, a combination of *quasi*-enantiomeric oxazolidinones (*R*)-**24b** and (*S*)-**24a** (in a lithiated form) was subjected to acylation with racemic pentafluorophenyl 2-methoxy-2-phenylacetate (*rac*-**33**) or pentafluorophenyl 2-phenoxy-propionate (*rac*-**36**) to form corresponding pairs of separable *quasi*-enantiomeric adducts (*syn*-**34** and *syn*-**35** or *syn*-**37** and *syn*-**38**) (Scheme **23**) [96]. Further transformations of isolated individual products *syn*-**34**, *syn*-**35**, *syn*-**37**, *syn*-**38** gave enantiomers of 2-methoxy-2-phenylacetic and 2-phenoxy-2-phenylpropionic acids [96].

The same approach was applied for parallel KR *via* diastereoselective acylation of a combination of *quasi*-enantiomeric oxazolidinones **24b** and **24j** with racemic pentafluorophenyl ester of Mosher's acid (2-methoxy-2-phenyl-2-trifluoromethyl-acetic acid) [100].

The above examples (Schemes **23** and **24**) demonstrate such a distinctive feature of the acylative KR as 'reciprocity'. Thus, the diastereoselective acylation of racemic oxazolidinone **24b** with a *quasi*-enantiomeric combination of active pentafluorophenyl esters (*R*)-**39** and (*S*)-**28b** (derived from (*S*)-ibuprofen) was also carried out to afford (*R*)- and (*S*)-enantiomers of both Mosher's acid and oxazolidinone **24b** after subsequent transformations [100].

In 2014, starting from (4a*R*,9a*S*)-bromohydroxamic acid derivative **42** and enantiomers of *O*-methylmandelic acid, Bode and coworkers synthesized the diastereoselective acylating agents (*R*,*S*,*S*)-**43** and (*R*,*S*,*R*)-**43**, in which the chiral centers are located in both the leaving group and acyl fragment (Scheme **25**) [101]. Acylation of racemic 3-benzylmorpholine (**44**) with reagents (*R*,*S*,*S*)-**43** and (*R*,*S*,*R*)-**43** was performed. It has been shown that configuration of acyl fragment plays a crucial role in stereoselectivity. Thus, acylation with compound (*R*,*S*,*S*)-**43** derived from (*S*)-*O*-methylmandelic acid resulted in (*S*,*S*)-amide **45** (*dr* 92:8, *s* 30), whereas the reaction of amine **44** with compound (*R*,*S*,*R*)-**43** derived from (*R*)-*O*-methylmandelic acid was unselective (*s* 1).

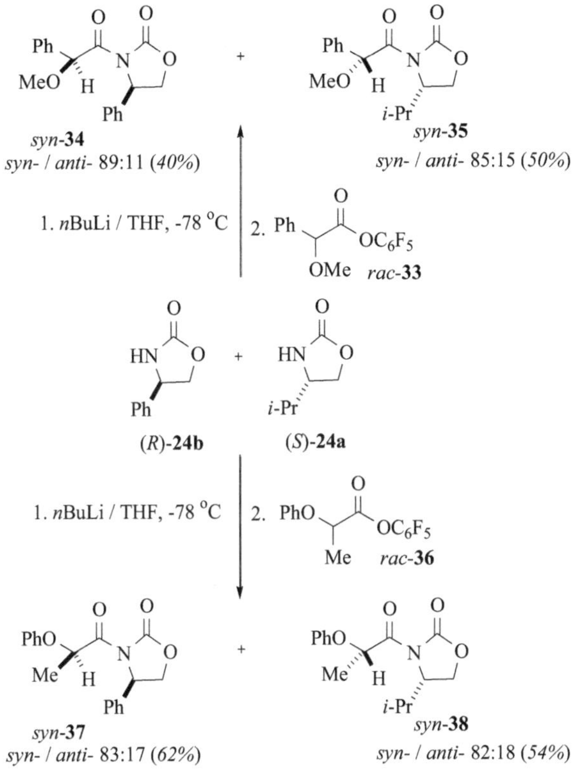

Scheme 23. Parallel KR of racemic active esters **33** and **36** using a combination of *quasi*-enantiomeric oxazolidinones **24a** and **24b**.

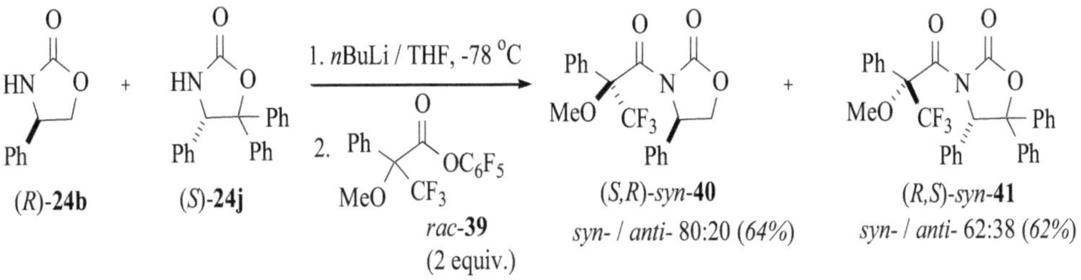

Scheme 24. Parallel KR of racemic active ester **39** using a combination of *quasi*-enantiomeric oxazolidinones **24b** and **24j**.

Diastereoselective acylating agents (*S*)-**46a-c** derived from achiral hydroxamic and (*S*)-*O*-methylmandelic acids exhibited rather high selectivity (*s* from 8 to 24) in acylation of racemic 3-benzylmorpholine (**44**) (Scheme **26**). However, it should be noted that in a number of cases the acylation with reagents **43** and **46a-c** was accompanied by epimerization of the chiral center in the mandelate fragment [101]. Reagents (*S*)-**46d,e** derived from *N*-hydroxysuccinimide and imidazole proved to be inefficient (Scheme **26**). The highest value of *dr* = 93:7 in the resulting amide (*S,S*)-**45** was observed in the acylation with (*S*)-**46b**. Other acylating agents **47a-f** obtained from achiral hydroxamic acid and various chiral 2-hydroxy acids turned out to be less selective than agent (*S*)-**46b** (Scheme **27**) [101].

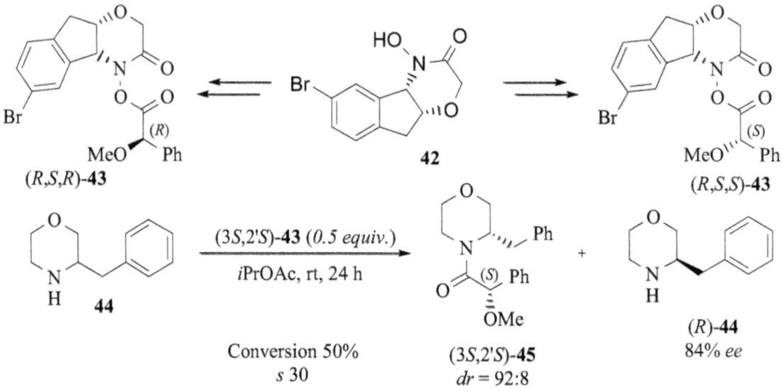

Scheme 25. Diastereoselective acylation of racemic 3-benzylmorpholine (**44**) with reagent (*R,S,S*)-**43**.

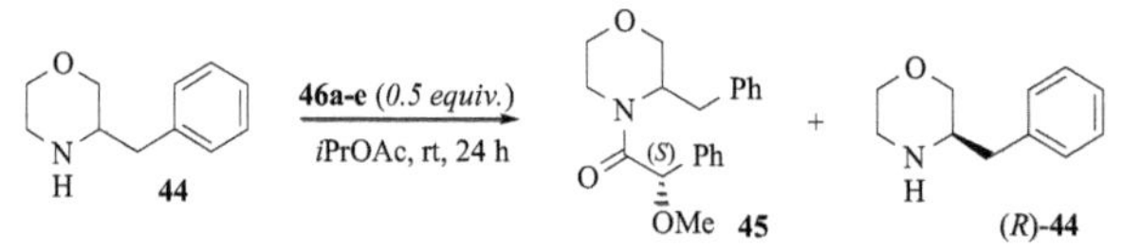

Acylating agent	s	Conversion, %	Amide **45** *dr*	(*R*)-**44**, *ee*
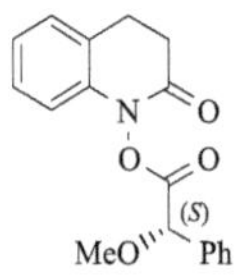 (*S*)-**46a**	9	32	(3*S*,2'*S*)-**45**, 86:14	33
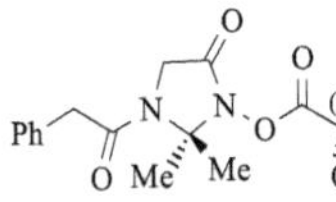 (*S*)-**46b**	24	40	(3*S*,2'*S*)-**45**, 93:7	58
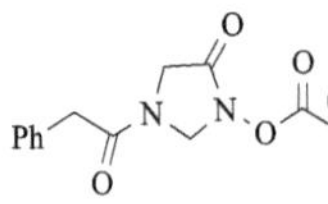 (*S*)-**46c**	8	40	(3*S*,2'*S*)-**45**, 84:16	46
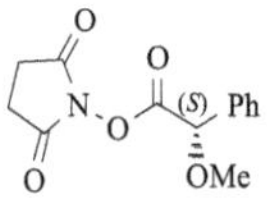 (*S*)-**46d**	2	36	(3*S*,2'*S*)-**45**, 66:34	18
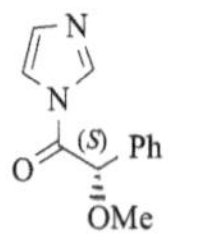 (*S*)-**46e**	1	50	(3*S*,2'*S*)-**45**, 50:50	0

Scheme 26. Diastereoselective acylation of racemic 3-benzylmorpholine (**44**) with reagents (*S*)-**46a-e**.

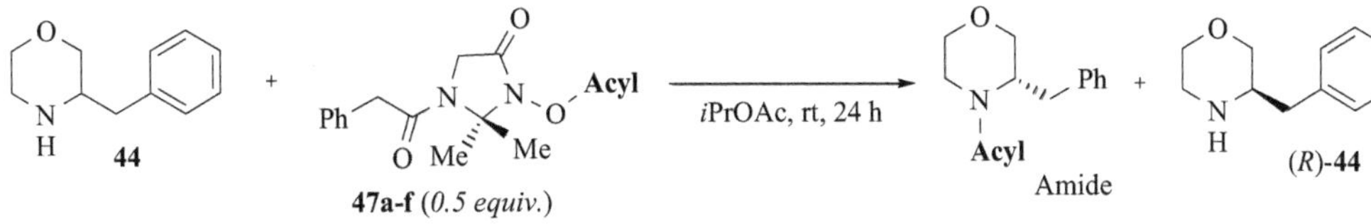

Acylating agent: **Acyl**		s	Conversion, %	Amide, dr	(R)-**44**, er
47a:	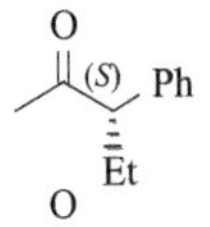 (S)-Ph, Et	3	24	69:31	56:44
47b:	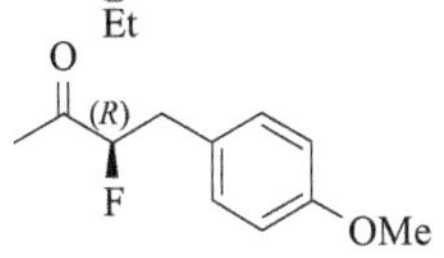 (R)-F, CH₂C₆H₄OMe	11	36	12:88	29:71
47c:	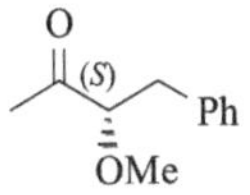 (S)-OMe, Ph	6	35	81:19	67:33
47d:	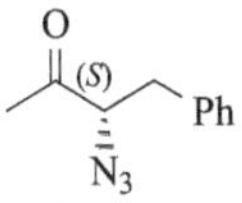 (S)-N₃, Ph	3	38	71:29	63:37
47e:	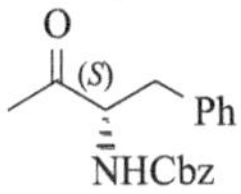 (S)-NHCbz, Ph	1	50	55:45	55:45
47f:	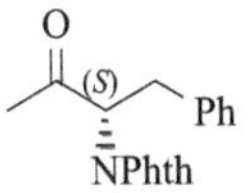 (S)-NPhth, Ph	–2*	18	41:59	48:52

*Opposite relative diastereoinduction was observed

Scheme 27. Diastereoselective acylation of racemic 3-benzylmorpholine (**44**) with reagents **47a-f**.

It has been shown that the (S)-mandelic acid-derived reagent (S)-**46b** is suitable for KR of several classes of racemic N-heterocycles (Scheme **28**). Acylation of amines with agent (S)-**46b** carried out in iPrOAc at room temperature for 24 h gave good conversions (36–43%) and acceptable selectivity (s 12–27) for the resolution of 2-substituted piperidine **48a**, morpholine **48b**, tetrahydroisoquinoline **48c**, and diazepanone **48d** [101]. Authors also proposed a model of transition states in the course of acylation of 3-benzylmorpholine (**44**) with reagent (S)-**46b**, which explains the observed stereoselectivity. A key role in the proposed model was assigned to the hydrogen bonding between the amino group of the amine and the carbonyl group in the hydroxamic acid residue, the steric hindrances created by the phenyl rings, and the preferred equatorial orientation of the substituent at the chiral center of the amine.

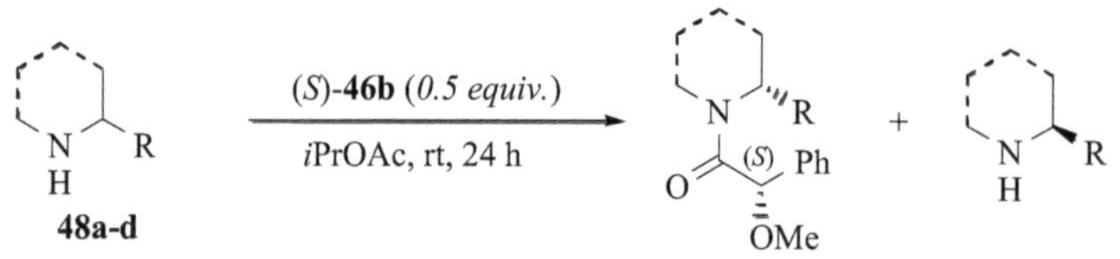

Scheme 28. Diastereoselective acylation of racemic amines **48a-d** with acylating agent (*S*)-**46b**.

Racemic amine	s	Conversion, %	Amide, *dr*	Unreacted amine, *er*
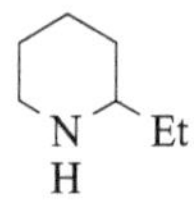 **48a**	12	39	88:12	74:26
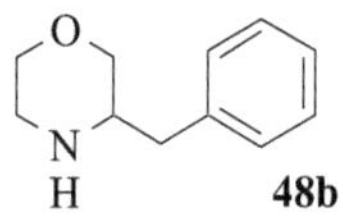 **48b**	24	40	93:7	79:21
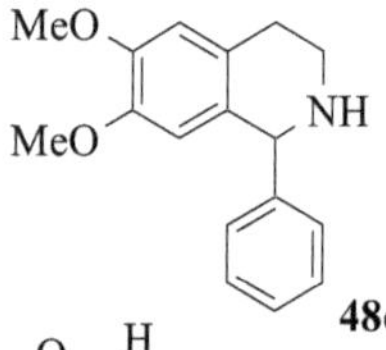 **48c**	18	36	92:8	74:26
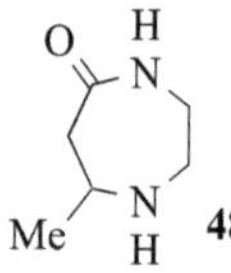 **48d**	27	38	94:6	77:23

In 2014, dynamic KR in the course of diastereoselective *N*-acylation with (*R*)-*O*-trimethylsilyl-4-fluoromandelic acid chloride ((*R*)-**49**) was applied for a scalable synthesis of a key intermediate of thrombin inhibitor, the active metabolite of propionate ester prodrug AZD8165 (Scheme **29**) [102]. Ethyl ester of 4,5-dihydro-1*H*-pyrazole-5-carboxylic acid (**50**) obtained *in situ* during the interaction of TMS-diazomethane and ethyl acrylate in the presence of trifluoroacetic acid (TFA) was used as a racemic substrate. Acylation of amino ester **50** (as a trifluoroacetate) with acyl chloride (*R*)-**49** in dichloromethane resulted a 70:30 mixture of diastereoisomeric amides (*R,S*)-**51** and (*R,R*)-**51** regardless of the amount of acylating agent and the reaction temperature. The presence of an auxiliary base was not necessary in this acylation reaction. The authors suggested that the N^2 atom of the dihydropyrazole ring of compound **50** serves as a proton acceptor, and racemization of the protonated amino ester **50** underlies the DKR. The target diastereoisomerically pure compound (*R,S*)-**51** (99% *ee*) was obtained after a single recrystallization. Further chemical transformations were developed to obtain the desired AZD8165 of high optical purity.

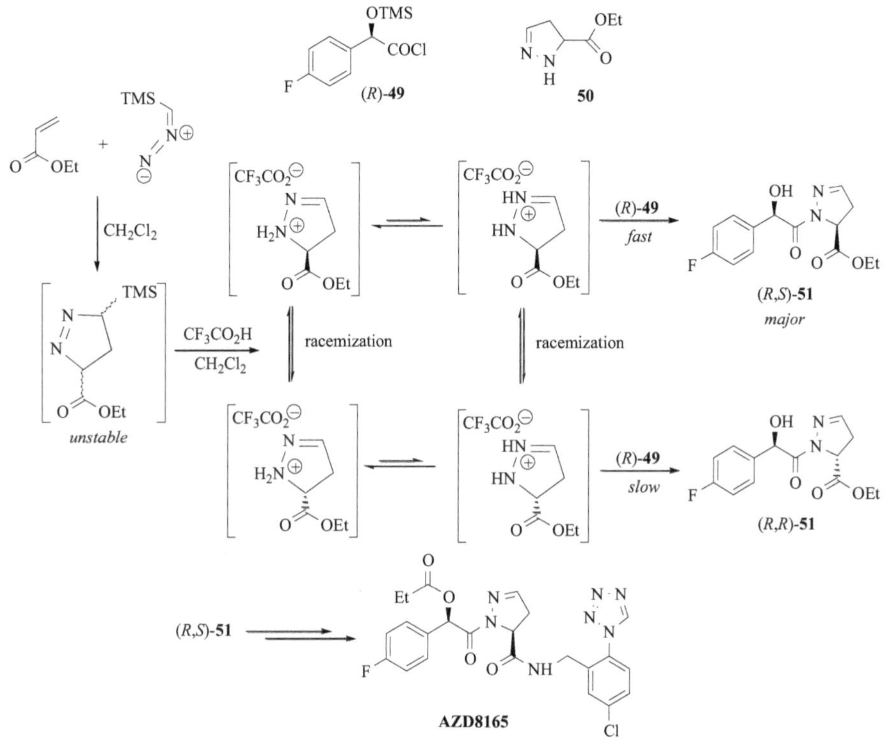

Scheme 29. A dynamic KR step in the synthesis of prodrug AZD8165.

In 2015, 2-phenoxy carboxylic acid chlorides (**52a-c**) were investigated as diastereoselective acylating agents for KR of racemic amines (Scheme **30**) [103]. It was found that acylation of 3-methylbenzoxazines **17a** and **17b** with acyl chlorides **52a-c** proceeded with higher stereoselectivity than that with 2-arylpropionyl chlorides **16**, **23a,b**. Acylation of racemic amines **17a,b** with racemic acyl chlorides **52a-c** (mutual KR) at a 2:1 molar ratio of reagents in toluene resulted in mixture of racemic (R,S)–(S,R) and (R,R)–(S,S) amides with significant predominance of (R,S)–(S,R)-diastereoisomers (according to GC/MS). 7,8-Difluoro-substituted amine **17b** was acylated with higher selectivity than its nonfluorinated analog **17a**. The highest stereoselectivity was observed when amines **17a,b** were subjected to acylation with 2-phenoxyisovaleric acid chloride (**52b**) with the bulky isopropyl substituent at the chiral center. The greatest value of selectivity factor s of about 500 was observed in the acylation reaction of amine **17b** with acyl chloride **52b** (in toluene at −20 °C) resulting in a mixture of amides with dr 99.8:0.2. This result is superior to the most up-to-date synthetic acylating reagents and catalysts for the kinetic resolution of racemic amines in stereoselectivity. The highest s value of 520 reported to date was obtained in the kinetic resolution of chiral oxazolidin-2-ones *via* catalytic enantioselective *N*-acylation [104].

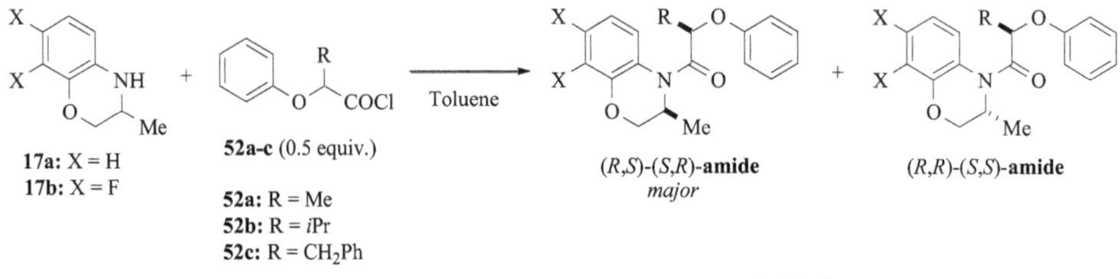

Racemic amine	Acylating agent	T, °C	Amide *dr* (*S,S*)-(*R,R*)/(*R,S*)-(*S,R*)
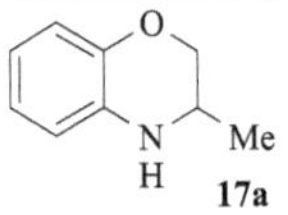 17a	52a	+20	97.2 : 2.8
	52a	-20	97.8 : 2.2
	52b	+20	99.0 : 1.0
	52b	-20	99.5 : 0.5
	52c	+20	98.2 : 1.8
	52c	-20	98.5 : 1.5
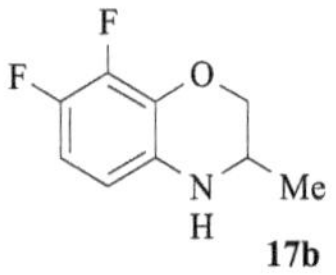 17b	52a	+20	98.2 : 1.8
	52a	-20	99.0 : 1.0
	52b	+20	99.5 : 0.5
	52b	-20	99.8 : 0.2
	52c	+20	99.2 : 0.8
	52c	-20	99.6 : 0.4

Scheme 30. Diastereoselective acylation of racemic amines **17a,b** with acyl chlorides **52a-c**.

Amino Acid Derivatives as Diastereoselective Acylating Agents

Amino acids being available in an optically pure form and very diverse in structure are very convenient building blocks to design a large range of chiral resolving agents.

In 2005, Toniolo *et al*. studied the C^{α}-methyl L-phenylglycine-based N^{α}-acetylated dipeptide 5(4*H*)-oxazolones **53a-f** as chiral acylating agents for KR of racemic 1-phenylethylamine (**1**) (Scheme 31) [105]. α-Amino acids with a quaternary C^{α} atom (for example, α-aminoisobutyric acid, L-2-methyl-2-phenylglycine (L-(αMe)Phg) or D-(αMe)Phg, 1-aminocycloalkane-1-carboxylic acids) were used to prepare oxazolones **53a-f**. Unlike 2,4-disubstituted 5(4*H*)-oxazolones derived from natural amino acids, oxazolones derived from optically pure 2-methyl-2-phenylglycine are configurationally stable and not prone to racemization. Each of the 5(4*H*)-oxazolones of *N*-acetyl-dipeptides **53a-f** was prepared *in situ* by treating the corresponding *N*-acetyl-dipeptide with EDC in acetonitrile. Diastereoselective acylation of racemic amine **1** with dipeptide 5(4*H*)-oxazolones **53a-f** was carried out at a reagent molar ratio of 8:1 in various solvents at +40 °C. In all cases, when oxazolone **53a** was used as acylating agent, (*R*)-enantiomer of amine **1** was predominantly acylated; the highest *de* value (36.3%) was observed in chloroform. In polar solvents (tetramethylurea (TMU), ethyl acetate, MeCN), the stereoselectivity was lower (10-19% *de*). A decrease in temperature to −10 °C resulted in an increase in *de* of the formed (*S,R*)-amide **54a** up to 51.7% (Scheme **31**).

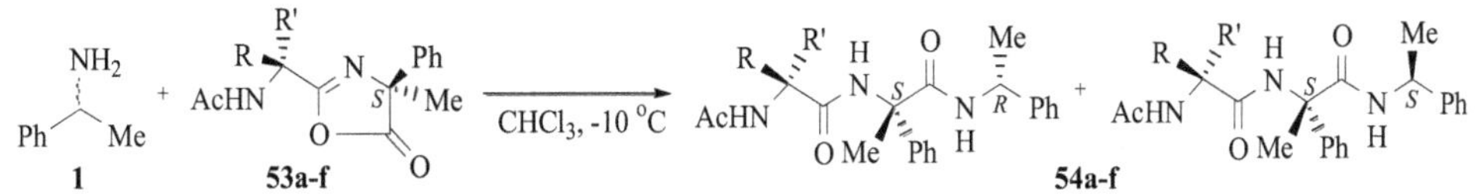

Acylating agent	R	R'	Amide 53	
			de, %	Configuration
53a	Me	Me	51.7	*S,R*
53b	(CH$_2$)$_3$		38.1	*S,R*
53c	(CH$_2$)$_6$		43.4	*S,R*
53d	(CH$_2$)$_{10}$		41.2	*S,R*
53e	Ph	Me	45.3	*S,S,R*
53f	Me	Ph	47.4	*R,S,R*

Scheme 31. Diastereoselective acylation of racemic 1-phenylethylamine (**1**) with 5(4*H*)-oxazolones **53a-f**.

N-Protected amino acyl chlorides were used in KR of racemic heterocyclic amines. A wide range of *N*-phthaloyl-(*S*)-amino acyl chlorides **55a-i** (Scheme **32**) was studied as diastereoselective agents in KR of racemic 3,4-dihydro-3- methyl-2*H*-[1,4]benzoxazine (**17a**) and 2-methyl-1,2,3,4- tetrahydroquinoline (**17c**) [106 - 111]. Just as in the case of KR with (*S*)-naproxen chloride ((*S*)-**16**), the products of acylation of racemic amines **17a,c** with acyl chlorides **55a-i** were enriched in (*S,S*)-diastereoisomers. The highest stereoselectivity was observed in the reaction carried out in dichloromethane at −20 °C.

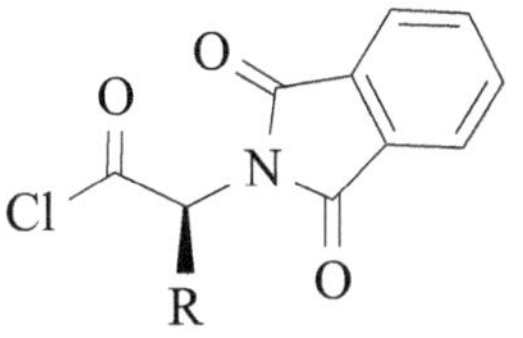

55a: R = Me (*Ala*)
55b: R = Ph (*Phg*)
55c: R = CH$_2$Ph (*Phe*)
55d: R = CH$_2$-4-NO$_2$-C$_6$H$_4$
55e: R = CH$_2$-4-OMe-C$_6$H$_4$
55f: R = *i*Pr (*Val*)
55g: R = *t*Bu (*tLeu*)
55h: R = CH$_2$*i*Pr (*Leu*)
55i: R = CH$_2$*cyclo*Hex

Scheme 32. Structures of *N*-phthaloyl amino acyl chlorides **55a-i**.

The effect of both aromatic interactions and spatial factors on the stereoselectivity of acylation with acyl chlorides **55a-i** has been demonstrated by the example of KR of racemic amines **17a** and **17c** (Scheme **33**). It was found that acyl chlorides **55b-e** containing aromatic substituents in the side chain of amino acid fragment were more selective acylating agents [107, 108] compared to the originally proposed *N*-phthaloyl-(*S*)-alanyl chloride (**55a**) [106]. Acylation with *N*-phthaloyl-(*S*)-phenylglycyl chloride (**55b**) proceeded with a high stereoselectivity (*s* up to 22 in the case of amine **17b**), but a low conversion of the starting amine (Scheme **33**) [107].

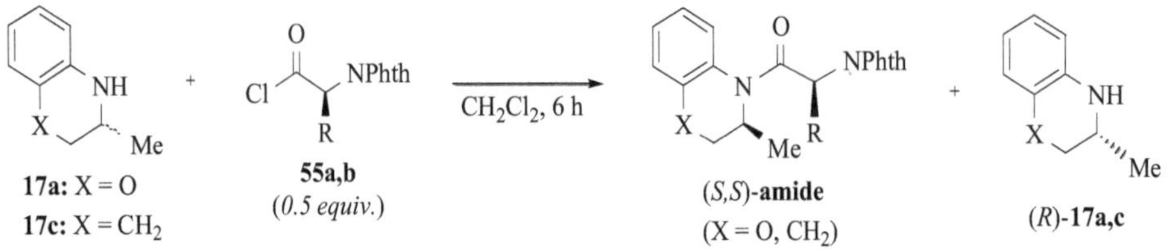

Scheme 33. Diastereoselective acylation of racemic amines **17a,c** with acyl chlorides **55a,b**.

	KR of amine **17a** (+20 °C)				KR of amine **17c** (-20 °C)			
Acyl chloride	(S,S)-**amide**, *de* (%)	(R)-**17a**, *ee* (%)	Conversion, %	*s*	(S,S)-**amide**, *de* (%)	(R)-**17c**, *ee* (%)	Conversion, %	*s*
55a: R = Me	49.3	33.1	40	4.0	63.3	41.5	40	6.6
55b: R = Ph	81.5	20.6	20	12	89.6	16.2	15	22

The electronic effects of side-chain substituents located distant from the chiral center on the stereochemical outcome were demonstrated when studying KR of amines **17a,c** with *N*-phthaloyl-(*S*)-phenylalanyl chloride (**55c**) and its *para*-substituted analogs **55d** and **55e** (Scheme **34**). In the transition from the electron-withdrawing to the electron-donating substituent in the aromatic ring of acylating agents **55c-e**, the selectivity factor *s* increases in the series: **55d** (*p*-NO$_2$) < **55c** (H) < **55e** (*p*-OMe) [108]. This indicates the important role of aromatic π-π interactions in the process of stereo discrimination.

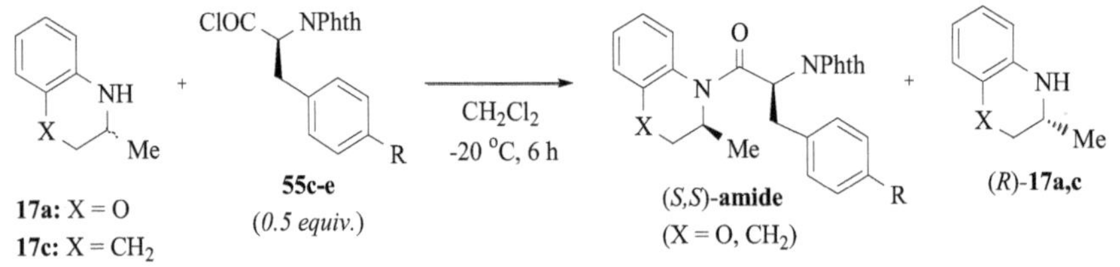

	KR of amine **17a**				KR of amine **17c**			
Acyl chloride	(S,S)-**amide**, *de* (%)	(R)-**17a**, *ee* (%)	Conversion, %	*s*	(S,S)-**amide**, *de* (%)	(R)-**17c**, *ee* (%)	Conversion, %	*s*
55c: R = H	64.3	46.4	42	7.2	74.4	60.4	45	12
55d: R = NO$_2$	63.0	44.0	41	6.7	69.0	59.2	46	9.9
55e: R = OMe	63.8	52.3	45	7.5	75.5	63.6	46	14

Scheme 34. Diastereoselective acylation of racemic amines **17a,c** with acyl chlorides **55c-e**.

A comparative study of KR of amines **17a** and **17c** with *N*-phthaloyl-(*S*)-amino acyl chlorides **55a,f,g** with alkyl side chains (Scheme **35**) showed that the size of the substituent directly linked to the chiral center of reagent **55** does not significantly affect the stereoselectivity: the acylation of racemic amines **17a** and **17c** proceeded approximately with the same selectivity (selectivity factor *s* was 4.0, 4.7 and 3.8 for acyl chlorides **55a** (R = Me), **55f** (R = *i*Pr) and **55g** (R = *t*Bu), respectively) [110]. At the same time, the presence of bulky *tert*-butyl substituent in the structure of *N*-phthaloyl-(*S*)-*tert*-leucyl chloride (**55g**) led to a significant decrease in the conversion. The stereoselectivity of acylation of racemic amines was significantly higher in the case of *N*-phthaloyl-(*S*)-leucyl (**55h**, R = CH_2*i*Pr; *s* = 14) and *N*-phthaloyl-3-cyclohexyl-(*S*)-alanyl chlorides (**55i**, R = CH_2*cyclo*Hex; *s* = 11) in which the branched alkyl substituent is separated from the chiral center by an additional methylene group. Apparently, in this case bulky alkyl substituents in the side chain of acyl chloride do not inhibit the acylation, while they create steric hindrances, which contributed to enhanced enantiomeric discrimination [110].

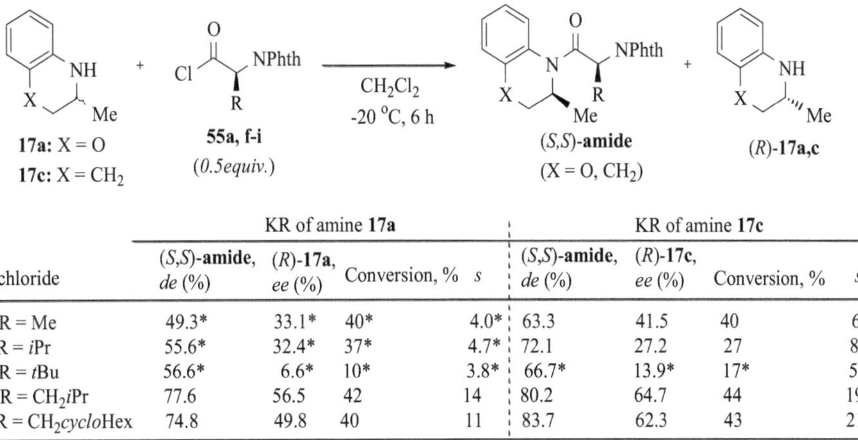

Acyl chloride	KR of amine **17a**				KR of amine **17c**			
	(*S,S*)-amide, *de* (%)	(*R*)-**17a**, *ee* (%)	Conversion, %	*s*	(*S,S*)-amide, *de* (%)	(*R*)-**17c**, *ee* (%)	Conversion, %	*s*
55a: R = Me	49.3*	33.1*	40*	4.0*	63.3	41.5	40	6.6
55f: R = *i*Pr	55.6*	32.4*	37*	4.7*	72.1	27.2	27	8.0
55g: R = *t*Bu	56.6*	6.6*	10*	3.8*	66.7*	13.9*	17*	5.7*
55h: R = CH_2*i*Pr	77.6	56.5	42	14	80.2	64.7	44	19
55i: R = CH_2*cyclo*Hex	74.8	49.8	40	11	83.7	62.3	43	21

* at +20 °C

Scheme 35. Diastereoselective acylation of racemic amines **17a,c** with acyl chlorides **55a,f-i**.

It should be noted that although *N*-phthaloyl-(*S*)-amino acyl chlorides (**55**) proved to be less efficient resolving agents in KR of the abovementioned racemic heterocyclic amines (*s* 5–10 in CH_2Cl_2 at +20 °C) compared to (*S*)-naproxen chloride (*s* 15–32 in CH_2Cl_2 at +20 °C) [84], they can also be used for the preparation of (*S*)-amines from racemates. Thus, the most efficient agents (derivatives of phenylalanine **55c** and leucine **55h**) in the series of *N*-phthaloyl-(*S*)-amino acyl chlorides were used to obtain (*S*)-enantiomers of amines **17a,c,d,f** (Scheme 36) in 20-25% overall yield (relative to racemate) [83, 107, 110, 111]. (*S*)-Enantiomers of 2-methyl-6-nitro-1,2,3,4-tetrahydroquinoline ((*S*)-**17h**) [109] and 3,4-dihydro-3-methyl-2*H*-[1,4]benzothiazine-1,1-dioxide ((*S*)-**17i**) [111] of high optical purity were obtained for the first time after subsequent transformations of (*S,S*)-amides **56** and **57** (*de* >99%), the products of KR of racemic 2-methyl-1,2,3,4-tetrahydroquinoline (**17c**) and 3,4-dihydro-3-methyl-2*H*-[1,4]benzothiazine (**17f**) with *N*-phthaloyl-(*S*)-leucyl chloride (**55h**) (Scheme 36).

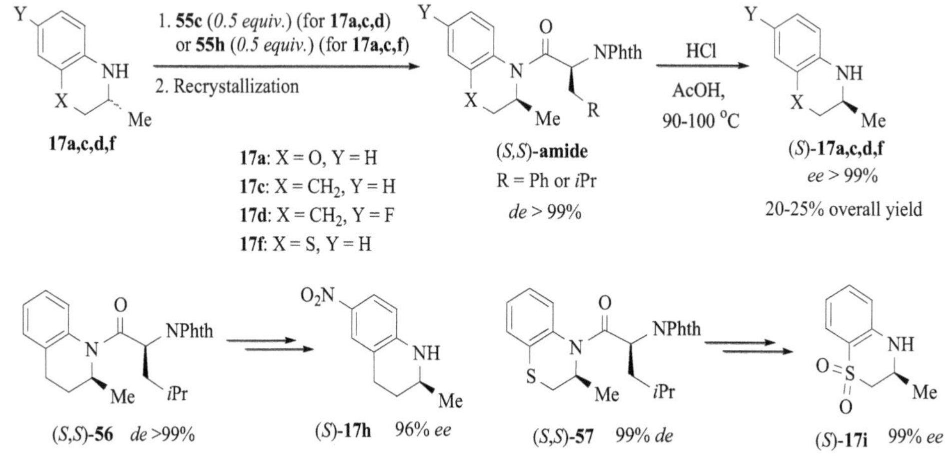

Scheme 36. Preparation of (*S*)-enantiomers of amines **17a,c,d,f-i** *via* acylative KR protocol.

In 2003, our research group proposed to use *N*-tosyl-(*S*)-prolyl chloride (**14a**) as a diastereoselective acylating agent (Scheme 37) in KR of racemic amines **17a,c,e** in order to prepare their (*R*)-enantiomers [112]. Later, the series of racemic substrates was extended [113]. So, the acylation of racemic amines **17a-g** and methylpiperidines **18a,b** with acyl chloride **14a** (0.5 equiv.) was carried out toluene (benzene) or dichloromethane, which resulted in the predominant formation of (*R,S*)-amides, and the unreacted amines were enriched in (*S*)-enantiomers (Scheme 37) [112, 113]. The highest stereoselectivity was observed in the case of acylation of 3-methyl-substituted benzoxazines **17a** and **17b** and benzothiazine **17f** (selectivity factor *s* up to 28 at +20 °C), while lowering the acylation temperature led to an increase in stereoselectivity (*s* up to 44 at –20 °C) [113]. The acylation of the tetrahydroquinoline derivatives **17c** (*s* 13 in toluene at –20 °C) and **17d** (*s* 6.5 in toluene at –20 °C) with acyl chloride **14a** proceeded with somewhat lower selectivity; stereoselectivity of acylation of 2-methylindoline **17e** was low (*s* 4.4 in toluene at –20 °C). It should be noted that an increase in the size of substituent at the chiral center of racemic amine (amine **17g**: Alk = *t*Bu) resulted in a sharp decrease in stereoselectivity, *s* 3.7 for amine **17g** as compared to *s* 28 for amine **17a** in toluene at +20 °C. The stereoselectivity of acylation of racemic amines, in the structure of which there is no fused aromatic ring, turned out to be significantly lower than in the acylation of tetrahydroquinoline derivatives and other amines. Thus, the acylation of racemic

2-methylpiperidine (**18a**) in both toluene and dichloromethane at +20 °C proceeded with low selectivity (s 1.7 and 3.0, respectively), and in the case of the isomeric amine **18b**, the acylation was nonselective (s 1.2 and 1.1).

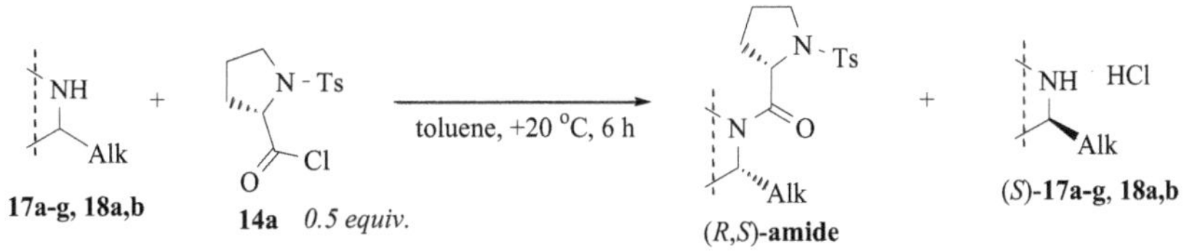

Racemic amine	(R,S)-Amide, de (%)	(S)-Amine, ee (%)	Conversion C, %	s
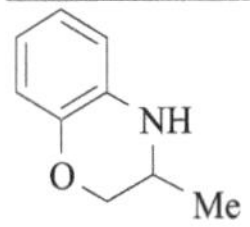 Me **17a**	84.6	75.0	47	28
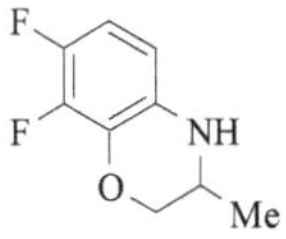 Me **17b**	76.7	59.0	44	14
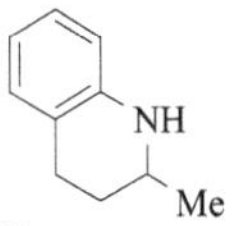 Me **17c**	70.4	58.8	46	10
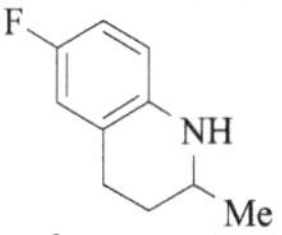 Me **17d**	56.0	40.5	42	5.2
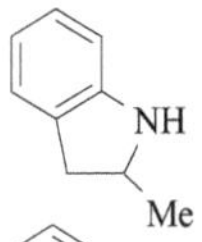 Me **17e**	39.4	37.8	49	3.3
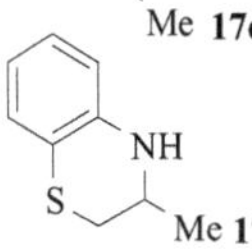 Me **17f**	89.4	35.9	29	25
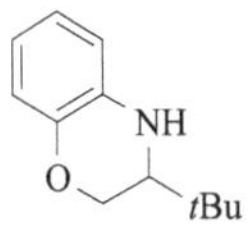 tBu **17g**	48.2	26.7	36	3.7
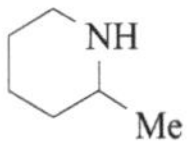 Me **18a**	18.0*	17.6*	49	1.7
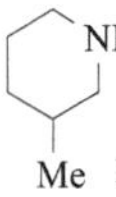 Me **18b**	8.6*	5.6*	39	1.2

* Configuration of the chiral centers was not assigned because of low selectivity of the acylation

Scheme 37. Diastereoselective acylation of racemic amines **17a-g** and **18a,b** with N-tosyl-(S)-prolyl chloride (**14a**).

The (*R*)-enantiomers of amines **17a-c,e,f** (*ee* ≥97%) were obtained from diastereomerically pure (*R,S*)-amides **58a-c,e** (*de* ≥99%) in the course of acidic hydrolysis (Scheme **38**) [112, 113]. Unlike the acidic hydrolysis of amides of (*S*)-naproxen or *N*-phthaloyl-(*S*)-amino acids, complete hydrolysis of the compounds (*R,S*)-**58a-c,e** required a longer time (up to 40 h), which resulted in a slight racemization of the chiral center in the amine moiety. Later, it was proposed to use alkaline conditions for the cleavage of the amide bond in compound (*R,S*)-**58b**; in this case, amine (*R*)-**17b** was obtained in high yield (93% relative to amide (*R,S*)-**58b**) without racemization (Scheme **38**) [113].

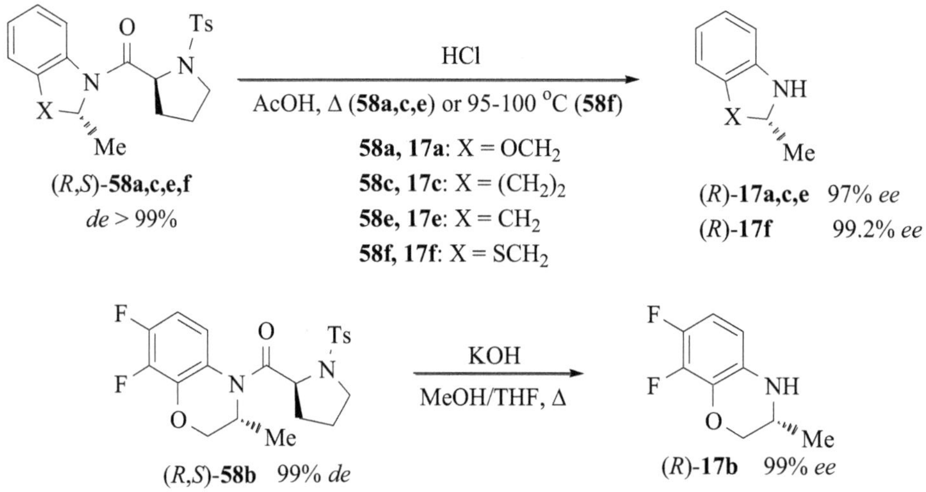

Scheme 38. Preparation of (*R*)-enantiomers of amines **17a,c,e,f** from (*R,S*)-amides of *N*-tosyl-(*S*)-proline.

It should be noted that it is this method that was used for preparation of (*R*)-enantiomer of 2-methylindoline (**17e**) for the first time [112]; it was later used by other researchers to synthesize chiral phosphamidite ligands based on enantiopure BINOL and (*R*)-**17e** [114].

We carried out a comparative study of diastereoselective acylation of heterocyclic amines **17a-c** with *N*-tosyl-(*S*)-prolyl chloride (**14a**) and its structural analogs **59a-g** (Scheme **39**) [115]. Acyl chlorides **14a** and **59a,b** differ in the structure of aryl substituent in the *N*-protecting group; reagents **59c,d** contain aliphatic sulfamide groups of different volumes; *N*-tosyl-(*S*)-indoline-2-carboxylic acid chloride (**59e**) has an additional fused aromatic moiety; and compound **59f** represents an acyclic analogue of acyl chloride **59e**. Acyl chloride **59g** may be considered as an oxygen-containing analog of proline derivatives. In general, it was found that high selectivity was observed when heterocyclic amines were acylated with reagents containing both a conformationally restricted pyrrolidine ring and an aromatic substituent in the *N*-acyl group (compounds **14a**, **59a,b,e,f**). The most selective acyl chloride in this series is *N*-tosyl-(*S*)-prolyl chloride (**14a**).

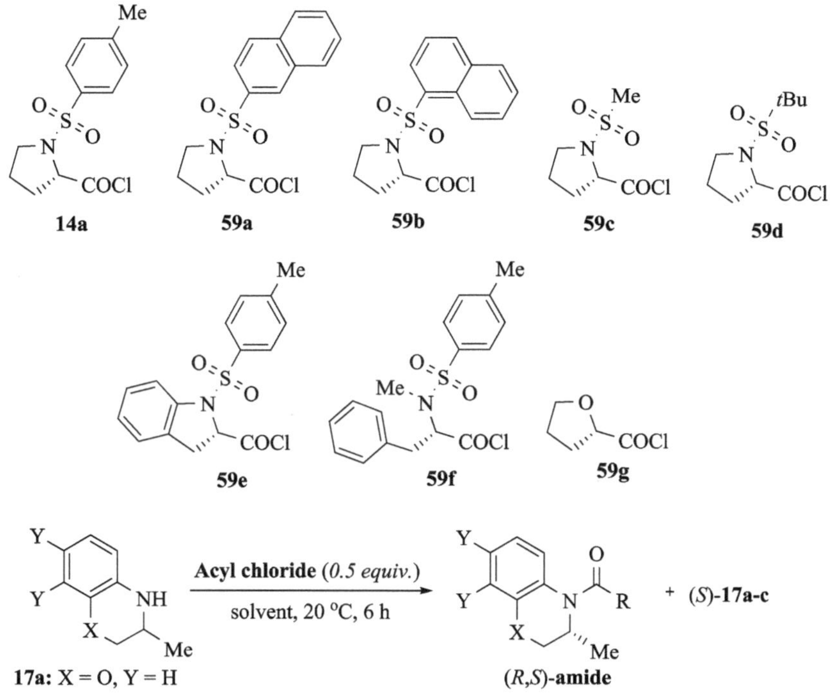

17a: X = O, Y = H
17b: X = O, Y = F
17c: X = CH$_2$, Y = H

Amine (solvent)	Acyl chloride	14a	59a	59b	59c	59d	59e	59f	59g
17a (toluene)	(*R,S*)-Amide, *de* (%)	85	80	72	54	74	86	59	-
	selectivity factor *s*	28	22	12	5.2	11	28	6.2	-
17b (CH$_2$Cl$_2$)	(*R,S*)-Amide, *de* (%)	83	78	79	70	81	80	41	21
	selectivity factor *s*	21	16	14	9.7	14	18	3.1	1.6
17c (toluene)	(*R,S*)-Amide, *de* (%)	70	54	45	26	-	67	36	-
	selectivity factor *s*	10	5.5	3.7	2.0	-	8.9	2.7	-

Scheme 39. Diastereoselective acylation of racemic amines **17a-c** with acyl chlorides **14a** and **59a-g**.

Thus, by the example of acylation of racemic heterocyclic aromatic amines, it has been shown that by selecting one or the other enantiomerically pure acyl chloride, it is possible to intentionally obtain mixtures with the predominance of (*R*)- or (*S*)-amines and the diastereomerically enriched reaction products, respectively. Despite the fact that the relatively high reactivity of acyl chlorides compared to other acylating agents [116] can cause nonselective acylation of highly nucleophilic primary and secondary amines, their use for KR of less reactive substrates (quinoline, benzoxazine and indoline derivatives) allows significant stereoselectivity to be achieved.

An attempt to obtain stereoisomers of trisubstituted pyrrolidine **60**, a precursor in the synthesis of thrombin inhibitors, using diastereoselective acylation with either (*S*)-naproxen chloride (**16**) or *N*-tosyl (**59f,h**) and *N*-phthaloyl-(*S*)-amino acyl chlorides (**55c,h**) (Scheme **40**) turned out to be unsuccessful [117]. We failed to select reagents and conditions that allow the enantiomers of compound **60** to be obtained with high optical purity. The highest stereoselectivity of acylation of racemic compound **60** was observed when using *N*-methyl-*N*-tosyl-(*S*)-phenylalanyl (**59f**) and *N*-phthaloyl-(S)-phenylalanyl (**55c**) chlorides; in toluene at +20 °C, the selectivity factor *s* was 5.0 and 4.8, respectively; *ee* of unreacted (−)-**60** of about 40%. The acylation of pyrrolidine *rac*-**60** with 0.75 equiv. of acyl chlorides **59f** or **55c** in the presence of 0.75 equiv. of *N,N*-diethylaniline (as HCl scavenger) in toluene at −20 °C resulted in an increase of *ee* of unreacted (−)-**60** up to 88 and 81%, respectively (selectivity factor *s* 6.5 and 8.3).

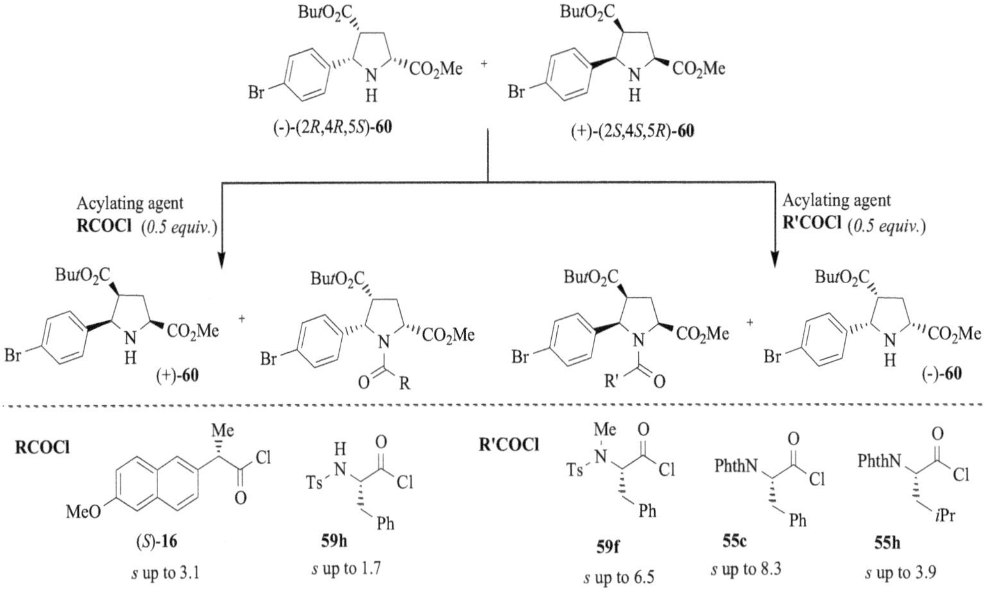

Scheme 40. Diastereoselective acylation of pyrrolidine derivative **60**.

However, the preparative isolation of scalemic compound (−)-**60** from the reaction mixture was laborious, the yield of enantioenriched (−)-**60** was only 13%. Nevertheless, the obtained results indicated the principal possibility of acylative KR of racemic heterocyclic amines, which do not have a condensed aromatic system in their structure, with chiral acyl chlorides. The enantiomers of compound

59 were obtained by preparative HPLC on a chiral stationary phase [117].

Acyl chlorides of *N*-protected (*S*)-amino acids were also used for KR of planar-chiral 1-substituted 3-amino-1,2-dicarba-*closo*-dode- caboranes **61a-c** (Scheme 41) [118 - 122]. The enantiomers of 3-amino-1-methyl-1,2-dicarba-*closo*-dodecaborane (**61a**) were first obtained by chromatographic separation of diastereoisomeric (*S*)-naproxen amides followed by acidic hydrolysis [118]. As a result of studying the KR of racemic 3-aminocarboranes, it was found that the process selectivity depended on both the structure of the resolving agent and 3-aminocarborane, and on the type of a solvent and an auxiliary tertiary amine [119]. *N*-Tosyl-(*S*)-prolyl chloride (**14a**) and *N*-phthaloyl-(*S*)-alanyl chloride (**55a**) proved to be more selective acylating agents than the (*S*)-naproxen chloride (**16**). The acylation of 1-methyl- and 1-phenyl-3-aminocarboranes **61a** and **61b** with acyl chloride **14a** led to the formation of diastereoisomeric mixtures enriched in the corresponding (*S,S*)-amide (unreacted aminocarborane was enriched in the (*R*)-enantiomer). During the acylation of aminocarboranes **61a** and **61b** with acyl chloride **55a**, the predominant formation of (*R,S*)-amides occurred. At the same time, the acylation of 1-isopropyl-3-aminocarborane (**61c**) with all the acyl chlorides studied under the KR conditions led to the enrichment of the reaction mixture in (*R,S*)-diastereoisomers.

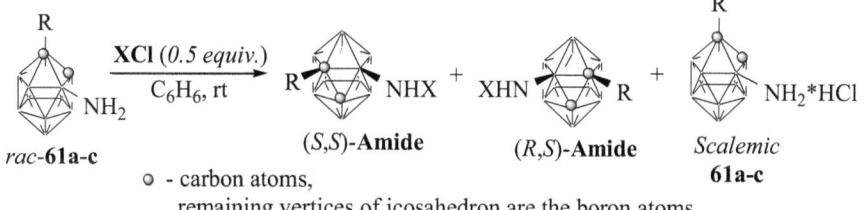

rac-Amine	Acylating agent XCl	Amide de, %	configuration
61a: R = Me	**14a**: X = *N*-Ts-(*S*)-Pro	28	(*S,S*)
61a: R = Me	**55a**: X = *N*-Phth-(*S*)-Ala	30	(*R,S*)
61a: R = Me	(*S*)-**16**: X = (*S*)-naproxen	28	(*S,S*)
61b: R = Ph	**14a**: X = *N*-Ts-(*S*)-Pro	36	(*S,S*)
61b: R = Ph	**55a**: X = *N*-Phth-(*S*)-Ala	39	(*R,S*)
61b: R = Ph	(*S*)-**16**: X = (*S*)-naproxen	16	(*S,S*)
61c: R = *i*Pr	**55a**: X = *N*-Phth-(*S*)-Ala	36	(*R,S*)

Scheme 41. Diastereoselective acylation of racemic 1-substituted 3-amino-1,2-dicarba-*closo*-dodecaboranes **61a-c** with acyl chlorides **14a**, **16**, and **55a**.

It was shown that, unlike the acidic hydrolysis of amides containing the 1-methyl-3-aminocarborane residue, the hydrolysis of *N*-acyl-1-phenyl- and 1-isopropyl-

3-acylaminocarboranes did not allow obtaining the enantiomers of carboranes **61b** and **61c**. In the case of amides containing a fragment of 1-phenylcarborane **61b**, the destruction of the carborane cage and the formation of *nido*-derivatives were observed simultaneously with the breakage of the amide bond [120]. Acidic hydrolysis of amides containing the 3-amino-1-isopropylcarborane residue **61c** led to the deboronation and racemization of the carborane fragment [121]. Scalemic 3-amino-carboranes **61b** and **61c** enriched in (*S*)- or (*R*)-enantiomers can be isolated after acylative KR. The optical purity of 3-aminocarboranes **61a-c** was determined by the methods of NMR with a chiral shift-reagent [122] or chiral HPLC [121].

Another example of the use of amino acid derivatives as diastereoselective acylating agents was reported [123]. *N*-Boc-Alanine (**62**) in the form of active ester with HOBt was applied in KR of racemic 1-(2,2-dimethoxyethyl)-1,2,3,4-tetrahydrocarboline (**63**) (Scheme **42**). The reagents were used in stoichiometric amounts, and the KR was achieved by stopping the reaction after 2 h. Only the (*S*)-enantiomer of amine **63** was subjected to acylation. (*S,S*)-Amide **64** and the unreacted (*R*)-**63** were isolated in optically pure form in 49% yields.

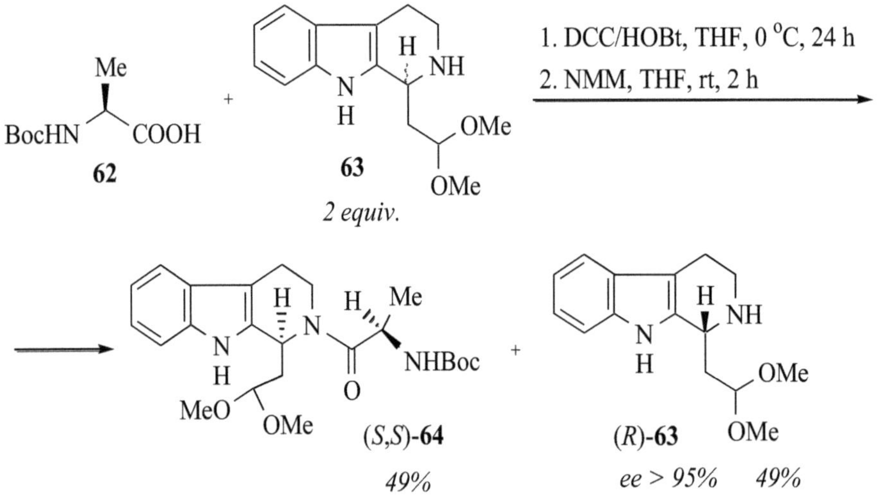

Scheme 42. Preparation of (*R*)-1-(2,2-dimethoxyethyl)-1,2,3,4-tetrahydrocarboline (**63**) *via* KR protocol.

In 2008, Chen *et al.* demonstrated the possibility of using the acylative KR with an *N*-protected amino acid for preparation of stereoisomers of amines of complex structure [124]. *N*-Boc-(*S*)-proline was used as the resolving agent for KR of racemic inherently chiral nitro-substituted aminocalix[4]arene **65** (Scheme **43**). The authors selected the reaction conditions so as to isolate the unreacted enantiomer of amine **65** with the highest optical purity. The reaction of *rac*-**65** with 2.0 equiv. N-Boc-(S)-proline in the presence of 2.0 equiv. DCC and 0.7 equiv. DMAP in dichloromethane for 20 h resulted in the acylation product (cS)-**66** (46% de, 66% yield) and unreacted amine (cR)-**65** (95% ee, 26% yield); the value of selectivity factor s was 9.9. It was found that the nature of solvent affected the process stereoselectivity. Thus, when rac-**65** was acylated with 2.0 equiv. N-Boc-(S)-proline in diethyl ether or THF for 20 h, unreacted (cR)-**65** could be isolated in high yields (70% and 76%, respectively), but its optical purity was low (13 and 17% ee, respectively). At the same time, when KR was carried out in toluene, compound (cR)-**65** was obtained with the greatest optical purity (98% ee), however, the yield was only 22%.

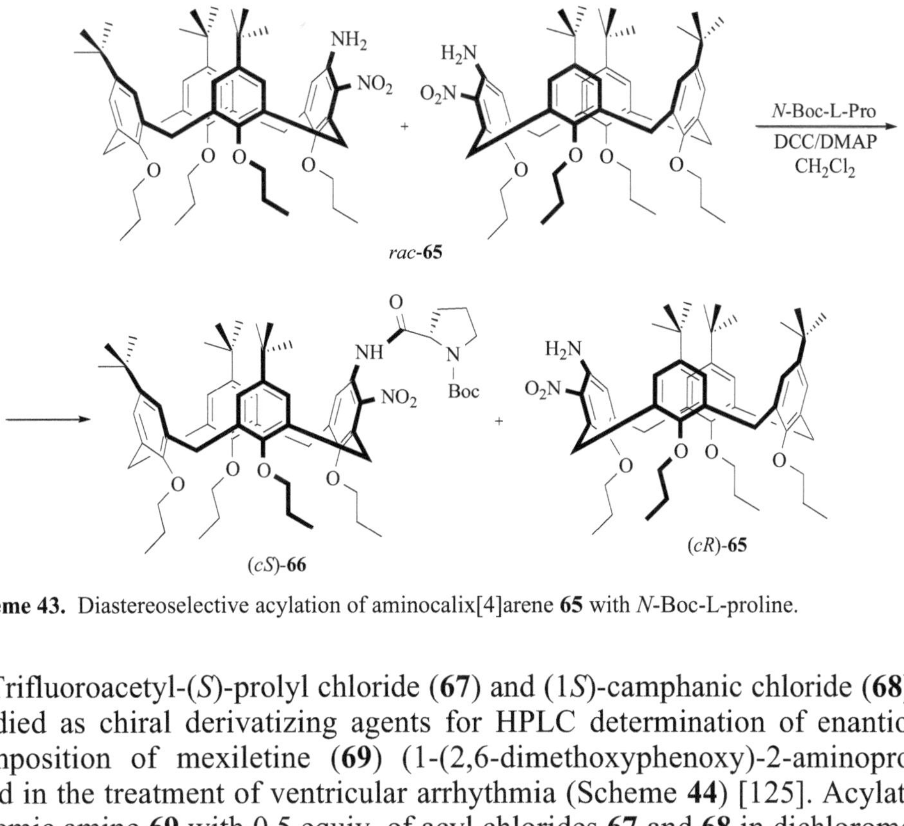

Scheme 43. Diastereoselective acylation of aminocalix[4]arene **65** with N-Boc-L-proline.

N-Trifluoroacetyl-(S)-prolyl chloride (**67**) and (1S)-camphanic chloride (**68**) were studied as chiral derivatizing agents for HPLC determination of enantiomeric composition of mexiletine (**69**) (1-(2,6-dimethoxyphenoxy)-2-aminopropane) used in the treatment of ventricular arrhythmia (Scheme **44**) [125]. Acylation of racemic amine **69** with 0.5 equiv. of acyl chlorides **67** and **68** in dichloromethane at room temperature proceeded with a noticeable stereoselectivity and led to the predominant formation of (R,S)-diastereoisomers of amides **70a** and **70b** (71-74% de). The individual (R,S)-amides **70a** and **70b** were isolated by preparative reversed-phase HPLC. Subsequent acidic hydrolysis of amides (R,S)-**70a** and (R,S)-**70b** led to (R)-mexiletine ((R)-**69**).

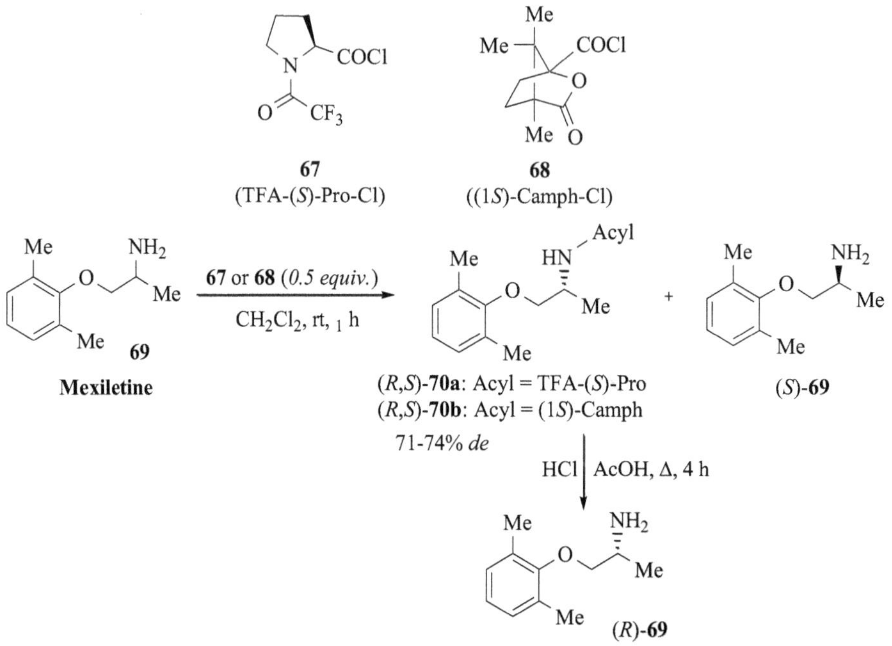

Scheme 44. KR of racemic mexiletine (**69**) *via* diastereoselective *N*-acylation with acyl chlorides **67** and **68**.

CONCLUDING REMARKS

In the last decade, significant progress has been achieved in the nonenzymatic KR of racemic amines and related compounds [48]. A considerable number of structurally diverse enantiopure amines were obtained as a result of acylative kinetic resolution. As a general rule, it can be noted that less reactive resolving agents are suitable for KR of racemic amines having a relatively high reactivity. At the same time, stereoselective acylation of less reactive amines is achieved by the action of highly reactive resolving agents.

A wide diversity of diastereoselective acylating agents made it possible to propose methods for the preparation of enantiomerically pure or enriched amines and their derivatives of various classes. In addition, the use of reactive reagents (chiral acyl chlorides) is an excellent method for KR of less nucleophilic heterocyclic aromatic amines. It is worth noting that it is not possible to formulate general regularities and mechanisms that explain stereoselectivity in each case, since the structures of substrates and resolving agents are very diverse. Despite the fact that in recent years some stereochemical results of acylative KR were substantiated by quantum chemical calculations [126–129], the selection of efficient substrate–resolving agent combinations can still be made only based on the experimental data.

The prevalence of KR processes using chiral acylating reagents is related, apparently, to the availability of a variety of reagents of this type, as well as the simplicity of the process implementation. When using a diastereoselective acylating agent, isomeric purity of diastereomerically enriched amide, the acylation product can be considerably increased by simple techniques (column chromatography or recrystallization), which makes it possible to obtain the target amine in the enantiomerically pure form (*ee* 99% and higher).

CONSENT FOR PUBLICATION

Not applicable.

CONFLICT OF INTEREST

The authors declare no conflict of interest, financial or otherwise.

ACKNOWLEDGEMENTS

Declared none.

REFERENCES

[1]   Pellissier, H. *Chirality from dynamic kinetic resolution*; RSC Publishing: Cambridge, **2011**, pp. xv-xvi.

[2]   Federsel, H-J. Chemical Process Research and Development in the 21st Century: Challenges, Strategies, and Solutions from a Pharmaceutical Industry Perspective. *Acc. Chem. Res.,* **2009**, *42*(5), 671-680.
      [http://dx.doi.org/10.1021/ar800257v]

[3]   Fogassy, E.; Nógrádi, M.; Pálovics, E.; Schindler, J. Resolution of Enantiomers by Non-Conventional Methods. *Synthesis,* **2005**, 1555-1568.
      [http://dx.doi.org/10.1055/s-2005-869903]

[4]   Breuer, M.; Ditrich, K.; Habicher, T. Hauer, B.; Keßeler, M.; Stürmer, R.; Zelinski, T. Industrial Methods for the Production of Optically Active Intermediates. *Angew. Chem. Int. Ed.,* **2004**, *43*, 788-824.
      [http://dx.doi.org/10.1002/anie.200300599]

[5]   Siedlecka, R. Recent developments in optical resolution. *Tetrahedron,* **2013**, *69*, 6331-6363.
      [http://dx.doi.org/10.1016/j.tet.2013.05.035]

[6]   Kagan, H.B.; Fiaud, J.C. Kinetic resolution. *Top. Stereochem.,* **1988**, *18*, 249-330.

[7]   Moss, G.P. Basic Terminology of Stereochemistry. *Pure Appl. Chem.,* **1996**, *68*, 2193-2222.
      [http://dx.doi.org/10.1351/pac199668122193]

[8]   Maddani, M.R.; Fiaud, J-C.; Kagan, H.B. *Separation of Enantiomers: Synthetic Methods*; Todd, M., Ed.; Wiley-VCH: Weinheim, **2014**, pp. 13-74.
      [http://dx.doi.org/10.1002/9783527650880.ch2]

[9]   Keith, J.M.; Larrow, J.F.; Jacobsen, E.N. Practical Considerations in Kinetic Resolution Reactions. *Adv. Synth. Catal.,* **2001**, *343*, 5-26.
      [http://dx.doi.org/10.1002/1615-4169(20010129)343:1<5::AID-ADSC5>3.0.CO;2-I]

[10] Vedejs, E.; Jure, M. Efficiency in Nonenzymatic Kinetic Resolution. *Angew. Chem. Int. Ed.,* **2005**, *44*, 3974-4001.
[http://dx.doi.org/10.1002/anie.200460842]

[11] Eliel, E.L.; Wilen, S.H. *Stereochemistry of Organic Compounds*; Wiley & Sons: New York, **1994**.

[12] Strauss, U.T.; Felfer, U.; Faber, K. Biocatalytic transformation of racemates into chiral building blocks in 100% chemical yield and 100% enantiomeric excess. *Tetrahedron Asymmetry,* **1999**, *10*, 107-117.
[http://dx.doi.org/10.1016/S0957-4166(98)00490-X]

[13] Spivey, A.C.; Arseniyadis, S. Amine, Alcohol and Phosphine Catalysts for Acyl Transfer Reactions. *Top. Curr. Chem.,* **2010**, *291*, 233-280.
[http://dx.doi.org/10.1007/978-3-642-02815-1_25]

[14] Vedejs, E.; Daugulis, O.; MacKay, J.A.; Rozners, E. Enantioselective Acyl Transfer Using Chiral Phosphine Catalysts. *Synlett,* **2001**, 1499-1505.
[http://dx.doi.org/10.1055/s-2001-17436]

[15] Müller, C.E.; Schreiner, P.E. Organocatalytic Enantioselective Acyl Transfer onto Racemic as well as *meso* Alcohols, Amines, and Thiols. *Angew. Chem. Int. Ed.,* **2011**, *50*, 6012-6042.
[http://dx.doi.org/10.1002/anie.201006128]

[16] Noyori, R.; Tokunaga, M.; Kitamura, M. Stereoselective Organic Synthesis *via* Dynamic Kinetic Resolution. *Bull. Chem. Soc. Jpn.,* **1995**, *68*, 36-56.
[http://dx.doi.org/10.1246/bcsj.68.36]

[17] Ward, R.S. Dynamic Kinetic Resolution. *Tetrahedron Asymmetry,* **1995**, *6*, 1475-1490.
[http://dx.doi.org/10.1016/0957-4166(95)00179-S]

[18] Nakano, K.; Kitamura, M. *Separation of Enantiomers: Synthetic Methods*; Todd, M., Ed.; Wiley-VCH: Weinheim, **2014**, pp. 161-215.
[http://dx.doi.org/10.1002/9783527650880.ch5]

[19] Pellissier, H. Recent developments in dynamic kinetic resolution. *Tetrahedron,* **2008**, *64*, 1563-1601.
[http://dx.doi.org/10.1016/j.tet.2007.10.080]

[20] Lee, J.H.; Han, K.; Kim, M-J.; Park, J. Chemoenzymatic Dynamic Kinetic Resolution of Alcohols and Amines. *Eur. J. Org. Chem.,* **2010**, 999-1015.
[http://dx.doi.org/10.1002/ejoc.200900935]

[21] Pellissier, H. Recent developments in dynamic kinetic resolution. *Tetrahedron,* **2011**, *67*, 3769-3802.
[http://dx.doi.org/10.1016/j.tet.2011.04.001]

[22] Marcos, R.; Martín-Matute, B. Combined Enzyme and Transition-Metal Catalysis for Dynamic Kinetic Resolutions. *Isr. J. Chem.,* **2012**, *52*, 639-652.
[http://dx.doi.org/10.1002/ijch.201200012]

[23] Verho, O.; Bäckvall, J-E. Chemoenzymatic Dynamic Kinetic Resolution: A Powerful Tool for the Preparation of Enantiomerically Pure Alcohols and Amines. *J. Am. Chem. Soc.,* **2015**, *137*, 3996-4009.
[http://dx.doi.org/10.1021/jacs.5b01031]

[24] Zhang, Q.; Curran, D.P. Quasienantiomers and Quasiracemates: New Tools for Identification, Analysis, Separation, and Synthesis of Enantiomers. *Chemistry,* **2005**, *11*, 4866-4880.
[http://dx.doi.org/10.1002/chem.200500076]

[25] Vedejs, E.; Chen, X. Parallel Kinetic Resolution. *J. Am. Chem. Soc.,* **1997**, *119*, 2584-2585.
[http://dx.doi.org/10.1021/ja963666v]

[26] Eames, J. Parallel Kinetic Resolutions. *Angew. Chem. Int. Ed.,* **2000**, *39*, 885-888.
[http://dx.doi.org/10.1002/(SICI)1521-3773(20000303)39:5<885::AID-ANIE885>3.0.CO;2-2]

[27] Dehli, J.R.; Gotor, V. Parallel kinetic resolution of racemic mixtures: a new strategy for the preparation of enantiopure compounds? *Chem. Soc. Rev.,* **2002**, *31*, 365-370.

[http://dx.doi.org/10.1039/b205280f]

[28] Russell, T.A.; Vedejs, E. *Separation of Enantiomers: Synthetic Methods*; Todd, M., Ed.; Wiley-VCH: Weinheim, **2014**, pp. 217-266.
[http://dx.doi.org/10.1002/9783527650880.ch6]

[29] Kreituss, I.; Bode, J.W. Flow chemistry and polymer-supported pseudoenantiomeric acylating agents enable parallel kinetic resolution of chiral saturated N-heterocycles. *Nat. Chem.*, **2017**, *9*, 446-452.
[http://dx.doi.org/10.1038/nchem.2681]

[30] Herlinger, H.; Kleimann, H.; Ugi, I. Stereoselektive Synthesen, II. Die stereoselektive Acylierung racemischer primärer Amine durch optisch aktive gemischte Anhydride. *Justus Liebigs Ann. Chem.*, **1967**, *706*, 37-46.
[http://dx.doi.org/10.1002/jlac.19677060105]

[31] Horeau, A. Method for obtaining an enantiomer containing less than 0.1% of its antipode. Determination of its maximum rotatory power. *Tetrahedron*, **1975**, *31*, 1307-1309.
[http://dx.doi.org/10.1016/0040-4020(75)80174-8]

[32] Hang, J.; Tian, S-K.; Tang, L.; Deng, L. Asymmetric Synthesis of α-Amino Acids *via* Cinchona Alkaloid-Catalyzed Kinetic Resolution of Urethane-Protected α-Amino Acid N-Carboxyanhydrides. *J. Am. Chem. Soc.*, **2001**, *123*, 12696-12697.
[http://dx.doi.org/10.1021/ja011936q]

[33] Hang, J.; Li, H.; Deng, L. Development of a Rapid, Room-Temperature Dynamic Kinetic Resolution for Efficient Asymmetric Synthesis of α-Aryl Amino Acids. *Org. Lett.*, **2002**, *4*, 3321-3324.
[http://dx.doi.org/10.1021/ol026660l]

[34] Tian, S-K.; Chen, Y.; Hang, J.; Tang, L.; McDaid, P.; Deng, L. Asymmetric Organic Catalysis with Modified Cinchona Alkaloids. *Acc. Chem. Res.*, **2004**, *37*, 621-631.
[http://dx.doi.org/10.1021/ar030048s]

[35] Hang, J.; Deng, L. Asymmetric synthesis of β,γ-unsaturated α-amino acids *via* efficient kinetic resolution with cinchona alkaloids. *Bioorg. Med. Chem. Lett.*, **2009**, *19*, 3856-3858.
[http://dx.doi.org/10.1016/j.bmcl.2009.03.152]

[36] Viso, A.; Lee, N.E.; Buchwald, S.L. Kinetic resolution of racemic disubstituted 1-pyrrolines *via* asymmetric reduction with a chiral titanocene catalyst. *J. Am. Chem. Soc.*, **1994**, *116*, 9373-9374.
[http://dx.doi.org/10.1021/ja00099a082]

[37] Yun, J.; Buchwald, S.L. Kinetic resolution and isomerization of 2,5-disubstituted pyrrolines. *Chirality*, **2000**, *12*, 476-478.
[http://dx.doi.org/10.1002/(SICI)1520-636X(2000)12:5/6<476::AID-CHIR31>3.0.CO;2-B]

[38] Yun, J.; Buchwald, S.L. Efficient Kinetic Resolution in the Asymmetric Hydrosilylation of Imines of 3-Substituted Indanones and 4-Substituted Tetralones. *J. Org. Chem.*, **2000**, *65*, 767-774.
[http://dx.doi.org/10.1021/jo991328h]

[39] Ros, A.; Magriz, A.; Dietrich, H.; Ford, M.; Fernández, R.; Lassaletta, J.M. Transfer Hydrogenation of α-Branched Ketimines: Enantioselective Synthesis of Cycloalkylamines *via* Dynamic Kinetic Resolution. *Adv. Synth. Catal.*, **2005**, *347*, 1917-1920.
[http://dx.doi.org/10.1002/adsc.200505291]

[40] Liang, J.; Ruble, J.C.; Fu, G.C. Dynamic Kinetic Resolutions Catalyzed by a Planar-Chiral Derivative of DMAP: Enantioselective Synthesis of Protected α-Amino Acids from Racemic Azlactones. *J. Org. Chem.*, **1998**, *63*, 3154-3155.
[http://dx.doi.org/10.1021/jo9803380]

[41] Krasnov, V.P.; Zhdanova, E.A.; Solieva, N.Z.; Sadretdinova, L.Sh.; Bukrina, I.M.; Demin, A.M.; Levit, G.L.; Ezhikova, M.A.; Kodess, M.I. Study of the effect of the nature of the side chain in esters of α-amino acids on the diastereoselectivity of condensation with 5(4H)-oxazolone in the synthesis of dipeptides with N-terminal N-acetylphenylalanine. *Russ. Chem. Bull.*, **2004**, *53*, 1331-1334.

[http://dx.doi.org/10.1023/B:RUCB.0000042296.13831.bd]

[42] Berkessel, A.; Cleemann, F.; Mukherjee, S.; Müller, T.N.; Lex, J. Highly Efficient Dynamic Kinetic Resolution of Azlactones by Urea-Based Bifunctional Organocatalysts. *Angew. Chem. Int. Ed.,* **2005**, *44*, 807-811.
[http://dx.doi.org/10.1002/anie.200461442]

[43] Berkessel, A.; Mukherjee, S.; Cleemann, F.; Müller, T.N.; Lex, J. Second-generation organocatalysts for the highly enantioselective dynamic kinetic resolution of azlactones. *Chem. Commun. (Camb.),* **2005**, 1898-1900.
[http://dx.doi.org/10.1039/B418666D]

[44] Zhdanova, E.A.; Solieva, N.Z.; Sadretdinova, L.Sh.; Ezhikova, M.A.; Kodess, M.I.; Krasnov, V.P. Study of the influence of the alkyl ester group in (*S*)-valinates on diastereoselectivity of their condensation with N-acetylphenylalanine by the mixed anhydride method. *Russ. Chem. Bull.,* **2006**, *55*, 925-927.
[http://dx.doi.org/10.1007/s11172-006-0353-5]

[45] Yang, X.; Lu, G.; Birman, V.B. Benzotetramisole-Catalyzed Dynamic Kinetic Resolution of Azlactones. *Org. Lett.,* **2010**, *12*, 892-895.
[http://dx.doi.org/10.1021/ol902969j]

[46] Miyano, S.; Lu, L.D-L.; Viti, S.M.; Sharpless, K.B. Kinetic resolution of racemic β-hydroxy amines by enantioselective N-oxide formation. *J. Org. Chem.,* **1983**, *48*, 3608-3611.
[http://dx.doi.org/10.1021/jo00168a064]

[47] Miyano, S.; Lu, L.D-L.; Viti, S.M.; Sharpless, K.B. Kinetic resolution of racemic β-hydroxy amines by enantioselective N-oxide formation. *J. Org. Chem.,* **1985**, *50*, 4350-4360.
[http://dx.doi.org/10.1021/jo00222a030]

[48] Krasnov, V.P.; Gruzdev, D.A.; Levit, G.L. Nonenzymatic Acylative Kinetic Resolution of Racemic Amines and Related Compounds. *Eur. J. Org. Chem.,* **2012**, 1471-1493.
[http://dx.doi.org/10.1002/ejoc.201101489]

[49] Wiesner, K.; Jay, E.W.K.; Poon-Jay, L. The Total Synthesis of Delphinine: Resolution of the Racemic Relay Compound into Optical Antipodes by an Asymmetric Reaction. *Experientia,* **1971**, *27*, 363.
[http://dx.doi.org/10.1007/BF02137247]

[50] Wiesner, K.; Jay, E.W.K.; Tsai, R. T.Y.; Demerson, C.; Jay, L.; Kanno, T.; Křepinský, J.; Vilím, A.; Wu, C.S. The Synthesis of Delphinine: A Stereoselective Total Synthesis of an Optically Active Advanced Relay Compound. *Can. J. Chem.,* **1972**, *50*, 1925-1943.
[http://dx.doi.org/10.1139/v72-308]

[51] Okamoto, K.; Minami, E.; Shingu, H. Mechanism of Asymmetric Reactions. II. Kinetical Resolution of Racemic Primary Amines by Means of Sulfonylation with (+)-Camphor-10-sulfonyl Chloride – Preferential Conservation of an Absolute Configuration in the Asymmetric Sulfonylation of Racemic Amines. *Bull. Chem. Soc. Jpn.,* **1968**, *41*, 1426-1432.
[http://dx.doi.org/10.1246/bcsj.41.1426]

[52] Bell, K.H. Regioselectivity and Stereoselectivity in the Reaction of *cis*-1,2,2-Trimethylcyclopentane-1,3-dicarboxylic Anhydride (*cis*-Camphoric Anhydride) with Chiral Primary Amines. *Aust. J. Chem.,* **1981**, *34*, 665-670.
[http://dx.doi.org/10.1071/CH9810665]

[53] Bell, K.H. Prediction of the Absolute Configurations of the Enantiomers of Racemic Primary and Secondary Alcohols and Primary Amines by Kinetic Resolution with (2*R*,3*R*)-2,3-Diacetoxysuccinic Anhydride. *Aust. J. Chem.,* **1979**, *32*, 2625-2629.
[http://dx.doi.org/10.1071/CH9792625]

[54] Kolomiets, V.F.; Gracheva, R.A.; Potapov, V.M. Kinetic Resolution of Alcohols and Amines using D-(+)-Dibenzoyltartaric Anhydride. *Zh. Org. Khim.,* **1980**, *16*, 322-324. [*in Russian*].

[55] Červinka, O.; Kroupová, E.; Bélovsky, O. Asymmetric Reactions. XXIX. Absolute Configuration of ω-Phenyl-2-alkylamines and Their N-Methyl Derivatives. *Collect. Czech. Chem. Commun.*, **1968**, *33*, 3551-3557.
[http://dx.doi.org/10.1135/cccc19683551]

[56] Červinka, O.; Fusek, J. Asymmetric reactions. XLII. Absolute configuration of alkylarylmethanols and alkylarylmethylamines. *Collect. Czech. Chem. Commun.*, **1973**, *38*, 441-446.
[http://dx.doi.org/10.1135/cccc19730441]

[57] Mukaiyama, T.; Onaka, M.; Shiono, M. Synthetic Control Based on Chiral 2-Halopyridinium Salts II. The Kinetic Resolution of Racemic Amines. *Chem. Lett.*, **1977**, *6*, 651-654.
[http://dx.doi.org/10.1246/cl.1977.651]

[58] Hiraki, Y.; Tai, A. Alkyl-Phenyl Interaction in Enantiomer Differentiating Acylation. *Chem. Lett.*, **1982**, *11*, 341-344.
[http://dx.doi.org/10.1246/cl.1982.341]

[59] Hiraki, Y.; Tai, A. A Basic Study of the Amino Acid Residue in Protein. The Role of Hydrocarbon Groups in Enantiomer-differentiating Acylation. *Bull. Chem. Soc. Jpn.*, **1984**, *57*, 1570-1575.
[http://dx.doi.org/10.1246/bcsj.57.1570]

[60] Deresa, L.I.; Shatskaya, V.A.; Popov, A.F. Medium Effect on the Enantioselectivity of Formation Reactions of Peptides and Amides. *Organic Reactivity (Tartu)*, **1988**, *25*, 189-194.

[61] Ötvös, L.; Tömösközi, I.; Mochácsi, T. Stereochemistry of the reactions of biopolymers II. Asymmetric synthesis III. Kinetic resolution in the acylation of DL-alanine ethyl ester with benzyloxycarbonylalanine peptide azides. *Tetrahedron Lett.*, **1970**, *11*, 1995-1998.
[http://dx.doi.org/10.1016/S0040-4039(01)98136-5]

[62] Sokołowska, T.; Bernat, J.F. Stereoselektywność w Syntezach Peptydów. II. Kinetycznie Kontrolowana Indukcja Asymetryczna w Syntesach Dwupeptydów. *Rocz. Chem.*, **1966**, *40*, 1665-1674.

[63] Teramoto, T.; Deguchi, M.; Kurosaki, T. Optically Active N-Hydroxytartrimides for Enantioselective Peptide Synthesis. *Tetrahedron Lett.*, **1981**, *22*, 1109-1112.
[http://dx.doi.org/10.1016/S0040-4039(01)90249-7]

[64] Takeda, K.; Tsuboyama, K.; Suzuki, A.; Ogura, H. Enantioselective Synthesis of Peptides Using (1R)-3-Hydroxy-1,8,8-trimethyl-3-azabicyclo[3.2.1]octane-2,4-dione ((+)-N-Hydroxy-camphorimide) an Active Ester Group. *Chem. Pharm. Bull. (Tokyo)*, **1985**, *33*, 2545-2548.
[http://dx.doi.org/10.1248/cpb.33.2545]

[65] Kostyanovsky, R.G.; Polyakov, A.E.; Shustov, G.V.; Zakharov, K.S.; Markov, V.I. Optically active diaziridines. *Dokl. Akad. Nauk SSSR*, **1974**, *219*, 873-876. [*in Russian*].

[66] Kostyanovsky, R.G.; Polyakov, A.E.; Markov, V.I. Optically active aziridines. *Russ. Chem. Bull.*, **1974**, *23*, 1601.
[http://dx.doi.org/10.1007/BF00929705]

[67] Kostyanovsky, R.G.; Polyakov, A.E.; Markov, V.I. Optically active symmetrically substituted diaziridine. *Russ. Chem. Bull.*, **1975**, *24*, 191.
[http://dx.doi.org/10.1007/BF00926331]

[68] Kostyanovsky, R.G.; Polyakov, A.E.; Shustov, G.V. Asymmetrical nonbridgehead nitrogen XII. The absolute configuration of chiral diaziridines. *Tetrahedron Lett.*, **1976**, *17*, 2059-2060.
[http://dx.doi.org/10.1016/S0040-4039(00)93817-6]

[69] Shustov, G.V.; Denisenko, S.N.; Shokhen, M.A.; Kostyanovsky, R.G. Asymmetric nitrogen 60. Acylation as a pathway to optically active 1,3,3-trisubstituted diaziridines. *Russ. Chem. Bull.*, **1988**, *37*, 1665-1871.
[http://dx.doi.org/10.1007/BF00961119]

[70] Konovalikhin, S.V.; Zolotoi, A.B.; Atovmyan, L.O.; Shustov, G.V.; Denisenko, S.N.; Kostyanovsky, R.G. Asymmetric nitrogen 77. Molecular structure of 3-methyl-1-((S)-1'-tosylprolyl)-1,2-diazacyclohex-2-ene. *Russ. Chem. Bull.*, **1995**, *44*, 483-486.
[http://dx.doi.org/10.1007/BF00702393]

[71] Korneev, V.A.; Shustov, G.V.; Chervin, I.I.; Kostyanovsky, R.G. (S)-1-(2,4-Dinitrophenyl)prolyl chloride – a novel reagent for kinetic optical resolution of diaziridines. *Russ. Chem. Bull.*, **1995**, *44*, 1346-1347.
[http://dx.doi.org/10.1007/BF00700919]

[72] Charushin, V.N.; Krasnov, V.P.; Levit, G.L.; Korolyova, M.A.; Kodess, M.I.; Chupakhin, O.N.; Kim, M.H.; Lee, H.S.; Park, Y.J.; Kim, K-C. Kinetic resolution of (±)-2,3-dihydro-3-methyl-4H-1,4-benzoxazines with (S)-naproxen. *Tetrahedron Asymmetry*, **1999**, *10*, 2691-2702.
[http://dx.doi.org/10.1016/S0957-4166(99)00276-1]

[73] Chupakhin, O.N.; Krasnov, V.P.; Levit, G.L.; Charushin, V.N.; Korolyova, M.A.; Tzoi, E.V.; Lee, H.S.; Park, Y.J.; Kim, M.H.; Kim, K-C. Preparation of (S)-benzoxazines and racemization of (R)-benzoxazines JP Patent 2000 178,265, June 27, 2000 [Chem. Abstrs., 2000, 133, 43530].

[74] Harrington, P.; Lodewijk, E. Twenty Years of Naproxen Technology. *Org. Process Res. Dev.*, **1997**, *1*, 72-76.
[http://dx.doi.org/10.1021/op960009e]

[75] Spahn, H. S-(+)-Naproxen chloride as acylating agent for separating the enantiomers of chiral amines and alcohols. *Arch. Pharm. (Weinheim)*, **1988**, *321*, 847-850.
[http://dx.doi.org/10.1002/ardp.19883211204]

[76] Büschges, R.; Linde, H.; Mutschler, E.; Spahn-Langguth, H. Chloroformates and isothiocyanates derived from 2-arylpropionic acids as chiral reagents: synthetic routes and chromatographic behaviour of the derivatives. *J. Chromatogr. A*, **1996**, *725*, 323-334.
[http://dx.doi.org/10.1016/0021-9673(95)00989-2]

[77] Solís, A.; Luna, H.; Pérez, I. H.; Manjarrez, N.; Sánchez, R. (S)-Naproxen® as a derivatizing agent to determine enantiomeric excess of cyanohydrins by HPLC. *Tetrahedron Lett.*, **1998**, *39*, 8759-8762.
[http://dx.doi.org/10.1016/S0040-4039(98)01990-X]

[78] Krasnov, V.P.; Levit, G.L.; Bukrina, I.M.; Demin, A.M.; Chupakhin, O.N.; Yoo, J.U. Efficient large (ca. 40 g) laboratory scale preparation of (S)- and (R)-valine *tert*-butyl esters. *Tetrahedron Asymmetry*, **2002**, *13*, 1911-1914.
[http://dx.doi.org/10.1016/S0957-4166(02)00532-3]

[79] Büyüktimkin, N.; Buschauer, A. Separation and determination of some amino acid ester enantiomers by thin-layer chromatography after derivatization with (S)-(+)-naproxen. *J. Chromatogr. A*, **1988**, *450*, 281-283.
[http://dx.doi.org/10.1016/S0021-9673(01)83582-2]

[80] Fauconnot, L.; Nugier-Chauvin, C.; Noiret, N.; Patin, H. Enantiomeric excess determination of some chiral sulfoxides by NMR: use of (S)-Ibuprofen® and (S)-Naproxen® as shift reagents. *Tetrahedron Lett.*, **1997**, *38*, 7875-7878.
[http://dx.doi.org/10.1016/S0040-4039(97)10153-8]

[81] Błażewska, K.; Gajda, T. (S)-Naproxen® and (S)-Ibuprofen® chlorides - convenient chemical derivatizing agents for the determination of the enantiomeric excess of hydroxy and aminophosphonates by ^{31}P NMR. *Tetrahedron Asymmetry*, **2002**, *13*, 671-674.
[http://dx.doi.org/10.1016/S0957-4166(02)00186-6]

[82] Krasnov, V.P.; Levit, G.L.; Andreyeva, I.N.; Grishakov, A.N.; Charushin, V.N.; Chupakhin, O.N. Kinetic resolution of (±)-2-methyl-1,2,3,4-tetrahydroquinoline and (±)-2-methylindoline. *Mendeleev Commun.*, **2002**, *12*, 27-28.
[http://dx.doi.org/10.1070/MC2002v012n01ABEH001545]

[83] Krasnov, V.P.; Levit, G.L.; Korolyova, M.A.; Bukrina, I.M.; Sadretdinova, L.Sh.; Andreeva, I.N.; Charushin, V.N.; Chupakhin, O.N. Kinetic resolution of heterocyclic amines by reaction with optically active acid chlorides. The effect of reaction conditions on the diastereoselectivity of acylation of (±)-3(+/-)-3-methyl-2,3-dihydro-4H-1,4-benzoxazine. *Russ. Chem. Bull.*, **2004**, *53*, 1253-1256.
[http://dx.doi.org/10.1023/B:RUCB.0000042282.29765.2f]

[84] Gruzdev, D.A.; Chulakov, E.N.; Levit, G.L.; Ezhikova, M.A.; Kodess, M.I.; Krasnov, V.P. A comparative study on the acylative kinetic resolution of racemic fluorinated and non-fluorinated 2-methyl-1,2,3,4-tetrahydroquinolines and 3,4-dihydro-3-methyl-2H-[1,4]benzoxazines. *Tetrahedron Asymmetry*, **2013**, *24*, 1240-1246.
[http://dx.doi.org/10.1016/j.tetasy.2013.07.024]

[85] Krasnov, V.P.; Levit, G.L.; Gruzdev, D.A.; Matveeva, T.V.; Chulakov, E.N.; Charushin, V.N. Method of producing (S)-7,8-difluoro-2,3-dihydro-3-methyl-4H-[1,4]benzoxazine RU Patent 2434004, November 20, 2011 [Chem. Abstrs., 2011, 155, 657015]

[86] Slepukhin, P.A.; Gruzdev, D.A.; Chulakov, E.N.; Levit, G.L.; Krasnov, V.P.; Charushin, V.N. Structures of the racemate and (S)-enantiomer of 7,8-difluoro-3-methyl-2,3-dihydro-4H-[1,4]benzoxazine. *Russ. Chem. Bull.*, **2011**, *60*, 955-960.
[http://dx.doi.org/10.1007/s11172-011-0150-7]

[87] Chulakov, E.N. *Kinetic Resolution of Racemic Amines in Acylation with Chiral 2-Arylalkanoyl Chlorides*, EkaterinburgJune;2013 [in Russian]

[88] Zhou, Y-G.; Yang, P-Y.; Han, X-W. Synthesis and Highly Enantioselective Hydrogenation of Exocyclic Enamides: (Z)-3-Arylidene-4-acetyl-3,4-dihydro-2H-1,4-benzoxazines. *J. Org. Chem.*, **2005**, *70*, 1679-1683.
[http://dx.doi.org/10.1021/jo048212s]

[89] Chulakov, E.N.; Gruzdev, D.A.; Levit, G.L.; Sadretdinova, L.Sh.; Krasnov, V.P.; Charushin, V.N. 2-Arylpropionyl chlorides in kinetic resolution of racemic 3-methyl-2,3-dihydro-4H-[1,4]benzoxazines. *Russ. Chem. Bull.*, **2011**, *60*, 948-954.
[http://dx.doi.org/10.1007/s11172-011-0149-0]

[90] Chulakov, E.N.; Levit, G.L.; Tumashov, A.A.; Sadretdinova, L.Sh.; Krasnov, V.P. Kinetic Resolution of Racemic 2-Methyl-1,2,3,4-tetrahydroquinoline and Its Structural Analogs by Using 2-Arylpropionyl Chlorides. *Chem. Heterocycl. Compd. (N. Y., NY, U. S.)*, **2012**, *48*, 724-732.
[http://dx.doi.org/10.1007/s10593-012-1051-x]

[91] Coumbarides, G.S.; Eames, J.; Flinn, A.; Northen, J.; Yohannes, Y. Probing the resolution of 2-phenylpropanoyl chloride using quasi-enantiomeric Evans' oxazolidinones. *Tetrahedron Lett.*, **2005**, *46*, 849-853.
[http://dx.doi.org/10.1016/j.tetlet.2004.11.142]

[92] Coumbarides, G.S.; Dingjan, M.; Eames, J.; Flinn, A.; Northen, J.; Yohannes, Y. Efficient parallel resolution of an active ester of 2-phenylpropionic acid using quasi-enantiomeric Evans' oxazolidinones. *Tetrahedron Lett.*, **2005**, *46*, 2897-2902.
[http://dx.doi.org/10.1016/j.tetlet.2005.02.139]

[93] Chavda, S.; Coulbeck, E.; Coumbarides, G.S.; Dingjan, M.; Eames, J.; Ghilagaber, S.; Yohannes, Y. Investigations into the parallel kinetic resolution of 2-phenylpropanoyl chloride using quasi-enantiomeric oxazolidinones. *Tetrahedron Asymmetry*, **2006**, *17*, 3386-3399.
[http://dx.doi.org/10.1016/j.tetasy.2007.01.002]

[94] Coumbarides, G.S.; Dingjan, M.; Eames, J.; Flinn, A.; Motevalli, M.; Northen, J.; Yohannes, Y. Efficient Parallel Resolution of Racemic Evans' Oxazolidinones Using quasi-Enantiomeric Profens. *Synlett*, **2006**, 101-105.

[95] Coumbarides, G.S.; Dingjan, M.; Eames, J.; Flinn, A.; Northen, J. Parallel Kinetic Resolution of an Oxazolidinone Using a Quasi-Enantiomeric Combination of [D,^{13}C]-Isotopomers of Pentafluorophenyl 2-Phenyl Propionate. *Chirality*, **2007**, *19*, 321-328.

[http://dx.doi.org/10.1002/chir.20367]

[96] Boyd, E.; Coulbeck, E.; Coumbarides, G.S.; Chavda, S.; Dingjan, M.; Eames, J.; Flinn, A.; Motevalli, M.; Northen, J.; Yohannes, Y. Parallel kinetic resolution of racemic oxazolidinones using quasi-enantiomeric active esters. *Tetrahedron Asymmetry,* **2007**, *18*, 2515-2530.
[http://dx.doi.org/10.1016/j.tetasy.2007.10.009]

[97] Chavda, S.; Coulbeck, E.; Dingjan, M.; Eames, J.; Flinn, A.; Northen, J. Parallel kinetic resolution of active esters using designer oxazolidin-2-ones derived from phenylglycine. *Tetrahedron Asymmetry,* **2008**, *19*, 1536-1548.
[http://dx.doi.org/10.1016/j.tetasy.2008.06.020]

[98] Coulbeck, E.; Eames, J. Parallel kinetic resolution of active esters using a quasi-enantiomeric combination of (R)-4-phenyl-oxazolidin-2-one and (S)-4,5,5-triphenyl-oxazolidin-2-one. *Tetrahedron Asymmetry,* **2008**, *19*, 2223-2233.
[http://dx.doi.org/10.1016/j.tetasy.2008.09.012]

[99] Andreou, A.; Al Shaye, N.; Brown, H.; Eames, J. Resolution of (4RS,5RS)-4,5-diphenylimidazolidine-2-thione using pentafluorophenyl active esters. *Tetrahedron Lett.,* **2010**, *51*, 6935-6938.
[http://dx.doi.org/10.1016/j.tetlet.2010.10.157]

[100] Chavda, S.; Coulbeck, E.; Eames, J. Investigations into the parallel kinetic resolution of acetyl mandelic acid. *Tetrahedron Lett.,* **2008**, *49*, 7398-7402.
[http://dx.doi.org/10.1016/j.tetlet.2008.10.062]

[101] Hsieh, S-Y.; Wanner, B.; Wheeler, P.; Beauchemin, A.M.; Rovis, T.; Bode, J.W. Stereoelectronic Basis for the Kinetic Resolution of N-Heterocycles with Chiral Acylating Reagents. *Chemistry,* **2014**, *20*, 7228-7231.
[http://dx.doi.org/10.1002/chem.201402818]

[102] Karlsson, S.; Brånalt, J.; Halvarsson, M.Ö.; Bergman, J. A One-Pot Asymmetric Synthesis of a N-Acylated 4,5-Dihydropyrazole, A Key Intermediate of Thrombin Inhibitor AZD8165. *Org. Process Res. Dev.,* **2014**, *18*, 969-975.
[http://dx.doi.org/10.1021/op500134e]

[103] Vakarov, S.A.; Gruzdev, D.A.; Sadretdinova, L.Sh.; Chulakov, E.N.; Pervova, M.G.; Ezhikova, M.A.; Kodess, M.I.; Levit, G.L.; Krasnov, V.P. Diastereoselective acylation of 3,4-dihydro-3-methyl-2H-[1,4]benzoxazines with 2-phenoxy carbonyl chlorides. *Tetrahedron Asymmetry,* **2015**, *26*, 312-319.
[http://dx.doi.org/10.1016/j.tetasy.2015.02.004]

[104] Birman, V.B.; Jiang, H.; Li, X.; Guo, L.; Uffman, E.W. Kinetic Resolution of 2-Oxazolidinones via Catalytic, Enantioselective N-Acylation. *J. Am. Chem. Soc.,* **2006**, *128*, 6536-6537.
[http://dx.doi.org/10.1021/ja061560m]

[105] Moretto, A.; Peggion, C.; Formaggio, F.; Crisma, M.; Kaptein, B.; Broxterman, Q.B.; Toniolo, C. Stereoselective Acylation of a Racemic Amine with C^{α}-Methyl Phenylglycine-Based Dipeptide 5(4H)-Oxazolones. *Chirality,* **2005**, *17*, 481-487.
[http://dx.doi.org/10.1002/chir.20182]

[106] Krasnov, V.P.; Levit, G.L.; Kodess, M.I.; Charushin, V.N.; Chupakhin, O.N. N-Phthaloyl-(S)-alanyl chloride as a chiral resolving agent for the kinetic resolution of heterocyclic amines. *Tetrahedron Asymmetry,* **2004**, *15*, 859-862.
[http://dx.doi.org/10.1016/j.tetasy.2004.01.025]

[107] Gruzdev, D.A.; Levit, G.L.; Krasnov, V.P.; Chulakov, E.N.; Sadretdinova, L.Sh.; Grishakov, A.N.; Ezhikova, M.A.; Kodess, M.I.; Charushin, V.N. Acylative kinetic resolution of racemic amines using N-phthaloyl-(S)-amino acyl chlorides. *Tetrahedron Asymmetry,* **2010**, *21*, 936-942.
[http://dx.doi.org/10.1016/j.tetasy.2010.05.013]

[108] Levit, G.L.; Gruzdev, D.A.; Krasnov, V.P.; Chulakov, E.N.; Sadretdinova, L.Sh.; Ezhikova, M.A.; Kodess, M.I.; Charushin, V.N. Substituent effect on the stereoselectivity of acylation of racemic heterocyclic amines with N-phthaloyl-3-aryl-(S)-alanyl chlorides. *Tetrahedron Asymmetry,* **2011**, *22*,

185-189.
[http://dx.doi.org/10.1016/j.tetasy.2010.12.017]

[109] Gruzdev, D.A.; Levit, G.L.; Kodess, M.I.; Krasnov, V.P. Synthesis of Enantiomers of 6-Nitro- and 6-Amino-2-methyl-1,2,3,4-tetrahydroquinolines. *Chem. Heterocycl. Compd. (N. Y., NY, U. S.)*, **2012**, *48*, 748-757.
[http://dx.doi.org/10.1007/s10593-012-1053-8]

[110] Gruzdev, D.A.; Levit, G.L.; Krasnov, V.P. Acylative kinetic resolution of racemic heterocyclic amines using N-phthaloyl-(*S*)-amino acyl chlorides with alkyl side chains. *Tetrahedron Asymmetry*, **2012**, *23*, 1640-1646.
[http://dx.doi.org/10.1016/j.tetasy.2012.11.001]

[111] Gruzdev, D.A.; Chulakov, E.N.; Sadretdinova, L.Sh.; Kodess, M.I.; Levit, G.L.; Krasnov, V.P. Synthesis of enantiomers of 3-methyl- and 3-phenyl-3,4-dihydro-2*H*-[1,4]benzothiazines and their 1,1-dioxides *via* an acylative kinetic resolution protocol. *Tetrahedron Asymmetry*, **2015**, *26*, 186-194.
[http://dx.doi.org/10.1016/j.tetasy.2015.01.010]

[112] Krasnov, V.P.; Levit, G.L.; Bukrina, I.M.; Andreeva, I.N.; Sadretdinova, L.Sh.; Korolyova, M.A.; Kodess, M.I.; Charushin, V.N.; Chupakhin, O.N. Kinetic resolution of (±)-2,3-dihydro-3-methl-4H-1,4-benzoxazine, (±)-2-methyl-1,2,3,4-tetrahydroquinoline and (±)-2-methylindoline using N-tosyl-(*S*)-prolyl chloride. *Tetrahedron Asymmetry*, **2003**, *14*, 1985-1988.
[http://dx.doi.org/10.1016/S0957-4166(03)00321-5]

[113] Gruzdev, D.A.; Vakarov, S.A.; Levit, G.L.; Krasnov, V.P. N-Tosyl-(*S*)-prolyl chloride in kinetic resolution of racemic heterocyclic amines. *Chem. Heterocycl. Compd. (N. Y., NY, U. S.)*, **2014**, *49*, 1795-1807.
[http://dx.doi.org/10.1007/s10593-014-1432-4]

[114] Liu, W-B.; He, H.; Dai, L-X.; You, S-L. Synthesis of 2-Methylindoline- and 2-Methyl-1,2,3,4-tetrahydroquinoline-Derived Phosphoramidites and Their Applications in Iridium-Catalyzed Allylic Alkylation of Indoles. *Synthesis*, **2009**, 2076-2082.

[115] Vakarov, S.A.; Gruzdev, D.A.; Chulakov, E.N.; Sadretdinova, L.Sh.; Ezhikova, M.A.; Kodess, M.I.; Levit, G.L.; Krasnov, V.P. Diastereoselective Acylation of Racemic Heterocyclic Amines with N-Tosyl-(*S*)-prolyl Chloride and Its Structural Analogs. *Chem. Heterocycl. Compd. (N. Y., NY, U. S.)*, **2014**, *50*, 838-855.
[http://dx.doi.org/10.1007/s10593-014-1538-8]

[116] Nigst, T.A.; Mayr, H. Comparison of the Electrophilic Reactivities of N-Acylpyridinium Ions and Other Acylating Agents. *Eur. J. Org. Chem.*, **2013**, 2155-2163.
[http://dx.doi.org/10.1002/ejoc.201201540]

[117] Chulakov, E.N.; Gruzdev, D.A.; Levit, G.L.; Kudryavtsev, K.V.; Krasnov, V.P. Enantiomers of all-*cis*-5-(4-bromophenyl)-4-tert-butoxycarbonyl-2-methoxycarbonylpyrrolidine: preparative HPLC separation and acylative kinetic resolution of the racemate. *Tetrahedron Asymmetry*, **2012**, *23*, 1683-1688.
[http://dx.doi.org/10.1016/j.tetasy.2012.10.020]

[118] Krasnov, V.P.; Levit, G.L.; Charushin, V.N.; Grishakov, A.N.; Kodess, M.I.; Kalinin, V.N.; Ol'shevskaya, V.A.; Chupakhin, O.N. Enantiomers of 3-amino-1-methyl-1,2-dicarba-*closo*-dodecaborane. *Tetrahedron Asymmetry*, **2002**, *13*, 1833-1835.
[http://dx.doi.org/10.1016/S0957-4166(02)00474-3]

[119] Levit, G.L.; Krasnov, V.P.; Demin, A.M.; Kodess, M.I.; Sadretdinova, L.Sh.; Matveeva, T.V.; Ol'shevskaya, V.A.; Kalinin, V.N.; Chupakhin, O.N.; Charushin, V.N. Kinetic resolution of 1-methyl- and 1-phenyl-3-amino-1,2-dicarba-*closo*-dodecaboranes *via* acylation with chiral acyl chlorides. *Mendeleev Commun.*, **2004**, *14*, 293-295.
[http://dx.doi.org/10.1070/MC2004v014n06ABEH002047]

[120] Levit, G.L.; Demin, A.M.; Kodess, M.I.; Ezhikova, M.A.; Sadretdinova, L.Sh.; Ol'shevskaya, V.A.;

Kalinin, V.N.; Krasnov, V.P.; Charushin, V.N. Acidic hydrolysis of N-acyl-1-substituted 3-amino-1,2-dicarba-*closo*-dodecaboranes. *J. Organomet. Chem.,* **2005**, *690*, 2783-2786.
[http://dx.doi.org/10.1016/j.jorganchem.2005.01.043]

[121] Krasnov, V.P.; Demin, A.M.; Levit, G.L.; Grishakov, A.N.; Sadretdinova, L.Sh.; Ol'shevskaya, V.A.; Glukhov, I.V.; Kalinin, V.N.; Charushin, V.N. Determination of enantiomeric purity of 1-substituted 3-amino-1,2-dicarba-*closo*-dodecaboranes by HPLC on chiral stationary phases. *Russ. Chem. Bull.,* **2008**, *57*, 2535-2539.
[http://dx.doi.org/10.1007/s11172-008-0364-5]

[122] Kodess, M.I.; Ezhikova, M.A.; Levit, G.L.; Krasnov, V.P.; Charushin, V.N. NMR determination of enantiomeric composition of 1-substituted 3-amino-1,2-dicarba-*closo*-dodecaboranes using Eu(hfc)$_3$. *J. Organomet. Chem.,* **2005**, *690*, 2766-2768.
[http://dx.doi.org/10.1016/j.jorganchem.2005.01.044]

[123] Zhao, M.; Wang, C.; Peng, S.; Winterfeldt, E. Easy generation of an enantiopure general indolalkaloid building block by kinetic resolution. *Tetrahedron Asymmetry,* **1999**, *10*, 3899-3905.
[http://dx.doi.org/10.1016/S0957-4166(99)00412-7]

[124] Xu, Z-X.; Zhang, C.; Yang, Y.; Chen, C-F.; Huang, Z-T. Effective Nonenzymatic Kinetic Resolution of Racemic *m*-Nitro-Substituted Inherently Chiral Aminocalix[4]arenes. *Org. Lett.,* **2008**, *10*, 477-479.
[http://dx.doi.org/10.1021/ol702884u]

[125] Dixit, S.; Dubey, R.; Bhushan, R. High-Performance Liquid Chromatography for Analytical and Small-Scale Preparative Separation of (*R,S*)-Mexiletine Using (–)-(*N*)-Trifluoroacetyl-Prolyl Chloride and (1*S*)-(–)-Camphanic Chloride and Recovery of Native Enantiomer by Detagging. *Acta Chromatogr.,* **2014**, *26*, 625-636.
[http://dx.doi.org/10.1556/AChrom.26.2014.4.5]

[126] Yang, X.; Bumbu, V.D.; Liu, P.; Li, X.; Jiang, H.; Uffman, E.W.; Guo, L.; Zhang, W.; Jiang, X.; Houk, K.N.; Birman, V.B. Catalytic, Enantioselective N-Acylation of Lactams and Thiolactams Using Amidine-Based Catalysts. *J. Am. Chem. Soc.,* **2012**, *134*, 17605-17612.
[http://dx.doi.org/10.1021/ja306766n]

[127] Allen, S.E.; Hsieh, S-Y.; Gutierrez, O.; Bode, J.W.; Kozlowski, M.C. Concerted Amidation of Activated Esters: Reaction Path and Origins of Selectivity in the Kinetic Resolution of Cyclic Amines *via* NHC and Hydroxamic Acid Co-Catalyzed Acyl Transfer. *J. Am. Chem. Soc.,* **2014**, *136*, 11783-11791.
[http://dx.doi.org/10.1021/ja505784w]

[128] Mittal, N.; Lippert, K.M.; De, C.K.; Klauber, E.G.; Emge, T.J.; Schreiner, P.R.; Seidel, D. A Dual-Catalysis Anion-Binding Approach to the Kinetic Resolution of Amines: Insights into the Mechanism *via* a Combined Experimental and Computational Study. *J. Am. Chem. Soc.,* **2015**, *137*, 5748-5758.
[http://dx.doi.org/10.1021/jacs.5b00190]

[129] Wanner, B.; Kreituss, I.; Gutierrez, O.; Kozlowski, M.C.; Bode, J.W. Catalytic Kinetic Resolution of Disubstituted Piperidines by Enantioselective Acylation: Synthetic Utility and Mechanistic Insights. *J. Am. Chem. Soc.,* **2015**, *137*, 11491-11497.
[http://dx.doi.org/10.1021/jacs.5b07201]

CHAPTER 5

Advances in the Synthesis of Functional α-Organyl *gem*-Bisphosphonates for Biomedical Applications

Vadim D. Romanenko[*]

Institute of Bioorganic Chemistry and Petrochemistry, National Academy of Sciences of Ukraine, Kiev-94, 02660, Ukraine

Abstract: This chapter deals with the synthesis of α-organyl-substituted methylenebisphosphonates and focuses on compounds not having the heteroatomic substituents at the geminal carbon. We present different methodologies on the basis of the reactivity of monophosphonate and methylene 1,1-bisphosphonate systems including the Michaelis-Arbuzov and Michaelis-Becker reactions, substitutions, Michael additions, cycloaddition, as well as more specific approaches. Together with the synthetic methods, the biomedical application of the compounds is also described.

Keywords: Alkenylidene-1,1-Bisphosphonates, Alkylation, Anti-Inflammatory Properties, Anti-Resorption Agents, Antitumor, Asymmetric Synthesis, Bone Targeting, Bisphosphonates, Bisphosphonate Conjugates, Cycloaddition, Enzyme Inhibitors, Michaelis-Arbuzov Reaction, Michael Additions, Phosphorylation, Pyrophosphate Bioisosteres, Phosphate Mimics, Phosphonate Carbanions, Three-Component Reactions.

INTRODUCTION

Geminal bisphosphonates represent a family of organophosphorus compounds structurally characterized by the presence of a P-C-P backbone. The structural similarity of the parent methylene bisphosphonic acid (**1**) to the naturally occurring endogenous pyrophosphate (**2**) is widely recognized and this relationship has been used for the synthesis of many pharmacologically active bisphosphonate derivatives (Fig. **1**) [1 - 3]. The central carbon (α-C) of methylene bisphosphonates imparts metabolic stability and provides a scaffold that can be modified with varied functional groups. The most studied subclasses of clinically important bisphosphonates, widely used in the treatment of several bone disorders such as Paget's disease, myeloma, skeletal metastases and osteoporosis, contain a

---

[*] Corresponding author **Vadim D. Romanenko**: Institute of Bioorganic Chemistry and Petrochemistry, National Academy of Sciences of Ukraine, 1-Murmanska Str., Kyiv-94, 02660, Ukraine; Tel: +38(044)5585388; Fax: 38(044)5732552; E-mail: romanenko@bpci.kiev.ua

combination of α-hydroxyl group and a nitrogen atom in the alkyl side chain or heterocyclic ring. These bisphosphonates bear a substantial negative charge at physiological pH and possess limited capacity to penetrate the cell membrane [4].

Fig. (1). Methylenebisphosphonate (1), pyrophosphate (2), and some representatives of α-organyl bisphosphonates of biomedical interest.

Since 2000, non-α-heteroatom containing bisphosphonates have gained significant attention. Methods for the synthesis, structure, molecular mechanism of action and biomedical properties of these compounds differ significantly from their analogues having a heteroatom bonded to the central carbon.

The introduction of lipophilic alkyl, aryl and heteroaryl groups into the P-C-P backbone increases the availability of bisphosphonate molecules in soft tissues and significantly changes a spectrum of their bioactivity. A number of publications dedicated to bisphosphonates lacking a heteroatom in position α-C have appeared and several sets of these compounds were synthesized and evaluated as potent inhibitors against the intracellular form of the parasites responsible for sleeping sickness, Chagas's disease, malaria and leishmaniosis. Thus, the work carried out by Rodriguez and co-workers has already resulted in some impressive examples of the efficiency of 2-alkylaminoethyl-1-1-bisphosphonates as growth inhibitors against *T. cruzi*, showing a high selectivity for the inhibition of the enzymatic activity of TcFPPS [5 - 7]. Other workers notably the group of Wiemer has described a new series of isoprenoid bisphosphonates as geranylgeranyl diphosphate synthase inhibitors [8]. Some other applications seem to be possible with α-organyl bisphosphonate derivatives [9]. Therefore, the development of new strategies for the synthesis and exploring the biological properties of functional α-organyl bisphosphonates became a promising area of medicinal chemistry.

This Chapter focuses solely on the recent developments in the synthesis of methylene bisphosphonates lacking heteroatom substituents on the geminal carbon. Since the early chemistry of *gem*-bisphosphonates has been partially reviewed, publications, which have appeared prior to 2000, are only mentioned when necessary for the context. Readers are referred to publications [10 - 14] for information on other types of *gem*-bisphosphonate systems.

SYNTHESIS FROM PHOSPHITES AND MONOPHOSPHONATES *VIA* C-P BOND FORMATION

New Variations in the Michaelis-Arbuzov and Michaelis-Becker Reactions

The longest known and widely used method for the synthesis of alkylphosphonates from haloalkanes and trialkyl phosphites (Michaelis-Arbuzov reaction) or metal dialkyl phosphites (Michaelis-Becker reaction) is also applicable for the preparation of 1-organyl-substituted bisphosphonates [4, 15].

Recently successful application of functionalized phosphonites for preparing bisphosphinates, which include pyridine moieties and steric hindered phenolic fragments, has been described by Russian chemists [16 - 18]. Trimethylsilyl phosphonites **3** readily react with benzalchloride in a methylene chloride solution to form bisphosphinates **4** in high yield. Transformation of trimethylsilyl esters **4** into functionalized acids **5** has been accomplished using excess methanol (Scheme **1**).

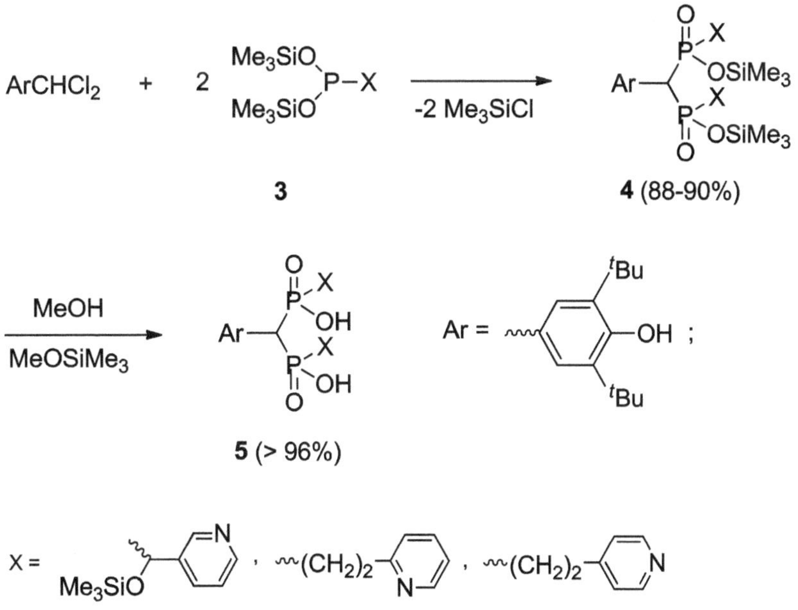

Scheme 1. Synthesis of *gem*-bisphosphinates *via* the Michaelis-Arbuzov reaction.

α-Monohalogenated benzylphosphonates ArCH(Hlg)P(O)(OR)$_2$, including those obtained by the direct bromination of the corresponding benzylphosphonates, are good electrophiles which are capable of reacting with trialkyl phosphites under Michaelis-Arbuzov conditions. This approach is illustrated by the synthesis of tetraethyl 4-aminophenylmethylene-1,1-bisphosphonate **8** (Scheme 2) [19].

Heteryl-substituted cyanomethylphosphonates, derived from diethyl cyanomethylphosphonate, were found to react with sodium dialkyl phosphites to give the corresponding symmetric and asymmetric bisphosphonates **9** and **10** in modest yield (Scheme 3) [20].

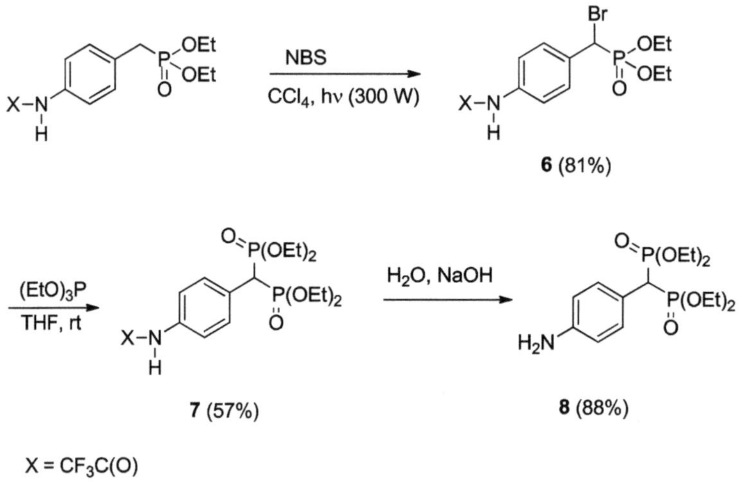

Scheme 2. Preparation of tetraethyl 4-aminophenylmethylene-1,1-bisphosphonate.

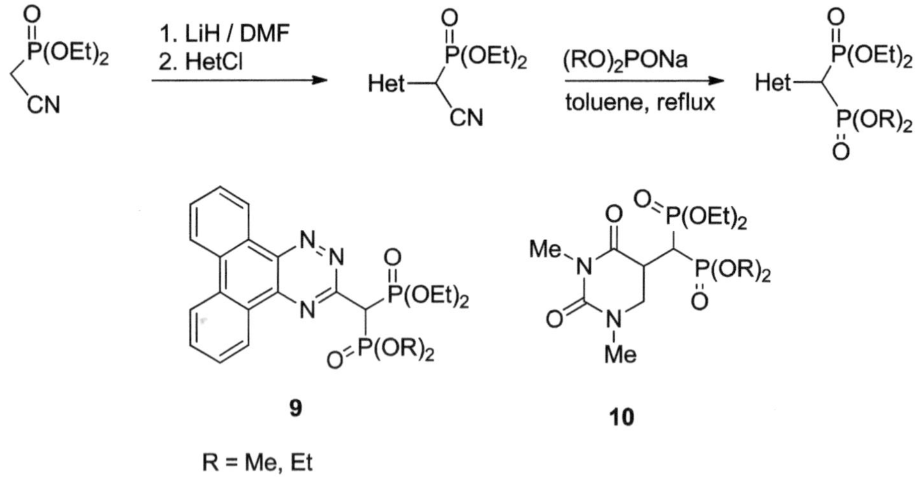

Scheme 3. Synthesis α-branched *N*-heterocycle-substituted bisphosphonates **9**, **10**.

Phosphorylation of α-Anionic Monophosphonates

The monophosphonates of the type $(RO)_2P(O)CHXY$, whose α-CH bond is activated toward deprotonation, are obvious precursors to bisphosphonates since α-phosphorylation of these compounds would be expected to give the corresponding bisphosphonate derivatives. A new example of this approach is outlined in Scheme 4. The diethyl 2-(1,3-dioxolane-2-yl)ethylidene 1,1-bisphosphonate **12**, the key intermediate in the synthesis of phosphonated tripodal ligand **14**, designed for actinides chelation therapy, was prepared from monophosphonate **11** *via* the lithiation with ^{n}BuLi and further coupling with diethyl chlorophosphate. The unprotected aldehyde **13** was then recovered by hydrolysis of acetal group and transformed into tripod **14** which exhibited excellent results in uranyl complexation studies [21].

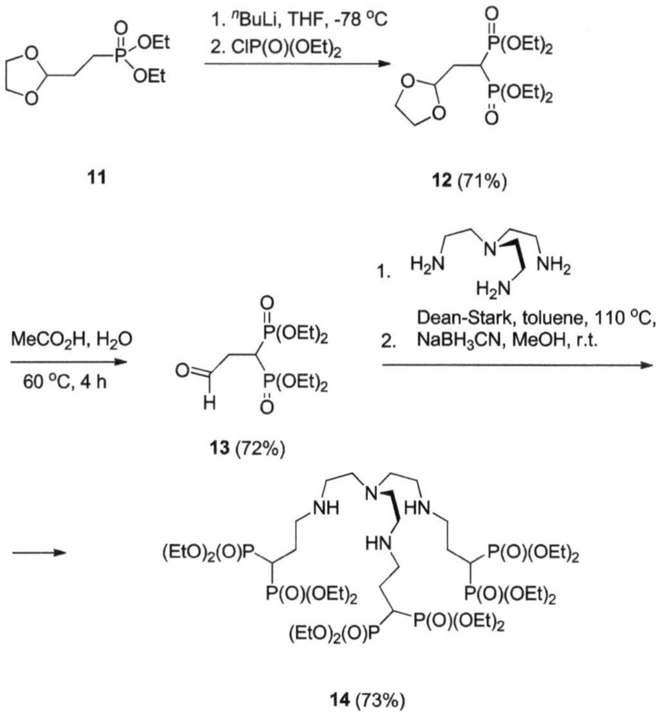

Scheme 4. Application of the deprotonation/phosphorylation methodology to the synthesis of chelating agent **14**.

Using the same approach, Bohacek and co-workers succeeded in preparing bisphosphonate precursor **16** for the synthesis of a high-affinity, bone-targeted Src SH2 inhibitor **18**. The phosphorylation step consists of initial deprotonation of the α-methylene group of the phosphonates framework of compound **15** with

LiHMDS, at -40 °C, followed by treatment with diethyl chlorophosphate (Scheme 5) [22].

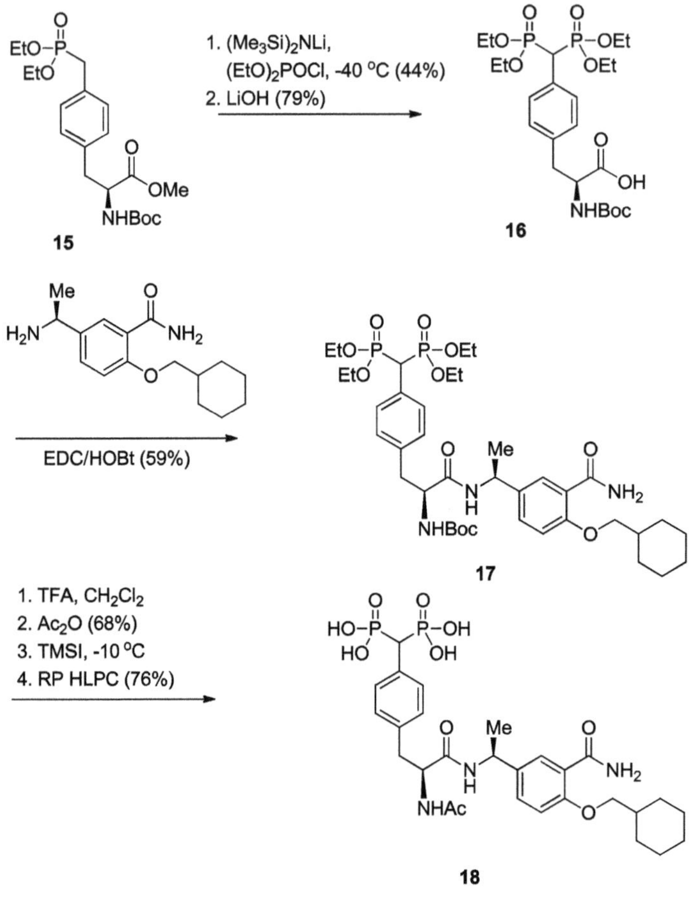

Scheme 5. Synthesis of bone-targeted Src SH2 inhibitor **18**.

Hoff and co-workers were able to prepare, isolate and characterize the ^{125}I and ^{131}I-labeled arylalkylidenebisphosphonates **22a,b** which have a potential for diagnosis and therapy of malignant osseous lesions from the *m*-trimethylsilylphenylethylidene-1,1-bisphosphonic acid **21** (Scheme 6) [23]. The phosphonate **19** was converted to the corresponding Li-carbanion using LDA, and the subsequent reaction with diethyl chlorophosphate gave the bisphosphonate **20** in 87% yield. The latter was labeled using an iododesilylation reaction. This was achieved by adding non carrier added (nca) Na^{125}I and Na^{131}I and *N*-chlorosuccinimide (NCS) as an oxidizing agent to a solution of the precursor **21** in a mixture of acetic acid and trifluoroacetic acid. The radioiodinated compounds were purified by HPLC and their structure were confirmed by co-elution with the corresponding non-radioactive compounds.

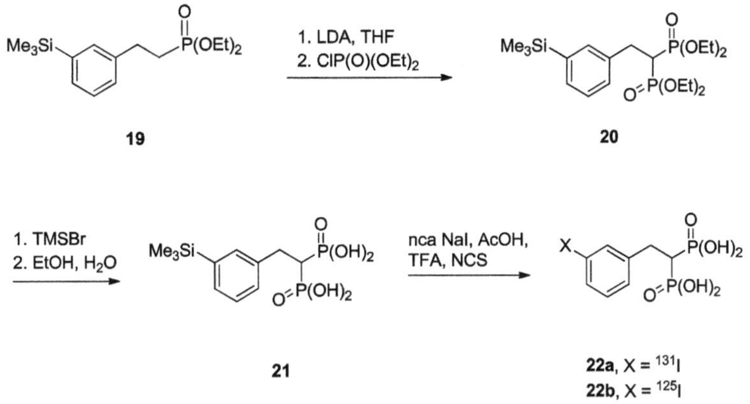

Scheme 6. Synthesis of the ^{125}I and ^{131}I-Labeled Arylalkylidenebisphosphonates **22**.

Since a large number of functionalized monophosphonate substrates are available, synthesis of bisphosphonates from monophosphonates *via* C-P bond formation is particularly useful for the preparation of some elaborated derivatives. Thus, Grison and co-workers reported on the preparation of trimethyl glycosyl alkylidenebisphosphonates through the phosphorylation of α-lithio alkylphosphonates and the subsequent direct introduction of the chloride bisphosphonate intermediate on a protected carbohydrate (Scheme **7**). These bisphosphonates represent potential mechanism-based transition state inhibitors of D-glycosyl phosphate transferase or DNA-polymerase. However, no antiviral activity was detected against a panel of RNA and DNA viruses [24].

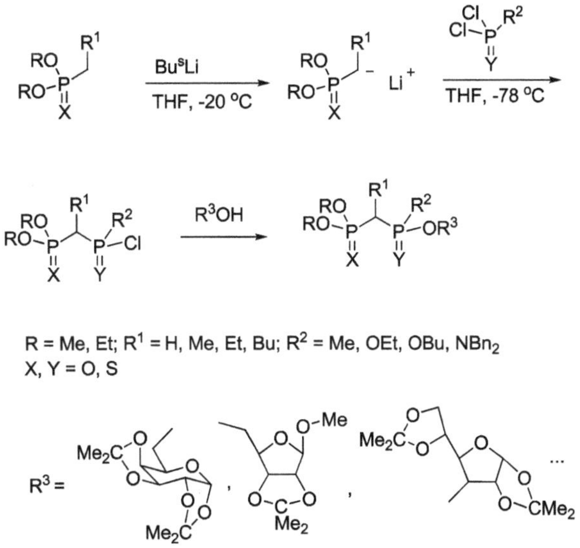

Scheme 7. Synthesis of monoglycosyl alkylidenebisphosphonates.

Synthesis *Via* Enolate Chemistry

Treatment of *N*-octylpyrrolidinone **23** with excess LDA and diethyl chlorophosphite followed by the oxidation of the reaction mixture with H_2O_2, results in the formation of the bisphosphonate **24** in a very good yield (Scheme **8**). These reaction conditions also were applied to the synthesis of *N*-farnesyl lactam bisphosphonates **25** and **26**. More varied results have been obtained with other carbonyl compounds (Table **1**). The simple acyclic amide, *N,N*-diethylacetamide **27**, reacted, under basic conditions, with $ClP(OEt)_2$, followed by oxidation of the reaction mixture with H_2O_2, to yield the bisphosphonate **28** in good yield. However, when the longer chain amide **29** was treated under the same reaction conditions, only the monophosphonate **30** was obtained. A distinct difference between the five and six membered imides **31** and **34** was also noticed: whereas the reaction afforded the vicinal bisphosphonates **32** with the former only the monophosphonate was found with the latter. In contrast, when step-by-step phosphorylation conditions with LHMDS as the base were applied to the imides **31** and **34**, the geminal bisphosphonates **33** and **35** were obtained in low yields accompanied by a large amount of the corresponding monophosphonates. With the 5-membered ring lactone **36**, application of the standard reaction conditions afforded the bisphosphonates **37** in good yield. The 6-membered ring lactone **38** on the other hand underwent transformation into the phosphonates-phosphate **39** [25]. Further studies on the synthesis of bisphosphonates *via* enolate chemistry will be necessary to better delineate the limits of this method.

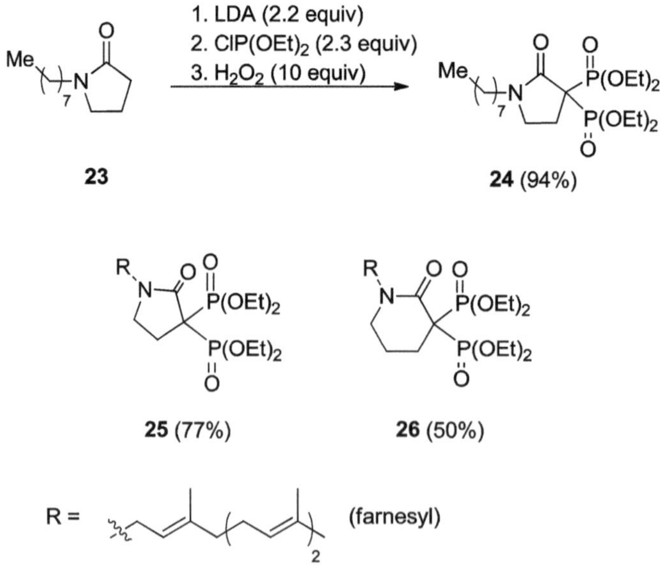

Scheme 8. Synthesis of lactam bisphosphonates **24-26** *via* enolate chemistry.

Table 1. Attempted synthesis of amide, lactam and lactone bisphosphonates [25].

Carbonyl compound[a]	Method[b]	Product	Yield bis (%) (+ yield mono)
27	A	28	72
29	A	30	95
31	A	32	46
31	B	33	30 (+43% mono)
34	B	35	18 (+48% mono)
36	A	37	67
38	A	39	55

[a]R = Farnesyl; [b]Method A: (i) LDA (2.2 equiv), (ii) ClP(OEt)$_2$ (2.3 equiv), (iii) H$_2$O$_2$ (10 equiv); Method B: (i) LHMDS, (ii) ClP(OEt)$_2$, (iii) LHMDS, (iv) ClP(OEt)$_2$, (v) H$_2$O$_2$ (20 equiv)

FUNCTIONALIZATION OF COMPOUNDS ALREADY CONTAINING THE P-C-P BACKBONE

Alkylation of Methylenebisphosphonate Carbanions

Deprotonation-alkylation of tetraalkyl methylenebisphosphonates provides a convenient and general method for producing 1-alkyl-substituted derivatives from readily available starting materials. Alkali metal hydrides (NaH, KH) are the most widely used reagents for the generation of the carbanions $[(RO)_2P(O)]_2CH^-$, although other strong bases such as ^nBuLi, ^tBuOK and dispersions of sodium or potassium in an ethereal or hydrocarbon solvent may also be utilized. In general, the stoichiometric reaction of sodium bisphosphonates with alkyl bromide, iodide, or tosylate in THF gives a good yield of C-monoalkylated bisphosphonate, but the product is often contaminated with a small amount of C,C-dialkylated derivative. Another possibility for the synthesis of C-alkylated methylenebisphosphonates involves the solid-liquid two-phase reaction in the presence of a base and phase transfer catalysis. For example, benzylation with benzyl bromide takes place efficiently in boiling acetonitrile in the presence of K_2CO_3 and triethylbenzylammonium chloride (TEBAC) [26]. The alkylation of tetraethyl methylenebisphosphonate was also investigated under solvent-free microwave (MW) conditions in the presence of Cs_2CO_3. The monoalkylated products were obtained in acceptable yields [27].

A conceptually similar chemistry was used by Moreau and Maffei to obtain 2-vinyl-1,1-cyclopropanediylbisphosphonate **40** from 1,4-dibromobut-2-ene. A process involving alkylation of the carbanion of $CH_2(PO_3Et_2)_2$ followed by the treatment of the isolated bisphosphonate **40** with dialkylamine in the presence of $Pd(PPh_3)_4$ resulted in the formation of bisphosphonates **41** in high yields and with high purity (Scheme **9**) [28].

Recently, the set of isoprenoid bisphosphonates bearing one [8, 29] or more isoprenoid substituents [8, 30] has been synthesized using deprotonation-alkylation methodology (Fig. **2**). The reaction of allylic bromides with bisphosphonate anions has been investigated in detail and has led to the conclusion that for the preparation of C,C-dialkylated bisphosphonates bearing isoprenoid chains of varying lengths, it is more convenient to add larger isoprenoid moiety to the template first and then install the smaller isoprenoid chain (Scheme **10**)[30]. Isoprenoid bisphosphonates **42-44** have been tested for *in vitro* inhibition of human recombinant geranylgeranyl diphosphate synthase (GGDPS) and cellular inhibition of protein geranylgeranylation. Bisphosphonates that contain at least one geranyl chain all are potent inhibitors of GGDPS. Isoprenoid bisphosphonates with *in vitro* GGDPS IC_{50} values of up to ~7μM were able to inhibit cellular geranylgeranylation with similar efficacy [8]. 2E,6E-farnesyl bisphosphonate inhibits protein geranylgeranylation in K562 leukemia cells [31].

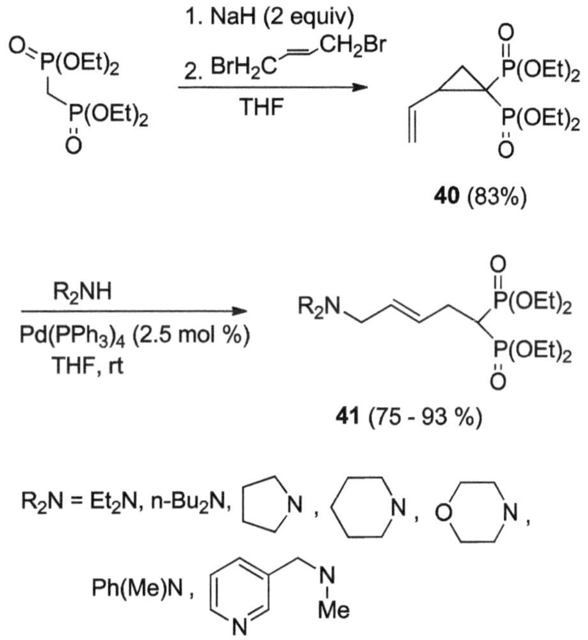

Scheme 9. Synthesis of functionalized methylenebisphosphonates involving the alkylation and Pd-catalyzed ring opening reactions.

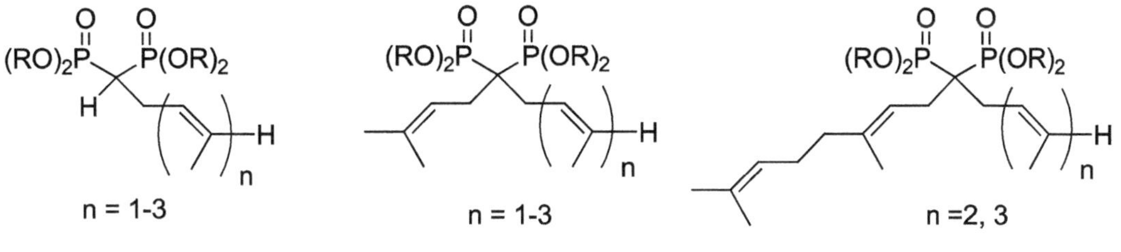

Fig. (2). Chemical structures of selected isoprenoid bisphosphonates [8].

Deprotonation-alkylation methodology was extended to fluorescently tagged isoprenoid bisphosphonates **47-49** that inhibit geranylgeranylation (Scheme **11**) [32]. The allylic bromide **45** was treated with the tetraethyl methylenebisphosphonate salt to afford the isoprenoid bisphosphonate **46** in modest yield. The *C,C*-dialkylated bisphosphonate derivative **48** was prepared through treatment of the allylic bromide **45** with the anion of geranyl bisphosphonate salt **42**.

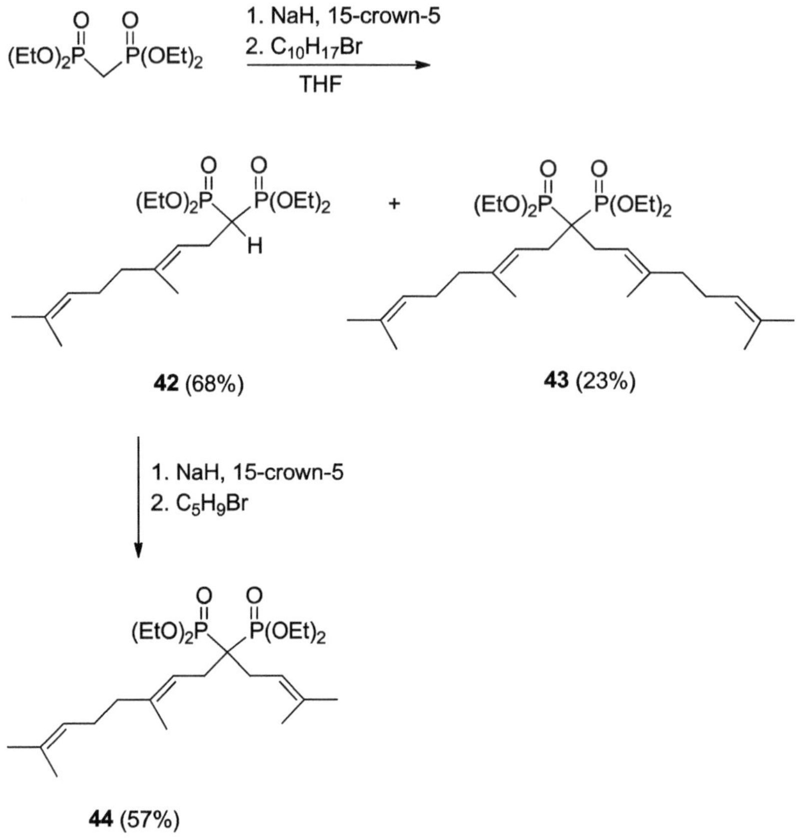

Scheme 10. Synthesis of bisphosphonates 42-44 with a C_{10} (geranyl) substituent.

A unique type of substrates that displays remarkably high reactivity with azides ("click" chemistry) and opens new route to potential anti-resorption bone drugs is α-propargyl-substituted bisphosphonates. The synthesis of the monopropargyl bisphosphonates has been reported by Li and Yuan *via* deprotonation of tetraethyl methylenebisphosphonate with NaH in toluene followed by alkylation with equimolar amount of propargyl bromide [33]. A disadvantage of this method is fact that reaction affords a mixture of the mono- and bis-propargyl substituted bisphosphonates **50, 51** (Scheme 12). Subsequently, an alternative and easy method for the preparation of **50** by selective addition of sodium acetylenide to ethenylidene-1,1-bisphosphonate has been developed [34, 35]. Nevertheless, the alkylation of methylenebisphosphonate with two equivalents of propargyl bromide under deprotonation with two equivalents of NaH in THF is a convenient route leading to dipropargyl derivative of bisphosphonate **51**.

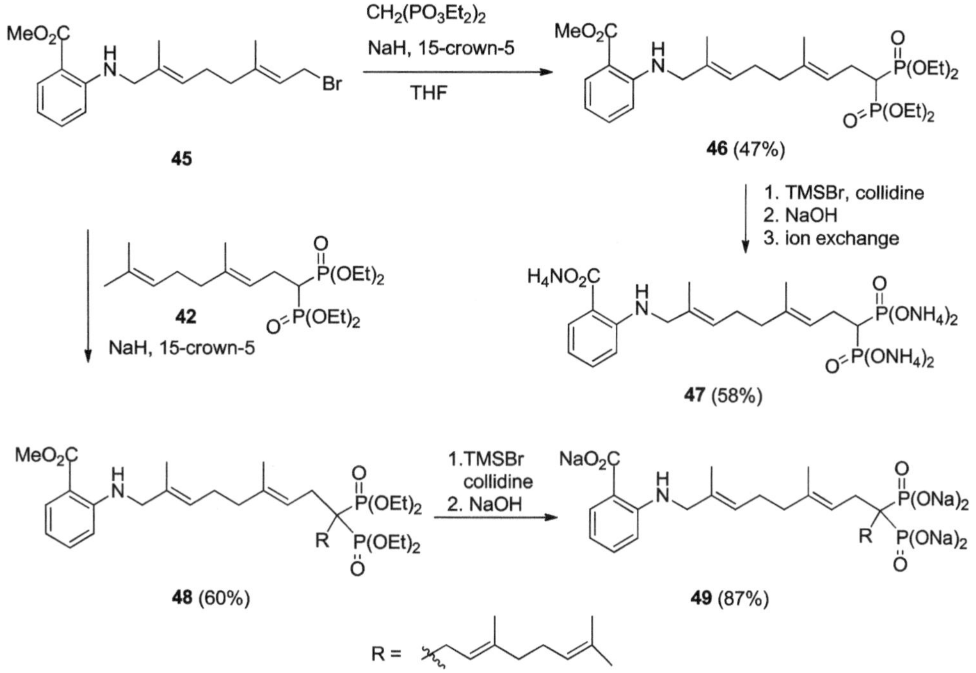

Scheme 11. Deprotonation-alkylation methodology used for the synthesis of isoprenoid bisphosphonates.

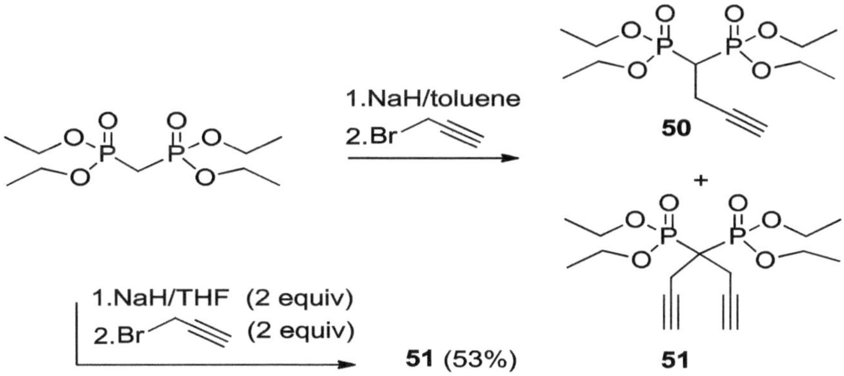

Scheme 12. Synthesis of α-propargyl-substituted bisphosphonates.

Only aromatic systems that are highly activated toward nucleophilic attack may be phosphonylated directly by nucleophilic aromatic substitution using methylenebisphosphonate carbanions. Pentafluoropyridine, for example, gave the corresponding bisphosphonate **52** when the reaction was carried out in DMF using NaH as a base in the presence of 18-crown-6 (Scheme **13**) [36].

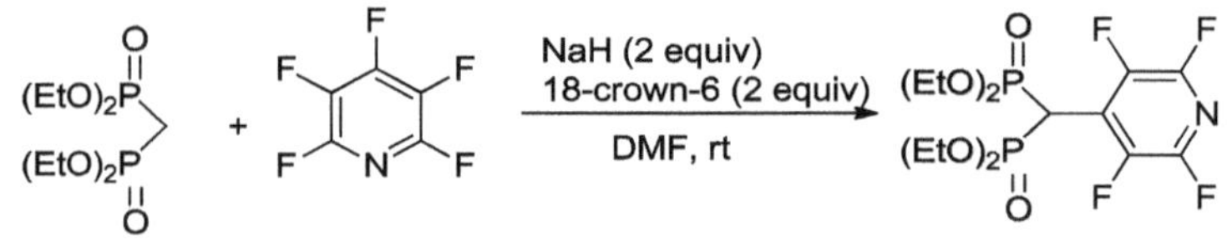

Scheme 13. Synthesis of bisphosphonate **52** using nucleophilic aromatic substitution.

Synthesis Based on "Click" Methodology

Roschenthaler's research group reported the synthesis of novel *N*-BPs based on the reaction of mono- and bis(propargyl)-substituted bisphosphonates with a variety of azides under Cu(I) catalysis ("click" chemistry) [34]. Thus, reaction of the α-propargyl-substituted methylenebisphosphonate with various azides leads to the formation of functionalized triazoles **53** in high yield (Scheme **14**). The catalyst can be directly introduced as a Cu(I) salt in an organic solvent (Method A), or generated *in situ* by reduction of Cu(II) salt with sodium ascorbate in a water-alcohol medium (Method B). A similar approach using bis(propargyl)-substituted bisphosphonate was developed to prepare bis(triazols) **54** and **55** in a good yield (Scheme **15**) [34]. This methodology was also applied for the synthesis of new conjugates of 3,5-bis(arylidene)-4-piperidone pharmacophore with bisphosphonate moiety **56** using a 1,2,3-triazole ring as a linker (Scheme **16**) [37, 38].

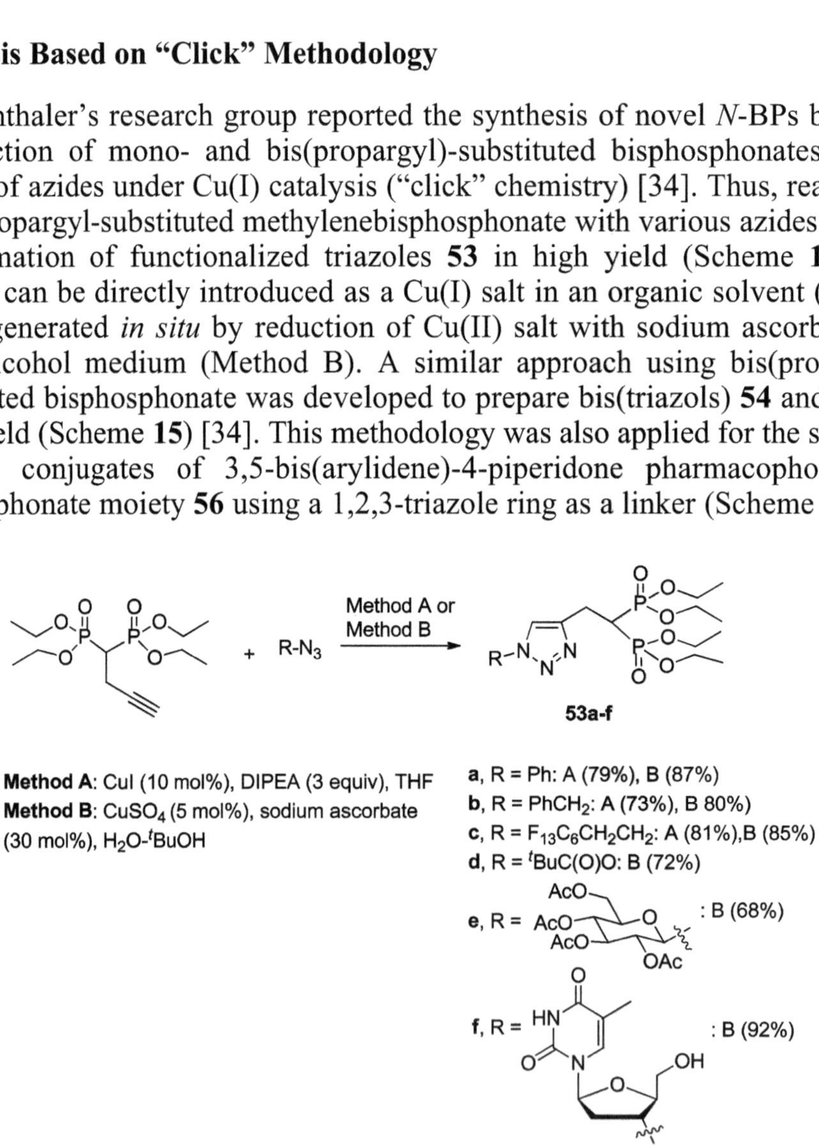

Method A: CuI (10 mol%), DIPEA (3 equiv), THF
Method B: CuSO$_4$ (5 mol%), sodium ascorbate (30 mol%), H$_2$O-^{t}BuOH

a, R = Ph: A (79%), B (87%)
b, R = PhCH$_2$: A (73%), B 80%)
c, R = F$_{13}$C$_6$CH$_2$CH$_2$: A (81%), B (85%)
d, R = ^{t}BuC(O)O: B (72%)
e, R = AcO-sugar (AcO, AcO, AcO, OAc) : B (68%)
f, R = thymidine : B (92%)

Scheme 14. Reaction of tetraethyl α-propargylmethylenebisphosphonate with azides.

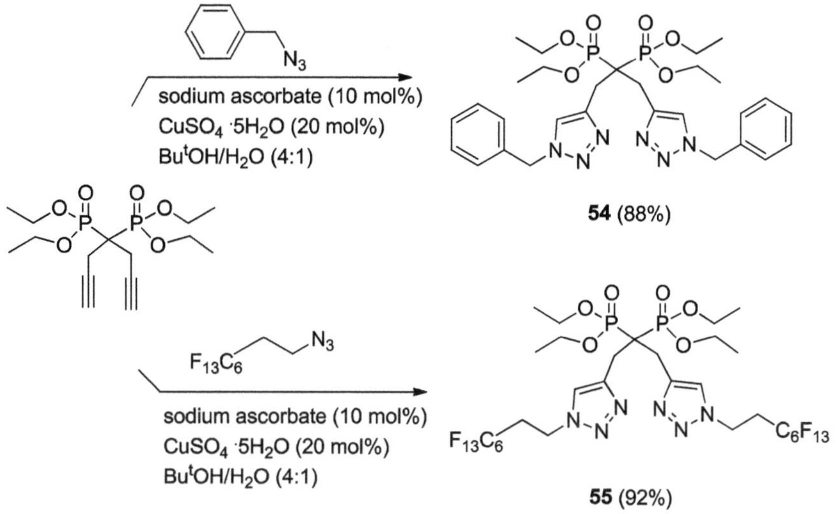

Scheme 15. Synthesis functionalized bisphosphonates **54** and **55** *via* click chemistry.

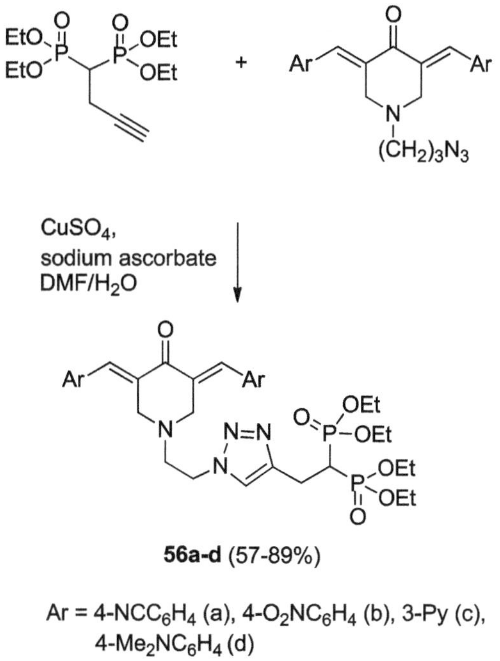

Scheme 16. Synthesis of 3,5-bis(arylidene)-4-piperidinones **56** modified with bisphosphonate groups.

Ar = 4-NCC$_6$H$_4$ (a), 4-O$_2$NC$_6$H$_4$ (b), 3-Py (c), 4-Me$_2$NC$_6$H$_4$ (d)

The antitumor activity of compounds **56a,b** was evaluated *in vitro* against the human cell lines HCT116, MCF7 and normal human embryonic fibroblasts HEF. It was shown that bisphosphonates **56a,b** possess moderate antitumor properties with the IC$_{50}$ values ranging from 5 to 7.5 μmol L^{-1} [37, 38].

Impressive examples of "click" methodology related to BPs were provided by Wiemer and co-workers [39 - 41]. They reported the synthesis of triazole-based inhibitors of geranylgeranyl diphosphate synthase **58** through click reaction between acetylenic bisphosphonate and isoprenoid azides (Scheme **17**). Interestingly, whether geranyl bromide or neryl bromide was employed as the starting material for the synthesis of the triazole bisphosphonates, the isomer ratio of the resulting product was approximately 2:1 in favor of the *E*-isomer [39]. Later, the same researchers have accomplished the stereoselective synthesis of the bisphosphonate triazoles **59** and **60** *via* the use of epoxy azides for cycloaddition reaction followed by regeneration of the desired olefin (Scheme **18**). The neryl triazole **59** was found to be a potent and selective inhibitor of geranylgeranyl diphosphate synthase (GGDPS) [40]. In parallel, synthesis of triazole bisphosphonate **62** has resulted in identification of even more potent inhibitor GGDPS (Scheme **19**). Compound **62** potently disrupts protein geranylgeranylation in cells at concentrations as low as 50 nM and induces cytotoxic effects at submicromolecular concentrations, suggesting potential clinical utility [41].

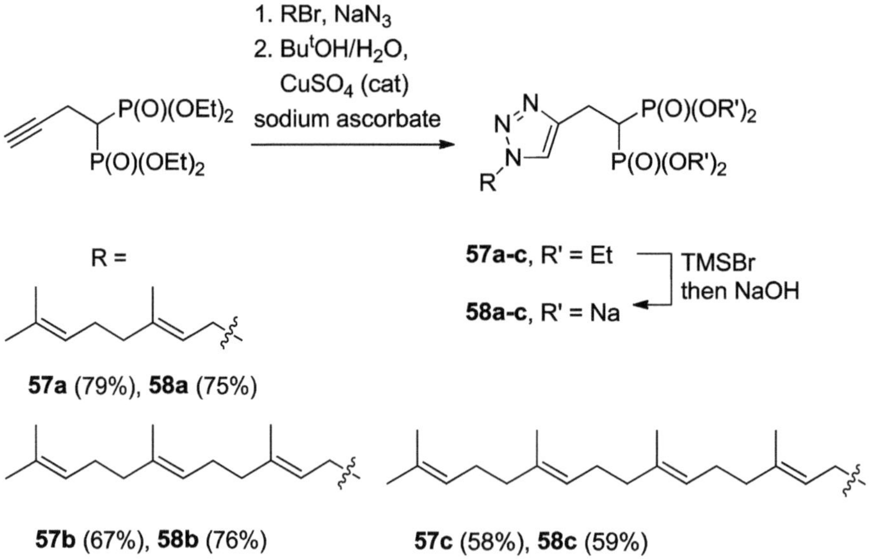

Scheme 17. Synthesis of the isoprenoid triazole bisphosphonates.

Radical Based Approach

In 2003, Gagosz and Zard showed that xanthate derivatives of bisphosphonates **63** add efficiently to various functionalized alkenes to give the corresponding adducts **64** *via* a radical chain reaction initiated by a small amount of lauroyl peroxide (Scheme **20**) [42]. The radical process tolerates many functional groups providing the expected products in yield ranging from 71% to 82%. Moreover, the synthetic utility of xanthate bisphosphonates can be extended through their further radical transformations. Thus, refluxing a solution of **64g** in 1,2-dichloroethane in the presence of peroxide induced ring-closure onto the benzene ring to give the indoline **65** in 62% yield (Scheme **21**) [42].

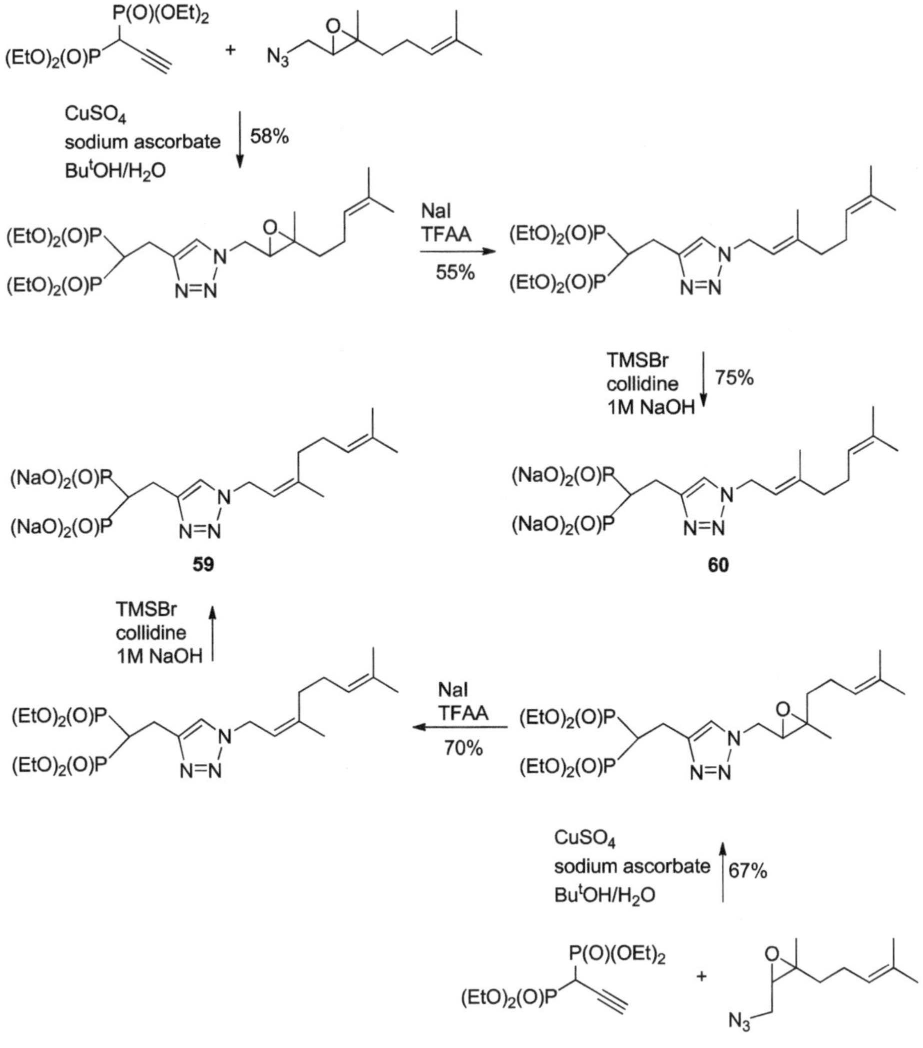

Scheme 18. Synthesis of neryl (**59**) and geranyl (**60**) triazole bisphosphonates.

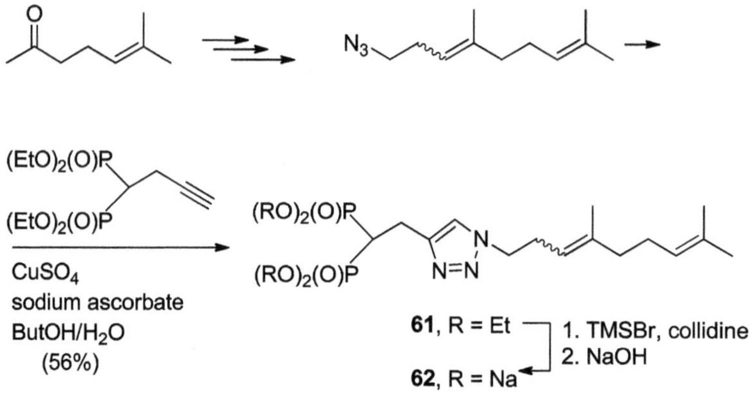

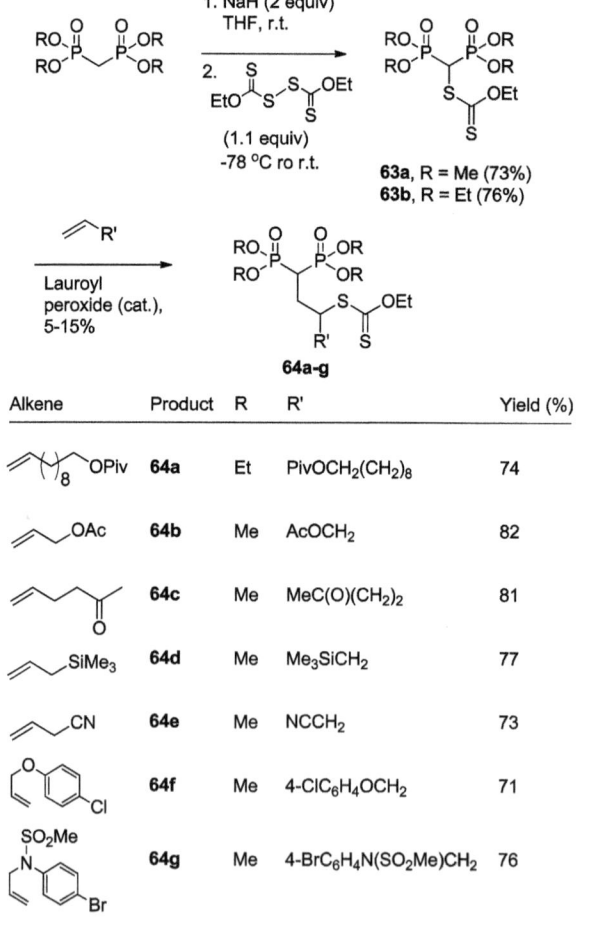

Scheme 19. Synthesis of triazole **62** as a ~3:1 mixture of *E*- and *Z*- olefin isomers.

Alkene	Product	R	R'	Yield (%)
	64a	Et	$PivOCH_2(CH_2)_8$	74
	64b	Me	$AcOCH_2$	82
	64c	Me	$MeC(O)(CH_2)_2$	81
	64d	Me	Me_3SiCH_2	77
	64e	Me	$NCCH_2$	73
	64f	Me	$4\text{-}ClC_6H_4OCH_2$	71
	64g	Me	$4\text{-}BrC_6H_4N(SO_2Me)CH_2$	76

Scheme 20. Radical additions of xanthate bisphosphonates to various alkenes.

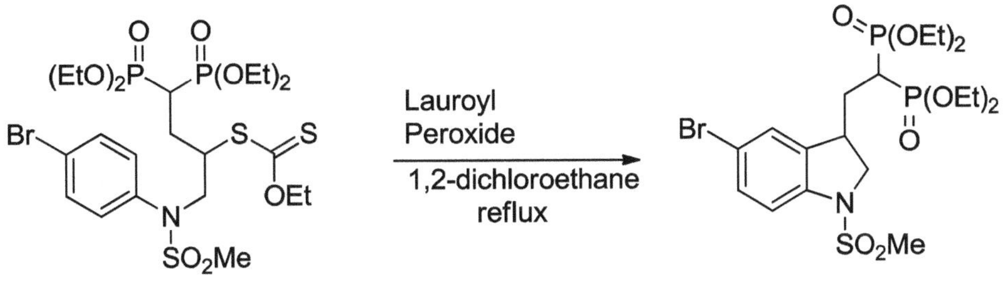

Scheme 21. Synthesis of bisphosphonate **65** *via* peroxide induced ring closure of **64g**.

Miscellaneous Reactions

Earlier work on the direct synthesis of 4-hydroxy-3,5-di-*tert*-butylphenyl-substituted bisphosphonate from 4-hydroxy-3,5-di-*tert*-butylbenzaldehyde and two equivalents of diethyl phosphite in the presence of diethylamine as a base has been extended recently. A slightly modified procedure (sodium diethyl phosphite in dioxane solution) was used for the preparation of 4-hydroxyphenyl derivative **66**. The standard dealkylation of tetraethyl ester **66** with Me$_3$SiBr and subsequent methanolysis of the silyl ester resulted in the corresponding bisphosphonic acid **67** in high yield (Scheme **22**) [43]. In addition, application of the method for synthesis of calix [4]arenes **68** and **69** bearing one or two methylenebisphosphonic acid fragments has been described. Both compounds displayed stronger inhibition of calf intestine alkaline phosphatase than simple methylenebisphosphonic acid and 4-hydroxyphenyl methylenebisphosphonic acids. Calixarene **69** bearing two methylenebisphosphonate motifs displays the strongest inhibition with $K_i = 0.38$ μM and $K'_i = 2.8$ μM [43 - 45].

In 2010, Jaffrès and co-workers reported an unexpected base-induced [1, 4]-phospha-Fries rearrangement that gives rise to the formation of a tetraethyl methylenebis(thiophosphonate) derivative (Scheme **23**) [46]. The phosphorylation of 2-methylresorcinol with *O,O*-diethylchlorothiophosphate in the presence of trimethylamine and subsequent treatment of thiophosphate **70** with ^{t}BuLi in THF afforded bisphosphonate **71**. The precise mechanism for the reaction is still unclear but should involve a double [1, 4]-phospha-Fries rearrangement which is only observed when the two thiophosphate groups are present in *ortho*-position of the methyl group.

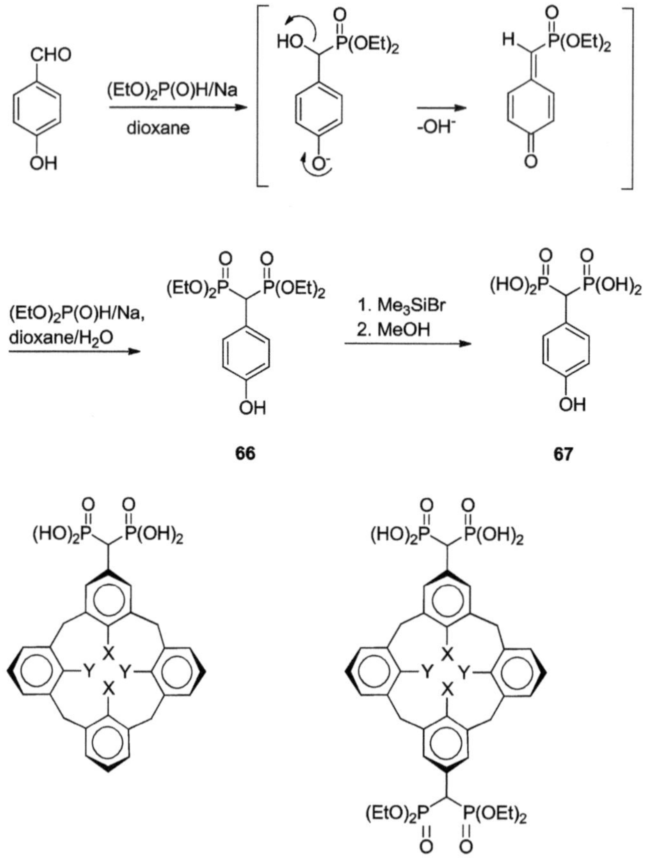

66 **67**

68 **69**

X = OH, Y = OPr

Scheme 22. Synthesis 4-hydroxyphenyl-substituted methylenebisphosphonates.

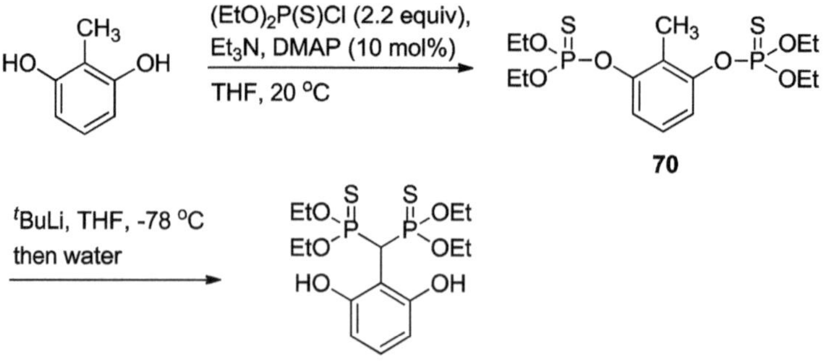

70

71 (29%)

Scheme 23. Base-induced [1, 4]-phospha-Fries rearrangement.

Alkylidene or arylidene substrates, whose double bonds are activated towards a nucleophilic attack, can undergo addition of bisphosphonate carbanions. This approach has been used for the preparation of a series of heteryl functionalized bisphosphonic acids. For example, 1,1-dicyano-2-(2-thienyl)ethylene reacted with tetraethyl methylenebisphosphonate (two-fold excess based on the alkene) in DMF containing excess NaH to give the corresponding bisphosphonate **72** (Scheme **24**) [47]. A variety of heteryl substituted electron-deficient alkenes were tested giving the desired bisphosphonates in reasonable yields [20, 48, 49]. Interestingly, the treatment of 5-arylidene-2-thioxo-4-thiazolidinones **73** with the bisphosphonate carbanion afforded the compound **75** as the major products (~47%) along with the monophosphonate **74** (~25% yield). However, when **73** and $CH_2(PO_3Et_2)_2$ were mixed in 1:1.5 ratio in DMF containing EtONa and heated in a microwave oven, the bisphosphonates **75** were exclusively obtained (> 85% yield) (Scheme **25**). The effects on bone resorption and the inflammatory joint diseases of the synthesized bisphosphonic acids **76** have been studied [48].

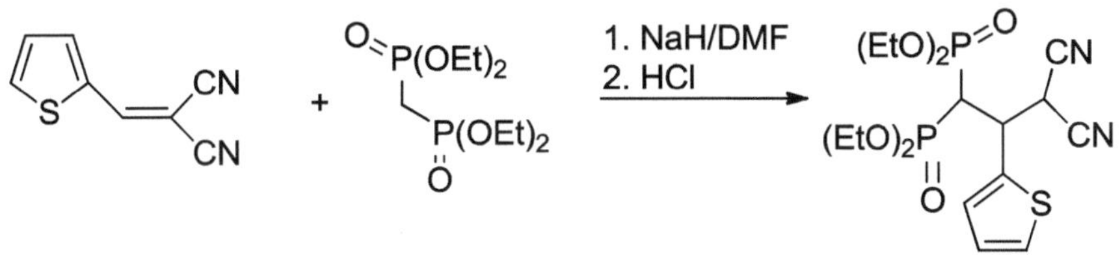

Scheme 24. The Michael addition of tetraethyl methylenebisphosphonate to 1,1-dicyano-2-(2-thienyl)-ethylene.

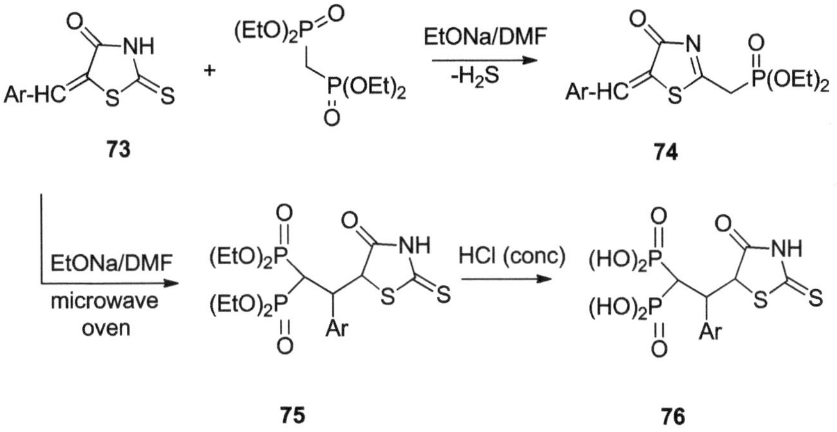

Ar = Ph, 4-Me$_2$NC$_6$H$_4$, 2-O$_2$NC$_6$H$_4$, 4-O$_2$NC$_6$H$_4$, 2-ClC$_6$H$_4$, 2-HOC$_6$H$_4$

Scheme 25. Synthesis of *gem*-diphosphono-substituted thiazolidinones.

En elegant route to medicinally useful 2-*gem*-bisphosphonate-substituted benzotriazoles *via* a copper-promoted domino condensation / S-arylation / heterocyclization has been developed be Chinese researches (Scheme **26**) [50]. Bis(diethoxyphosphoryl)methanide, generated *in situ* in this cascade three-component process, reacts with CS_2 and 2-iodoanilines in the presence of $CuCl_2$ to give the compounds **77** as a tautomeric mixture of imine form **77′** and enamine form **77″**.

SYNTHESIS *VIA* ALKENYLIDENE-1,1-BISPHOSPHONATES

Alkenylidene-1,1-bisphosphonates are definitely one of the most important types of electron-deficient alkene substrates for the synthesis of elaborated bisphosphonate structures [51, 52]. Vast number publications describe the Michael addition reactions of these compounds with carbon, nitrogen, oxygen, sulfur, and phosphorus nucleophiles. Another important feature of alkenylidene-1,1-bisphosphonates is their ability to undergo polar cycloadditions giving rise to carbocyclic and heterocyclic compounds containing the bisphosphonate moiety. This section presents the potential of alkenylidene-1,1-bisphosphonates as Michael acceptors in the synthesis of organyl-substituted bisphosphonates.

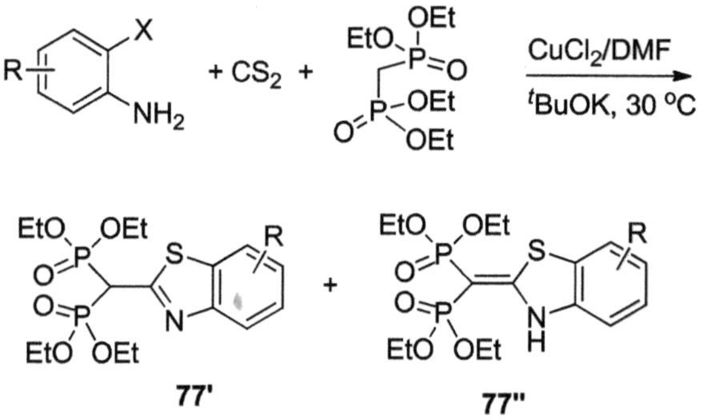

2-Haloaniline	Product*	Yield, % (77′/77″)
1-H_2N-2-IC_6H_4	**77a**	75 (1/3.3)
1-H_2N-2-I-4-MeC_6H_3	**77b**	73 (1/2.4)
1-H_2N-2-I-5-MeC_6H_2	**77c**	65 (1/4.0)
1-H_2N-2-I-4-$MeOCH_6H_3$	**77d**	45 (1/0.94)
1-H_2N-2-I-4-FC_6H_3	**77e**	44 (1/1.8)

*Optimized conditions: ^{t}BuOK (3 equiv), CS_2 (1.2 equiv), $CuCl_2$ (1 equiv)

Scheme 26. Selected examples of an approach to bisphosphonate-substituted benzotriazoles *via* a copper-promoted three-component reaction.

Addition Reactions

The Michael-type additions of various nucleophiles to the double bond of either non-esterified ethenylidenebisphosphonic acid, $H_2C=C(PO_3H_2)_2$, or tetraalkyl ethenylidenebisphosphonates, $H_2C=C(PO_3R_2)_2$, proved to be a versatile method for preparation of simple and functionalized 1-alkyl-substituted methylenebisphosphonic acids and their derivatives.

Reactions with Organometallic Reagents

Addition of Grignard reagents RMgHlg to ethenylidenebisphosphonate esters is the starting point of a highly effective synthesis of substituted methylenebisphosphonates of the general formula $RCH_2CH(PO_3R^1{}_2)_2$. Thus, reactions of ethenylidenebisphosphonate ester **78a** with organomagnesium reagents $Me(CH_2)_nMgHlg$ (n = 1-10) provides an easy route to the corresponding bisphosphonates $Me(CH_2)_nCH_2CH(PO_3Et_2)_2$ derived from fatty acids. Some of the thus-obtained bisphosphonic acids proved to be potent inhibitors against the protozoan *Trypanosoma cruzi*, which causes Chagas disease [53]. The addition reactions easily occur even if functionalized Grignard reagents are used (Scheme 27) [54 - 56]. Another illustrative example is the synthesis of nitrooxy NO-donor bisphosphonate **80** *via* the reaction of **78a** with the 3-butenylmagnesium bromide followed by the action of $AgNO_3$ and I_2 on unsaturated ester **79** (Scheme 28) [57]. The corresponding bisphosphonic acid displays affinity for the bone and inhibits the differentiation of pre-osteoclastic cells to functional osteoclasts by a prevalent NO-mediated action [57].

When the reaction of **78a** with tribromomethyllithium was performed in THF at -78 °C, the adduct **81** was obtain in 58% yield. The later cyclized under basic conditions to give the 1,1-dibromo-2,2-bis(diethoxyphosphoryl)cyclopropane **82** (Scheme **29**) [58].

Water enhanced synthesis of β-aryl-substituted bisphosphonates **83** *via* Rh(I) mediated conjugate addition of arylboronic acids to ethenylidenebisphosphonate esters was devised by Bianchini *et al*. (Scheme 30) [59]. The optimized catalytic system was applied to ethenylidenebisphosphonate **78a** on a 0.2 mmol scale with arylboronic acid catalyzed by 4.8 mol % of $[Rh(COD)Cl]_2$ in water-dioxane (9:1). The reaction well tolerates the presence of alkoxy, nitro, and aldehyde functional groups or halogen atoms in the phenyl and naphthyl rings. The isolated new bisphosphonate tetraesters were subjected to deprotection by treatment with $TMSBr/H_2O$.

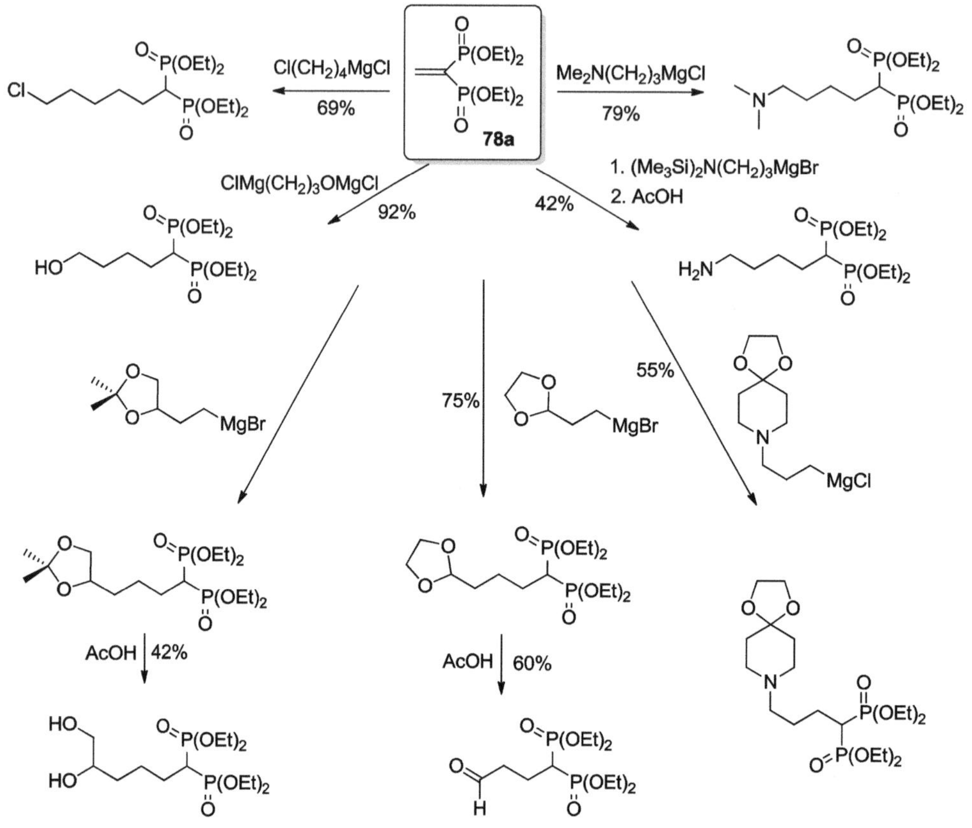

Scheme 27. Addition reactions of tetraethyl ethenylidenebisphosphonate with organometallic reagents.

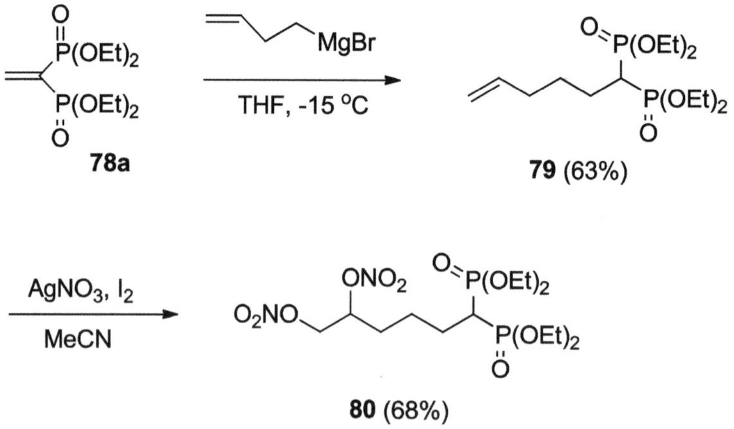

Scheme 28. Synthesis of bisphosphonate containing nitrooxy NO-donor function.

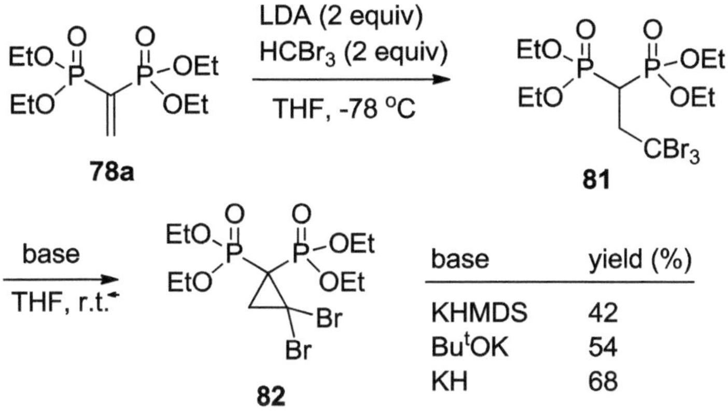

base	yield (%)
KHMDS	42
Bu^{t}OK	54
KH	68

Scheme 29. Synthesis of 1,1-dibromo-2,2-bis(diethoxyphosphoryl)cyclopropane.

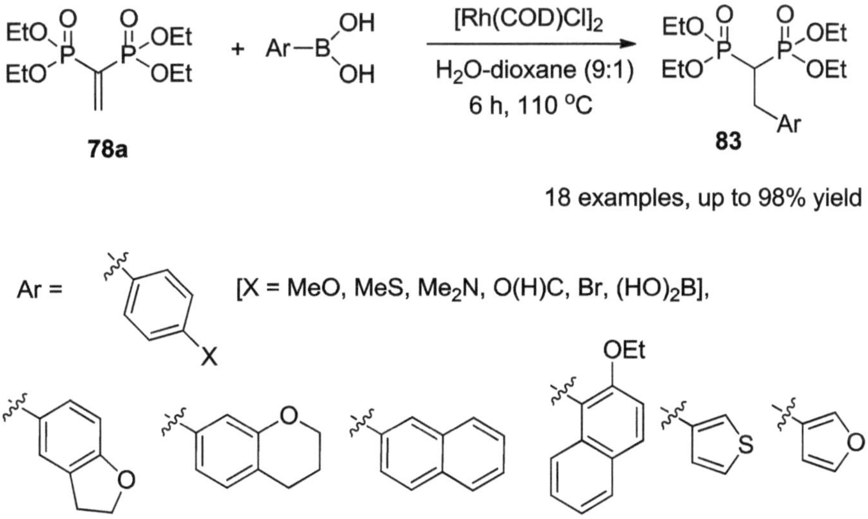

Scheme 30. Conjugate addition of arylboronic acids to tetraethyl ethenylidenebisphosphonate mediated by Rh(I) catalyst.

Base-Promoted Addition of Carbon Nucleophiles

Sturtz and Guervenou have reported the synthesis of cortisone-bisphosphonate conjugates for the inhibition of bone resorption from butanoic acid derivative $HO_2C(CH_2)_2CH(PO_3Et_2)_2$ using standard peptide coupling technique. They derived the above mentioned bisphosphonate from ethenylidenebisphosphonate ester **78a** and diethyl malonate anion followed by basic hydrolysis and decarboxylation [60]. Later on, this reaction sequence was shorten using the Meldrum's acid as a preactivated carboxylic acid equivalent. Addition of Meldrum's acid to **78a** using DBU as the basic mediator provided essentially pure bisphosphonate derivative

84. Coupling **84** with pregnenolone by heating in toluene gave steroidal bisphosphonate conjugate **85**. The transformation was also successful for other steroids such as oestrone and *trans*-androsterone, to give compounds **86** and **87**, respectively (Scheme **31**) [61, 62].

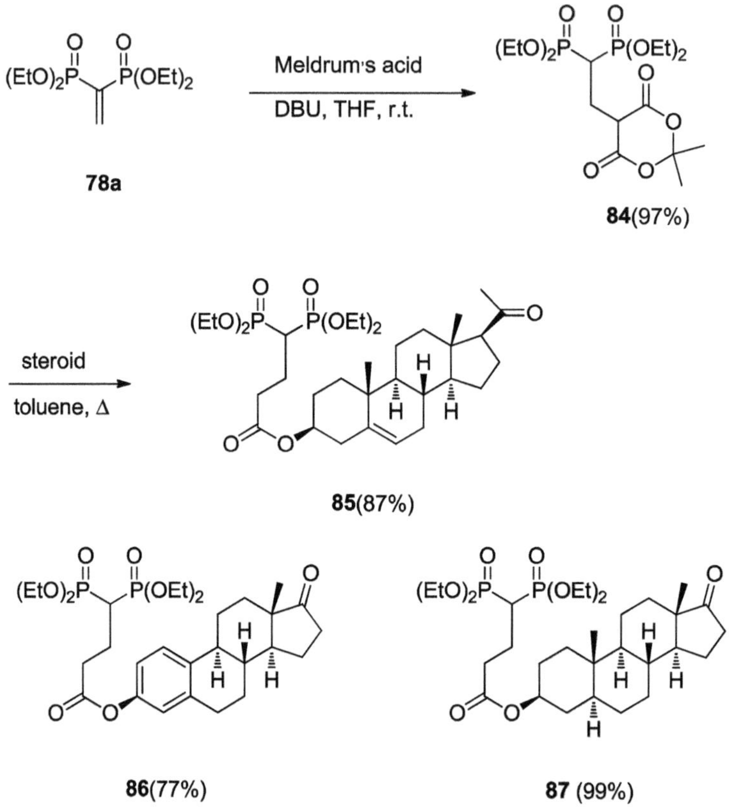

Scheme 31. Synthesis of steroid-bisphosphonate conjugates.

An interesting application of the Michael addition to the preparation of quaternary amino acids **89** containing a bisphosphonate moiety was reported by Albrecht and co-workers [63]. They developed two-step reaction sequence consisting of Michael addition of α-substituted azlactones to tetraethyl ethenylidenebisphosphonate followed by azlactone ring opening under acidic conditions (Scheme **32**). Initial attempts to develop an enantioselective version of the reaction employing different chiral Brønsted bases have been demonstrated.

Recently, synthesis of new indole-based bisphosphonates using tetraethyl ethylidenebisphosphonate as an enophile in *ene*-type reactions has been reported [64]. Thus, heating an equimolar mixture of the suitable Boc-protected 3-methyleneindoline and **78a** at 100 °C for 2-4 h in solvent-free conditions affords indolylalkylbisphosphonates **90** in 50-70% yield. The protecting *tert*-butoxycarbonyl group was removed by rising the temperature to 150-190 °C to give final indoles **91** (Scheme **33**). Evaluation of their chelating ability in PE/CA-PJ15 cells showed that the indole moiety plays an important role in cell permeability and metabolism properties.

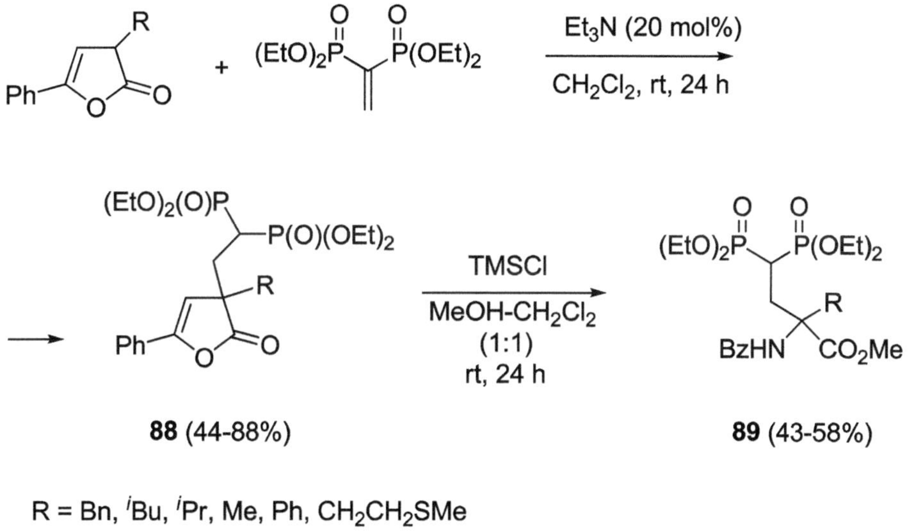
Et3N (20 mol%)
CH2Cl2, rt, 24 h
TMSCl
MeOH-CH2Cl2
(1:1)
rt, 24 h
88 (44-88%)
89 (43-58%)
R = Bn, iBu, iPr, Me, Ph, CH2CH2SMe

Scheme 32. Michael addition of α-substituted azlactones to tetraethyl ethenylidenebisphosphonate.

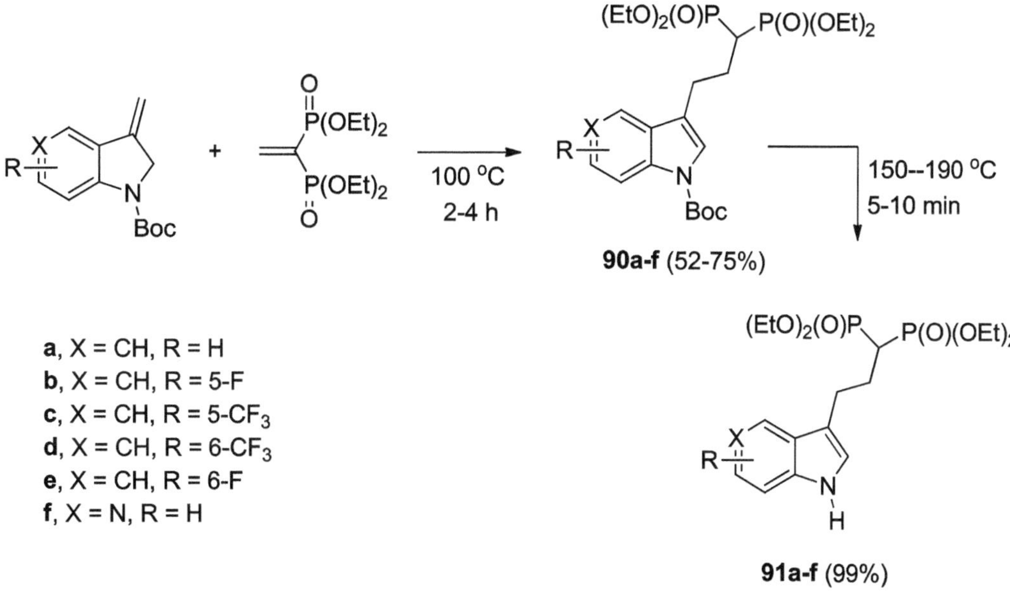
100 °C
2-4 h
150--190 °C
5-10 min
90a-f (52-75%)
a, X = CH, R = H
b, X = CH, R = 5-F
c, X = CH, R = 5-CF3
d, X = CH, R = 6-CF3
e, X = CH, R = 6-F
f, X = N, R = H
91a-f (99%)

Scheme 33. Synthesis of regiosubstituted indole-based bisphosphonates by "ene" reactions.

The Michael addition to ethenylidenebisphosphonate esters has also been conducted stereoselectively. Alexakis and co-workers described the asymmetric conjugate addition of aldehydes to bisphosphonate **78a** catalyzed by chiral amines [65]. The stereochemical outcome was examined by screening several pyrrolidine-core organocatalysts among which (*S*)-diphenylprolinol silyl ester **93** was found to induce particularly high stereocontrol. The stereoselectivity was excellent in cases of branched aldehydes but dropped when linear aldehydes such as propionaldehyde and valeraldehyde were employed (Scheme **34**).

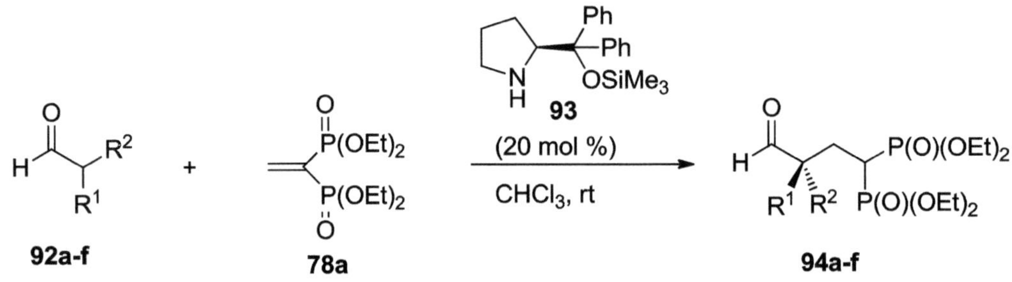

Aldehyde	R^1, R^2	Product 94 time (h)	yield (%)[a]	ee (%)
92a	^{i}Pr, H	12	80	90 (S) (+)
92b	^{t}Bu, H	12	85	97 (+)
92c	^{n}Pr, H	12	75	86 (+)
92d	Me, H	8	75	75 (+)
92e	allyl, H	12	65	46 (+)

[a] Performed with **92a-e** (3.33 mmol), **78a** (0.333 mmol), and **93** (0.066 mmol).

Scheme 34. Enantioselective organocatalytic conjugate addition of aldehydes to ethenylidenebisphosphonate **78a**.

Jørgensen and co-workers have developed the asymmetric addition of β-ketoesters **95** to ethenylidenebisphosphonates catalyzed by cheap and commercially available cinchona alkaloid derivatives [66]. During their study of the addition of 1-oxoindan-2-carboxylate (**95a**) to the bisphosphonates **78a,b** these authors identified dihydroquinine **98** as the most effective catalyst. They subsequently investigated the scope of the reaction performing a systematic study with variously functionalized indanone structutes (**95a-i**), benzofuranone (**95g**), benzothiophenone (**95h**), cyclopentenones (**95j-l**) and α-fluorinated β-ketoesters (**95m**). High yields and enantioselectivities (up to 90% ee) were achieved for a wide range of indanone-based β-ketoesters and 5-*tert*-butyloxycarbonyl cyclopentenones (Scheme 35).

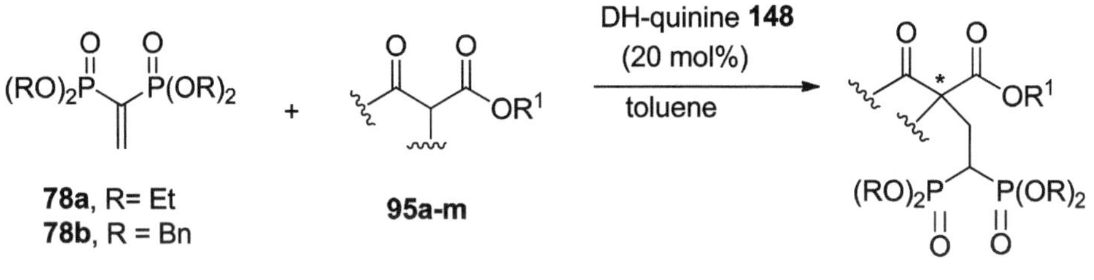

78a, R = Et
78b, R = Bn

95a-m

DH-quinine **148** (20 mol%), toluene

up to 99% ee

Selected nucleophiles (**95**) and cinchona alkaloids (**96-100**):

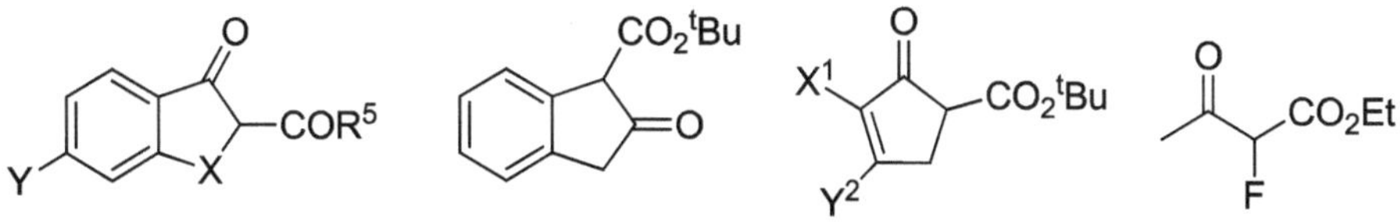

95a-h

95i

95j-l

95m

95a: R^5 = Me, X = CH_2, Y = H
95b: R^5 = Bn, X = CH_2, Y = H
95c: R^5 = ^{i}Pr, X = CH_2, Y = H
95d: R^5 = ^{t}Bu, CH_2, Y = H
95e: R^5 = ^{t}Bu, X = CH_2, Y = Cl
95f: R^5 = ^{t}Bu, X = CH_2, Y = OMe

95g: R^5 = ^{t}Bu, X = O, Y = H
95h: R = ^{t}Bu, X = S, Y = H
95j: X^1 = H, Y^2 = OMe
95k: X^1 = Me, Y2 = H
95l: X^1 = OMe, Y^2 = Me

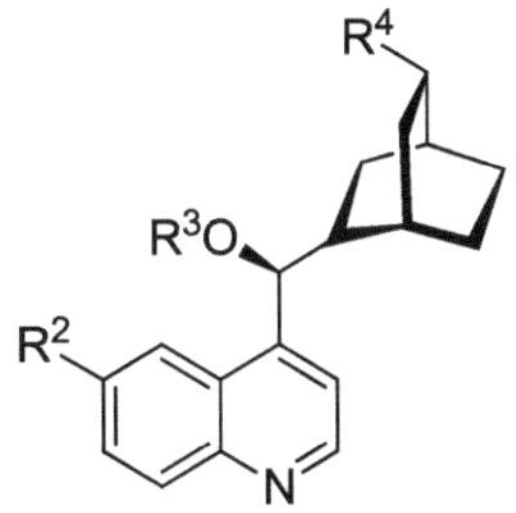

96, R^2 = OMe, R^3 = H, R^4 = vinyl
97, R^2 = H, R^3 = H, R^4 = vinyl
98, R^2 = OMe, R^3 = H, R^4 = Et
99, R^2 = OMe, R^3 = 4-Cl-Bz, R^4 = vinyl
100, R^2 = OH, R^3 = H, R^4 = vinyl

Scheme 35. Organocatalyzed Michael-type addition of β-ketoesters.

The asymmetric synthesis of γ-keto bisphosphonates by Michael addition was reported by Barros and Faisca Phillips [67]. Addition of cyclohexanones to **78a** catalyzed by (S)-(+)-1-(2-pyrrolidinylmethyl)pyrrolidine **101** and benzoic acid gave the products **102** in yields of up to 86%, dr (*cis/trans*) > 1:99 and ee of up to 83%. With cyclopentanone a *C2*-symmetric disubstituted product **103** was obtained (Scheme **36**).

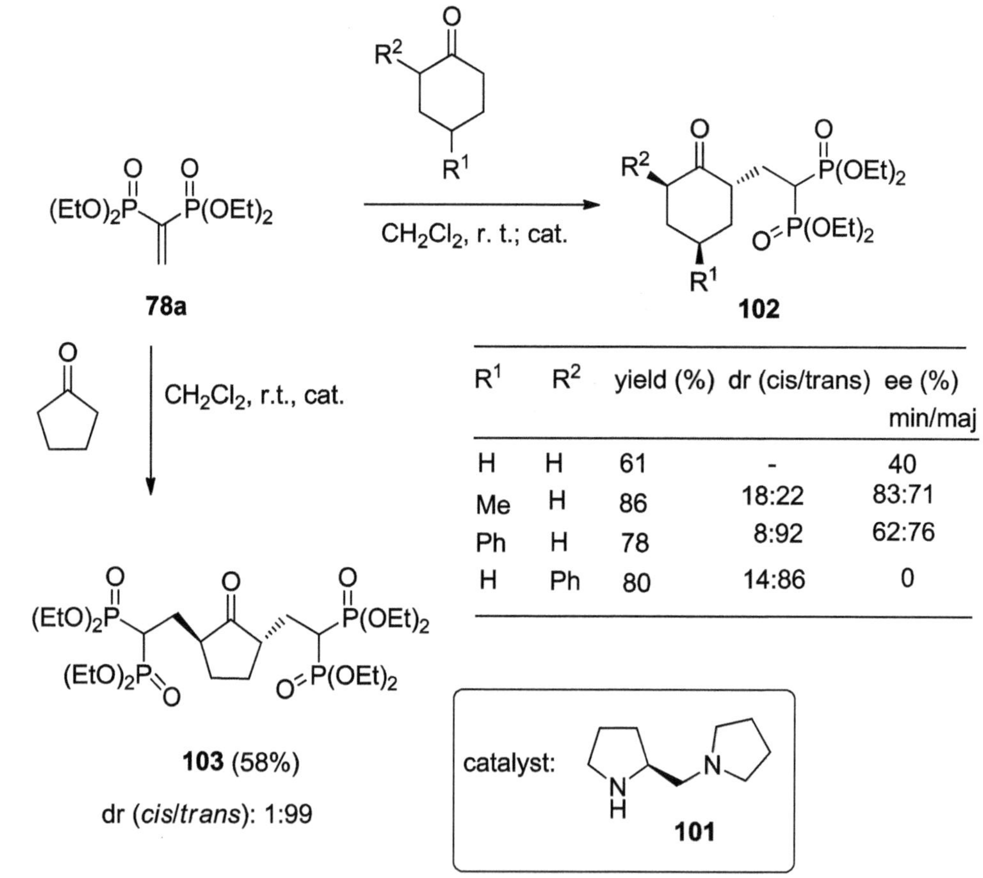

R^1	R^2	yield (%)	dr (cis/trans)	ee (%) min/maj
H	H	61	-	40
Me	H	86	18:22	83:71
Ph	H	78	8:92	62:76
H	Ph	80	14:86	0

Scheme 36. Enamine catalysis in the synthesis of chiral bisphosphonates.

Shibasaki and co-workers [68] reported the catalytic asymmetric synthesis of nitrogen-containing bisphosphonates **106** using α-substituted nitroacetates **104** as nitrogen-containing reagents and dinuclear Ni$_2$-Shiff base complex **105** as catalyst in good to excellent yields (65 - 94%) and high enantioselectivities (76 - 93%) (Scheme **37**). The nitro group in **106a** was reduced using NiCl$_2$ and NaBH$_4$ to afford bisphosphonates with an α-amino ester group in 92% yield.

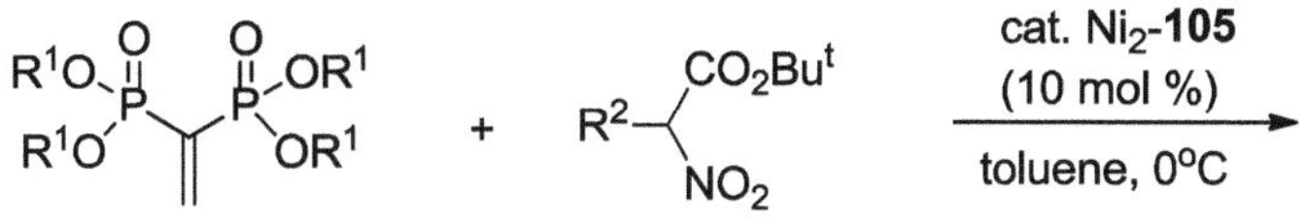

106	R¹	R²	yield (%)	ee (%)
106a	Et	Me	83	82
106b	Et	Et	94	82
106c	Et	Pr	81	84
106d	Et	Bn	88	82
106e	Et	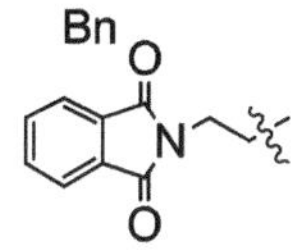 (phthalimidomethyl)	81	93
106f	Bn	Me	69	81
106g	allyl	Me	65	76

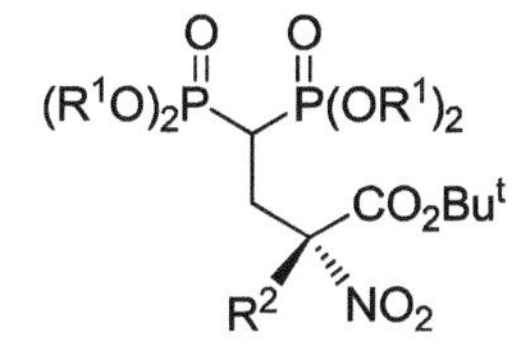

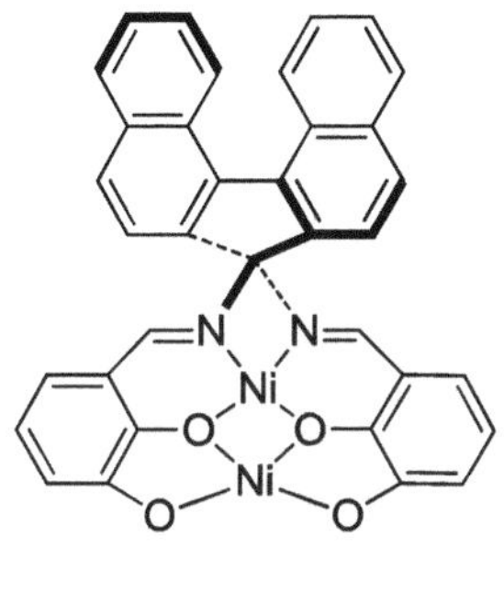

Scheme 37. Catalytic asymmetric 1,4-addition of nitroacetates to ethenylidenebisphosphonates.

Wang and co-workers were able to prepare unnatural amino acids **109** bearing bisphosphonate moiety in good yields with excellent diastereoselectivity (>99:1) and high enantioselectivities (89-99% ee) through an asymmetric reaction between *N*-benzylidene glycine methyl ester **107** and a variety of alkylidenebisphosphonates catalyzed by Cu(MeCN)$_4$/TF-BiphamPhos [(*R*)-**108**] system. Representative examples of this approach are illustrated in (Scheme 38)[69].

Inspired by the success of Wang's work, Fukuzawa and co-workers applied the Ag(I)/TCF complex to the catalysis of the asymmetric Michael addition of a glycine imino ester **110** to alkenylidenebisphosphonates. Under optimized conditions (base Cs$_2$CO$_3$, temperature -20 °C, THF solvent), the corresponding adducts **111** were prepared in good yields with high enantioselectivities (Scheme 39) [70].

107

109 (syn/anti > 99:1)

109		yield (%)	% ee
a	R = H	75	89
b	R = Ph	90	94
c	R = 4-FC_6H_4	90	99
d	R = 4-ClC_6H_4	85	95
e	R = 4-$O_2NC_6H_4$	91	93
f	R = 2-naphthyl	85	96
g	R = 2-furyl	82	96
h	R = $CyCH_2$	72	89
i	R = H	75	89

(*R*)-**108**

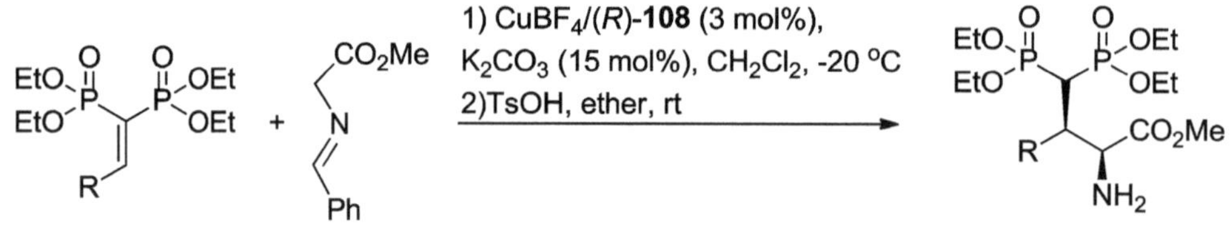

Scheme 38. Cu(I)/TF-BiphamPhos-catalyzed asymmetric Michael additions of imino ester **107** and various alkenylidenebisphosphonates.

111a-l

TCF:

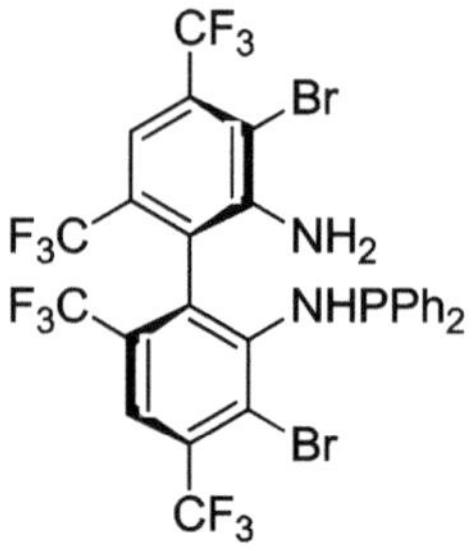

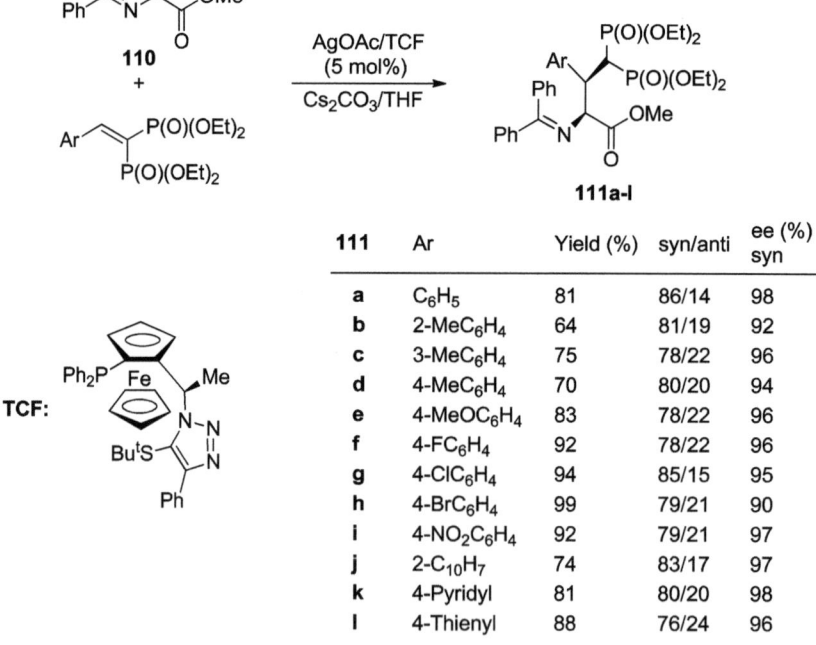

111	Ar	Yield (%)	syn/anti	ee (%) syn
a	C_6H_5	81	86/14	98
b	2-MeC_6H_4	64	81/19	92
c	3-MeC_6H_4	75	78/22	96
d	4-MeC_6H_4	70	80/20	94
e	4-$MeOC_6H_4$	83	78/22	96
f	4-FC_6H_4	92	78/22	96
g	4-ClC_6H_4	94	85/15	95
h	4-BrC_6H_4	99	79/21	90
i	4-$NO_2C_6H_4$	92	79/21	97
j	2-$C_{10}H_7$	74	83/17	97
k	4-Pyridyl	81	80/20	98
l	4-Thienyl	88	76/24	96

Scheme 39. The asymmetric Michael addition of a glycine imino ester to alkenylidenebisphosphonates.

Reactions with Alcohols, Thiols and Amines

Several 2-alkoxyethylene-1,1-bisphosphonates have been prepared by the addition of alcohols in the presence of triethylamine or sodium alkoxides to the compound $H_2C=CH(PO_3Et_2)_2$ (**78a**) [71]. Reaction conditions were found to be critical; best results were obtained by the low temperature addition of sodium alkoxide to ethenylidenebisphosphonates in alcoholic solution. Thus, reaction of **78a** with RONa (R = Me, Et, ^{i}Pr) at 0° C gave the corresponding 2-alkoxyethylene-1-1-bisphosphonates in good yields. However, when bisphosphonate **78a** and RONa were allowed to react at room temperature for an extended period, the main product was identified as 2-alkoxyethylphosphonate derivatives. The postulated mechanism for the formation of the latter is depicted in Scheme **40** [71].

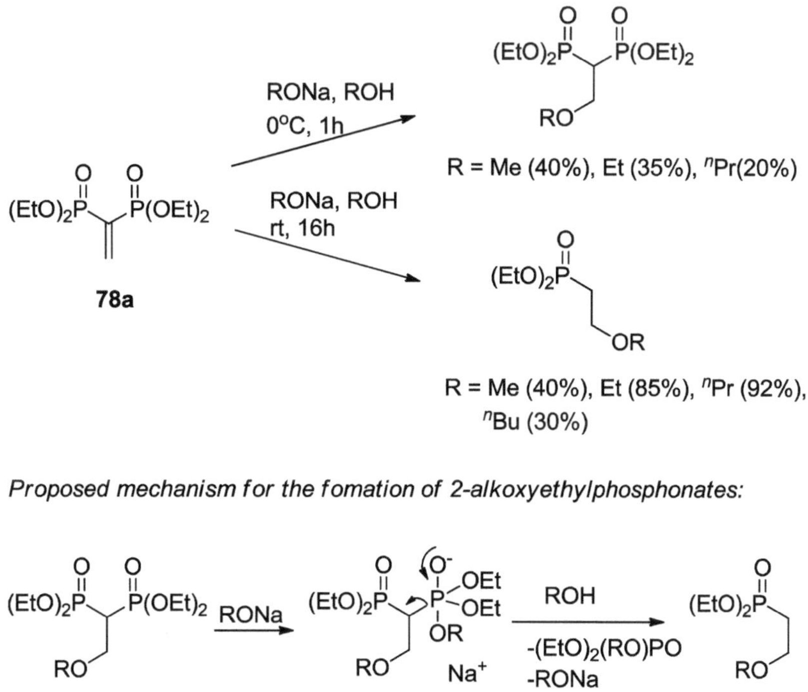

Scheme 40. Addition reaction of tetraethyl ethenylidenebisphosphonate with alcohols.

When tetraethyl ester **78a** is used in the reaction with primary, secondary and heterocyclic amines, the expected adducts are formed in good or excellent yields (Scheme **41**) [6, 72 - 78]. The following hydrolysis with concentrated hydrochloric acid gives the corresponding bisphosphonic acids. Oldfield and co-workers reported the inhibition of a recombinant vacuolar pyrophosphatase, *Tb*VSP1, expressed in *Escherichia coli* by a panel of synthesized bisphosphonates. The IC_{50} values were found to vary from ~2 to 850 µM [72].

Effects of these compound on the growth of *Entamoeba histolytica* and *Plasmodium* species *in vitro* and *in vivo* have also been studied [73].

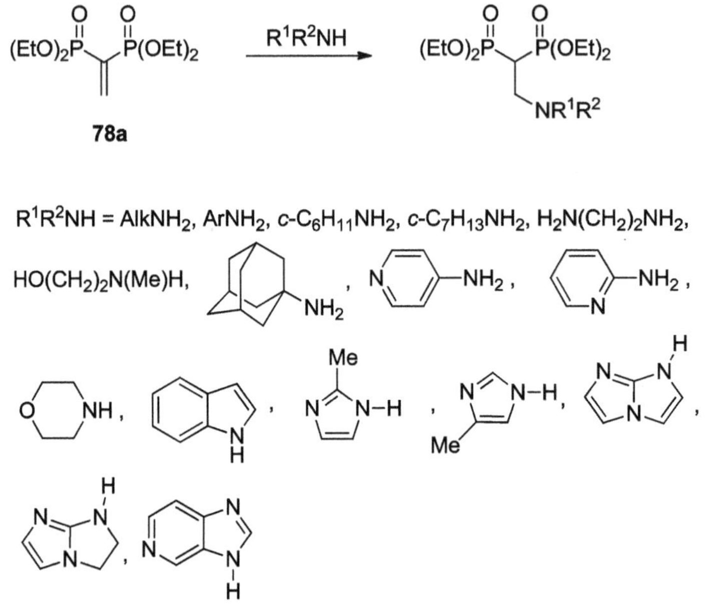

Scheme 41. Michael-type addition reaction of **78a** with alkyl-, aryl- and hererylamines.

Kukhar and co-workers have reported a mild and convenient synthesis of 2-aminoethylene-1,1-bisphosphonic acids *via* the Michael addition of amines to easily available tetrakis(trimethylsilyl) ethenylidenebisphosphonate **78c** (Scheme **42**) [79]. This approach avoided hydrolysis of the addition products in boiling hydrochloric acid, bromo- or iodotrimethylsilane alcoholysis, or catalytic hydrogenolysis of tetrabenzyl bisphosphonate esters.

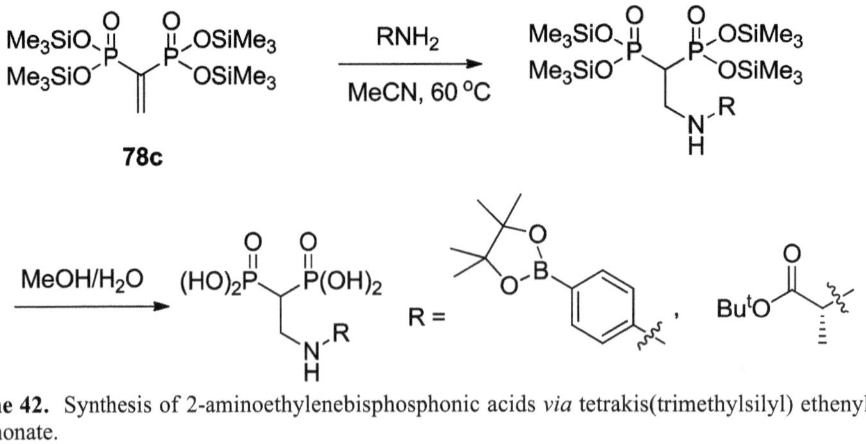

Scheme 42. Synthesis of 2-aminoethylenebisphosphonic acids *via* tetrakis(trimethylsilyl) ethenylidenebisphosphonate.

A wide variety of dipodal and tripodal amines (*e.g.***112-114**) on treatment with **78a** give the corresponding addition products. The latter were quenched by addition of acetic anhydride to avoid the retro-Michael reaction and the resulting bisphosphonate esters were then cleaved by treatment with TMSBr affording the bisphosphonates-based ligands **115** (Scheme **43**). Tripodal bisphosphonates were tested for their uranium-binding properties and were found to display high UO_2^{2+}-binding capabilities with binding constants up to $K_{cond} = 10^{19.5}$ at physiological pH [80].

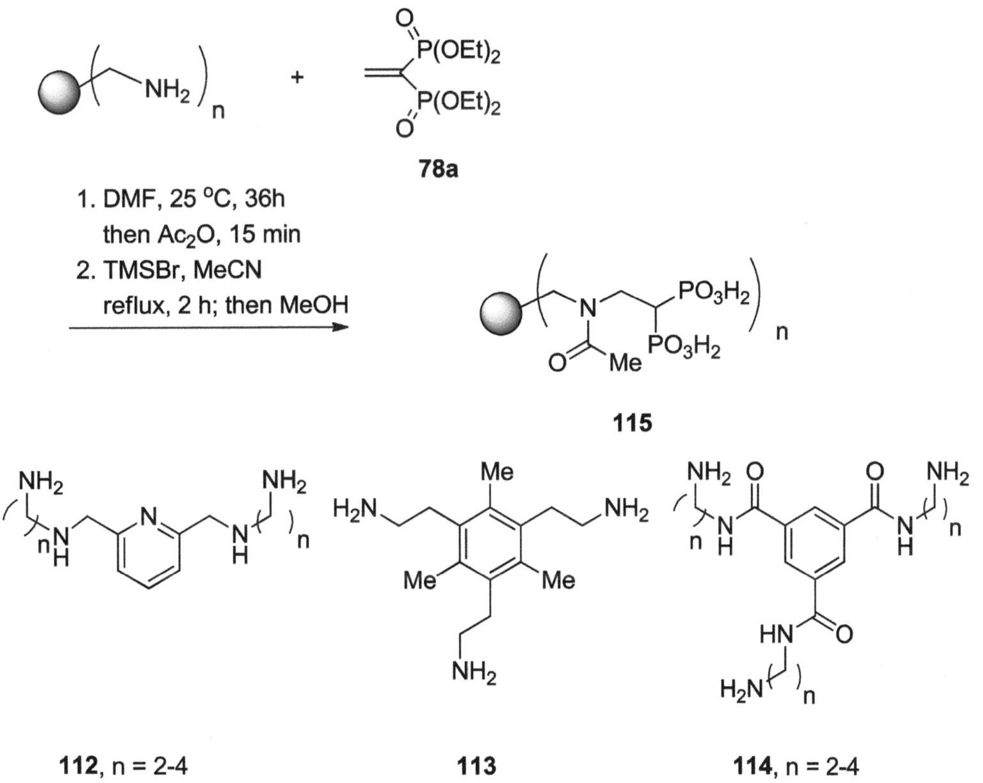

Scheme 43. Synthesis of library bisphosphonate ligands.

Bisphosphonate derivatives of three fluoroquinolone antibacterials, norfloxacin (**116**), enoxacin (**117**), and ciprofloxacin (**118**), were prepared by the reaction of ethenylidenebisphosphonate **78a** with compounds **116-118** in the presence of triethylamine. Bromotrimethylsilane-mediated hydrolysis of ester groups in the compounds **119-121** resulted in the free bisphosphonic acid derivatives **122-124** (Scheme **44**). These compounds have the ability to bind to bone and to inhibit the growth of Gram-negative bacteria [81].

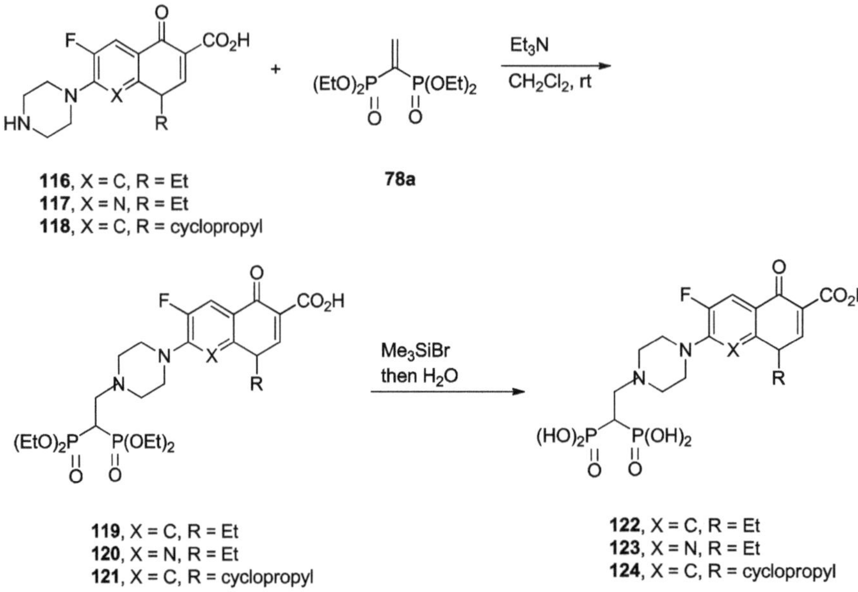

Scheme 44. Synthesis of bisphosphonate derivatives of fluoroquinolone antibacterials.

2-(Methylamino)ethanol undergoes addition reaction to the ethenylidenebisphosphonate **78a** exclusively at the nitrogen atom to give the compound HO(CH$_2$)$_2$N(Me)CH$_2$CH(PO$_3$Et$_2$)$_2$ [82]. However, in the reaction of **78a** with cystamine the thiol-containing bisphosphonate **125** is a side product (< 2% yield), while the main reaction is the addition of SH group to the C=C double bond giving the isomeric 2-(2-aminoethylthio)-substituted bisphosphonate **126**. 2-(2-Mercaptoethylamino)-derivative **129** has been made by addition of cystamine to **78a** followed by reduction of disulfide bonds with Me$_3$P (Scheme **45**). This compound is representative of novel thiol-containing bisphosphonates that are capable of efficiently reducing the *in vivo* calcification of triglycidyl-amine-treated bioprosthetic tissues [83].

Among organyl-substituted bisphosphonates sulfur-containing compounds are very interesting because of their low toxicity and high biological activity. In particular, Italian chemists have demonstrated unprecedented anti-osteoclast activity of compounds RSCH$_2$CH$_2$(PO$_3$H$_2$)$_2$, either containing an aromatic ring or an aliphatic moiety on position R [84]. An improved synthesis of these compounds relies on the use of **78a** as an electron-deficient substrate and thiols in the presence of trimethylamine as the catalyst. The free bisphosphonic acids are prepared in almost quantitative yield *via* transesterification of bisphosphonate esters with bromotrimethylsilane followed by hydrolysis with methanol.

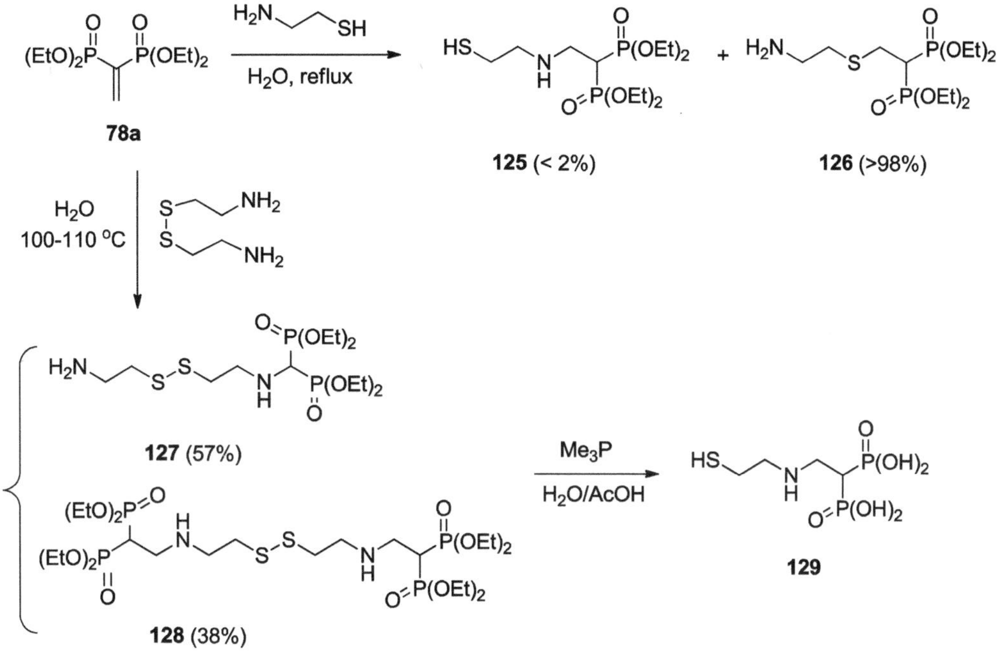

Scheme 45. Rections of ethenylidenebisphosphonate **78a** with amino thiols.

As part of efforts aimed at searching for new antiparasitic agents, Rodriguez and co-workers have synthesized several series of linear sulfur-containing methylenebisphosphonic acids **130-133** (Scheme **46**) [85]. Many of these bisphosphonate derivatives were potent inhibitors against the intracellular form of *T. cruzi* and also against *T. gondii*. Bisphosphonic acids bearing a sulfoxide unit at C-3 were also potent anti-*Toxoplasma* agents.

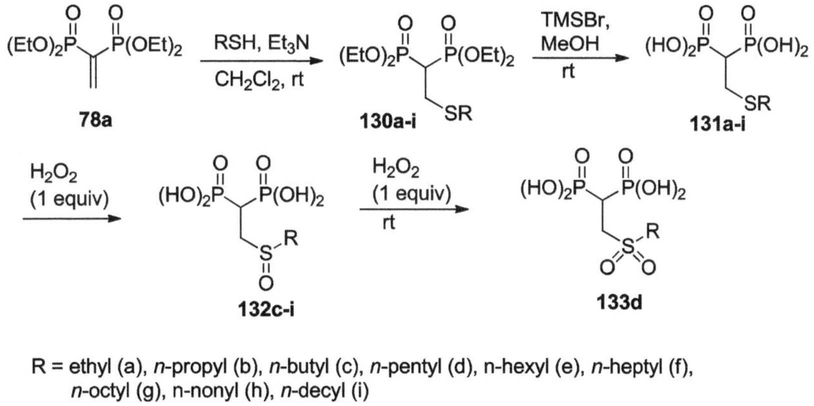

R = ethyl (a), *n*-propyl (b), *n*-butyl (c), *n*-pentyl (d), n-hexyl (e), *n*-heptyl (f), *n*-octyl (g), n-nonyl (h), *n*-decyl (i)

Scheme 46. Synthetic approach for the preparation of sulfur-containing bisphosphonic acids.

A rare example of sulfur-containing boronated bisphosphonates has been described by Shore and co-workers [86]. Organic thioethers are known to form sulfonium salts upon reaction with activated alkenes. Similarly, ethenylidenebisphosphonate **78a** is excellent Michael acceptor for methyl thioether $[MeSB_{12}H_{11}]^{2-} [Me_4N]^{+2}$ (**134**). Addition of **78a** to a solution of thioether **134** in the presence of HCl affords boronated bisphosphonate **135** that is potentially attractive boron entities for the design of therapeutics for boron neutron capture therapy of cancer (Scheme **47**).

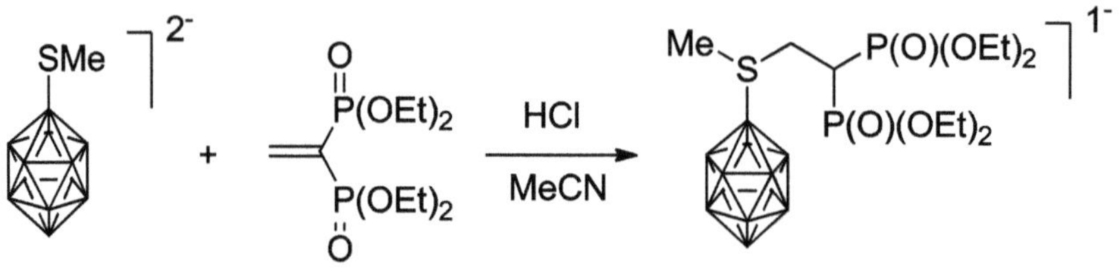

Scheme 47. Michael-type addition as a route to the boronated *gem*-bisphosphonates derived from *closo*-$[B_{12}H_{12}]^{2-}$.

A special case in the chemistry of ethenylidenebisphosphonates is the preparation of various dialkylphosphono-substituted heterocycles *via* a tandem Michael/Horner-Wadworth-Emmons reaction [87, 88]. Thus, when the sodium salt of 2-hydroxycyclohexanone is used in the reaction, the addition product undergoes cyclization to form a cyclic bisphosphonates intermediate **136**. Loss of sodium diethyl phosphate affords the monophosphonate heterocycle **137** (Scheme **48**) [88]. Similarly, treatment of the bisphosphonates **78a** with sodium phthalimide gives phosphono-substituted heterocycle **138** (27%) along with the uncyclized addition compound **139** (30%) (Scheme **49**) [87].

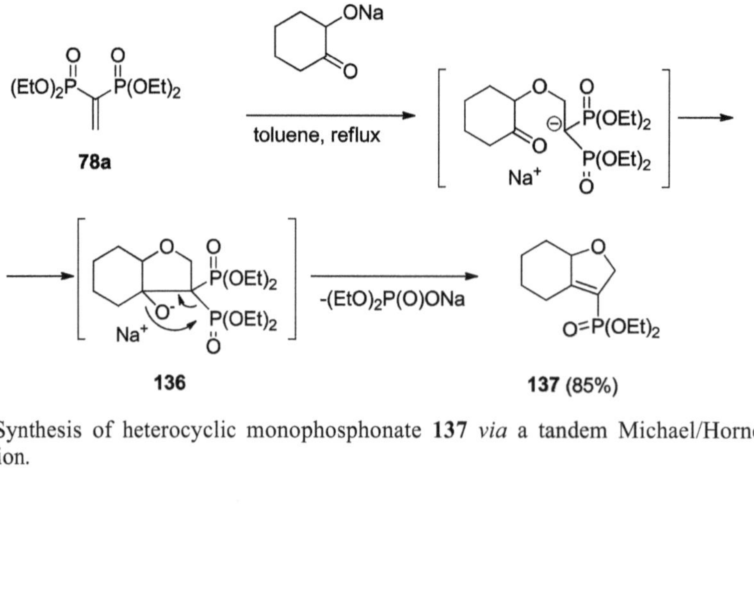

Scheme 48. Synthesis of heterocyclic monophosphonate **137** *via* a tandem Michael/Horner-Wadwort--Emmons reaction.

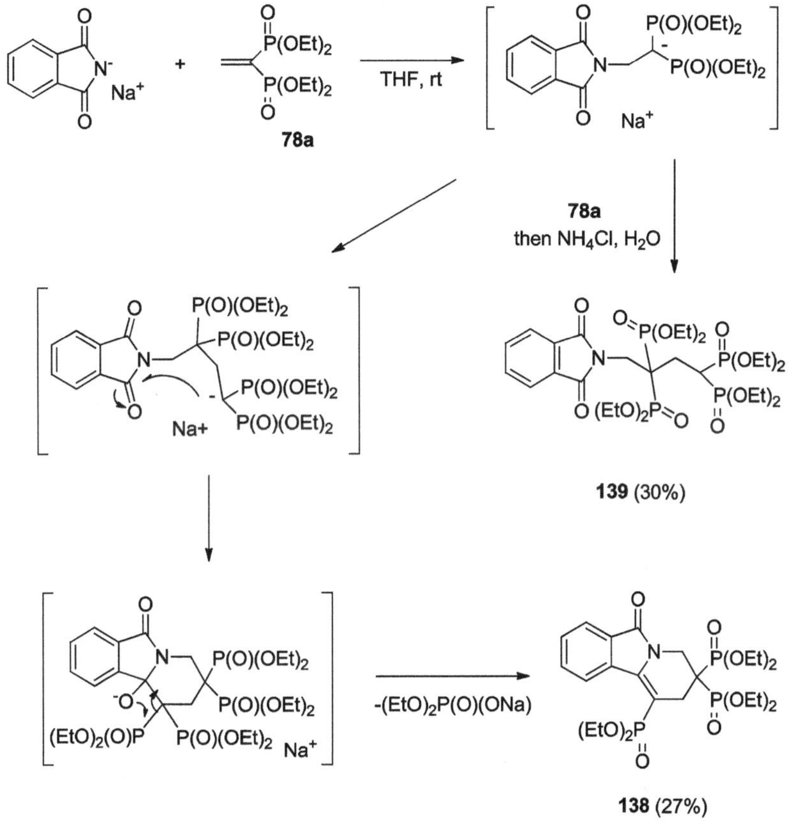

Scheme 49. Reaction of ethenylidenebisphosphonate **78a** with sodium phthalimide.

Reactions with Phosphorus Nucleophiles

Diethyl and diisopropyl phosphite react smoothly with ethenylidenebisphosphonate ester **78a** in the presence of one equivalent of ^{i}Pr$_2$NH as base to give the corresponding addition products. Other nucleophilic phosphorus reagents such as HP(O)Ph(OEt), HP(S)(OEt)$_2$, HP(BH$_3$)Ph$_2$, and HPPh$_2$ also react with **78a** in protic solvents (ethanol or methanol) in the presence of a catalytic amount of sodium alkoxide to produce compounds **140-145** (Scheme **50**) [89]. The authors have stated that the retro-Michael reactions do not occur at a room temperature, but the reverse reaction is observed at a temperature upper than 160 °C.

Cycloaddition reactions

Examples, where tetraethyl ethenylidenebisphosphonate **78a** undergoes [2+4]-cycloaddition, behaving as a dienophile, are illustrated in Scheme **51** [90]. It was noted that as a dienophile, **78a** is less reactive than α,β-unsaturated carbonyl derivatives and nitriles. However, in the presence of Lewis acids, **78a** is able to undergo cycloaddition reactions with a variety of simple dienes affording biologically important cyclohex-3-ene-1,1-bisphosphonates.

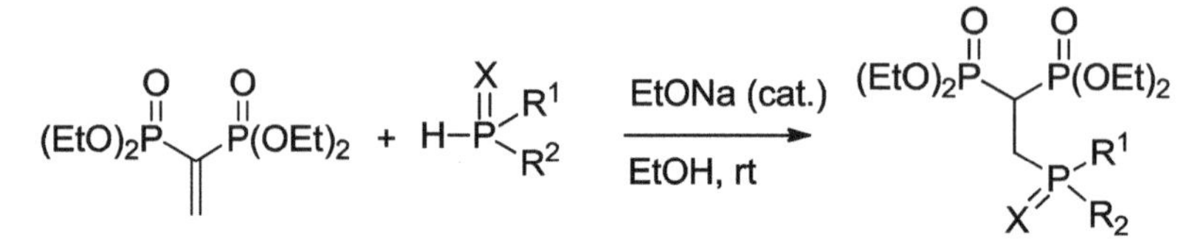

Compd	X	R^1	R^2	Yield (%)
140	O	MeO	MeO	89
141	O	EtO	EtO	93
142	O	Ph	EtO	67
143	S	EtO	EtO	75
144	2e	Ph	Ph	77
145	BH$_3$	Ph	Ph	74

Scheme 50. Michael addition of nucleophilic phosphorus derivatives on tetraethyl ethenylidene-1,1-bisphosphonate.

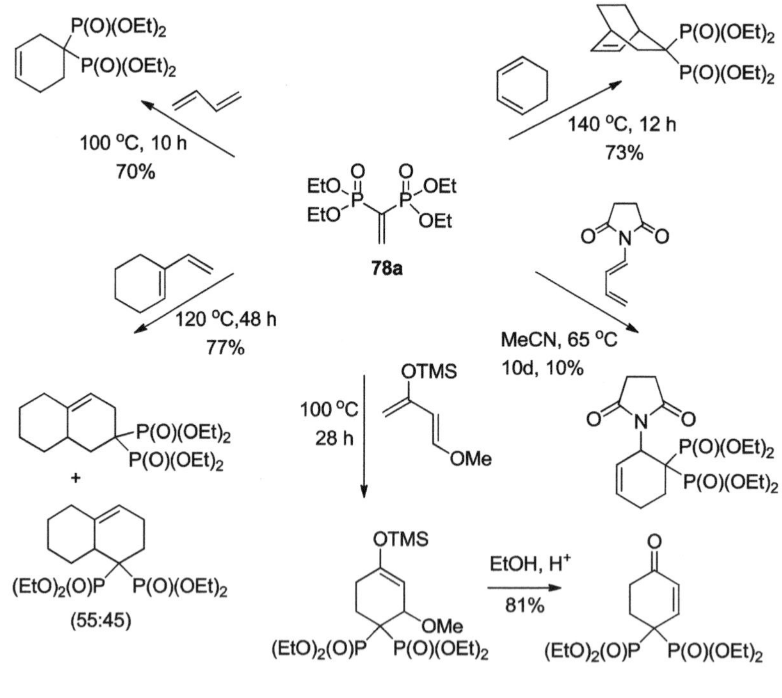

Scheme 51. [4+2]-Cycloaddition reactions of tetraethyl ethylidene-1,1-bisphosphonate with 1,3-dienes.

Cycloadditions to tetraethyl ethenylidene-1,1-bisphosphonate are also widely used in the synthesis of heterocyclic compounds containing the bisphosphonate unit. Thus, pyrazoline derivatives **146**, bisphosphonates bearing a substituted isoxazolidine ring **147**, and a series of 4,5-dihydroisoxazoles **148** have been conveniently prepared *via* reactions of **78a** with diazo compounds [91], nitrones [92], and nitrile oxides [93, 94] (Scheme **52**). Pyrazoline bisphosphonate esters **146** exhibit anti-inflammatory activity and are capable of inhibiting chronic arthritis and inflammation in animals [91].

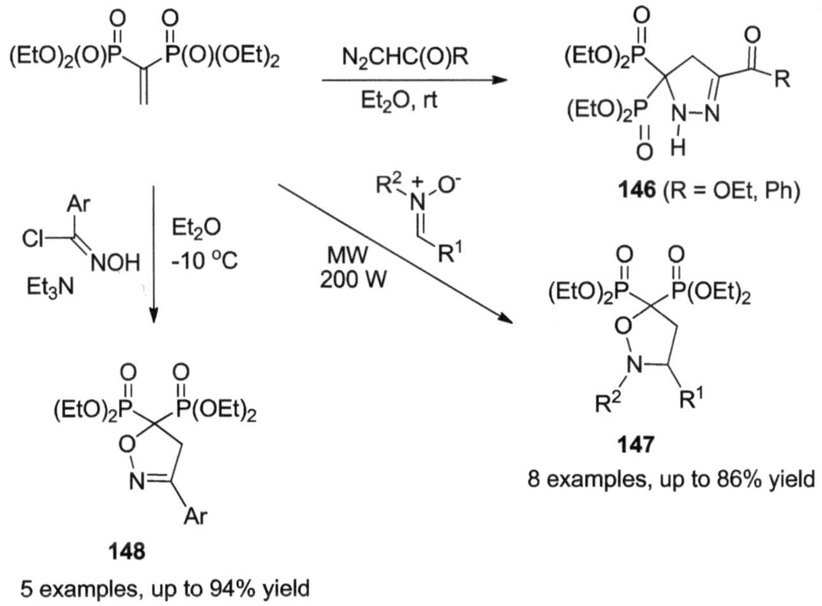

Scheme 52. 1,3-Dipolar cycloaddition reactions of **78a** with diazo compounds, nitrones, and nitrile oxides.

A successful method for triazole bisphosphonate synthesis involving ethenylidene-1,1-bisphosphonate **78a** utilizes azide-containing compounds. Indeed, the reaction between **78a** and benzyl azide in acetonitrile at room temperature afforded cycloadduct **149** in 83% yield. Similarly, the azide **150** also gave rise to a 1,2,3-triazole **151** in 68% yield (Scheme **53**) [94].

Recent work by Chen and co-workers has explored a tandem Michael addition reaction and 1,3-dipolar cycloaddition for the construction of *N*-attached 1,2,3-triazole containing bisphosphonates **152** (Scheme **54**) [95]. Both reactions were integrated into one-pot process in the presence of ultrasound irradiation. At least two pathways are conceivable for this reaction: addition reaction of NaN_3 with ethenylidene-1,1-bisphosphonate giving the adduct $N_3CH_2CH(PO_3Et_2)_2$, which can then react with alkyne, or alternatively, 1,3-dipolar cycloaddition between NaN_3

and alkyne followed by a Michael-type reaction of thus formed 1,2,3-triazole with ethenylidene-1,1-bisphosphonate. Interestingly, contrary to what was published [96], the reaction between **78a** and sodium azide did not lead to the Michael adduct but to **153** instead in excellent yield (Scheme 55) [94].

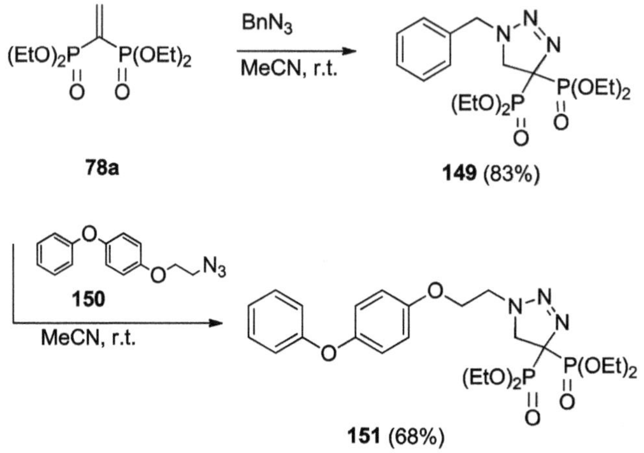

Scheme 53. 1,3-Dipolar addition reactions between **78a** and azides.

Recently, research group in China has reported strategy in which ethenylidenebisphosphonate **78a** is used as a dipolarophile to prepare spiro[indole-pyrrolizine], spiro[indole-indolizine], and spiro[indole-pyrrolidine] bisphosphonates by one-pot multicomponent reactions between compound **78a**, isatins, and amino acids in the presence of montmorillonite as the catalyst [97]. The experimental data are consistent with a 1,3-dipolar cycloaddition mechanism involving the azomethine ylide formed by decarboxylation, as in Scheme 56.

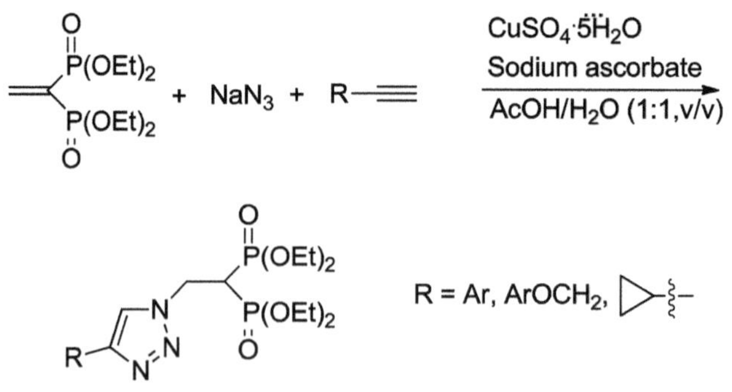

14 examples, up to 85% yield

Scheme 54. One-pot preparation of *N*-attached 1,2,3-triazole containing bisphosphonates.

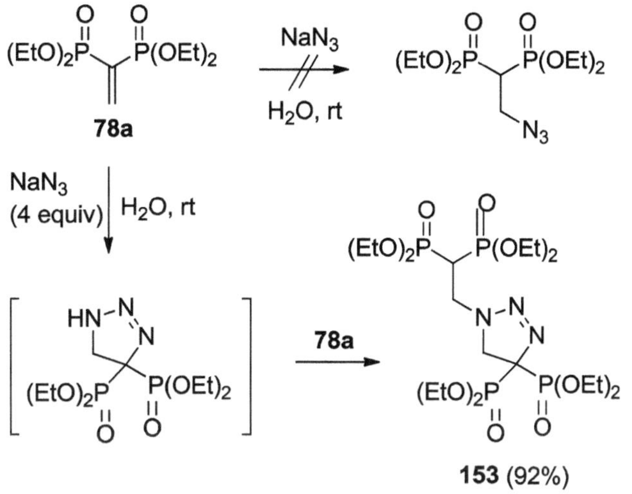

Scheme 55. 1,3-Dipolar cycloaddition between **78a** and NaN_3 in water.

Another interesting example for the use of ethenylidene-1,1-bisphosphonate as a dipolarophile is the Grigg azomethine ylide cascade cycloaddition (Scheme **57**) [94]. The probable reaction pathway, which was suggested following experiments, involves *in situ* imine formation from an aldehyde and α-amino acid derivative. This imine behaves as an azomethyne ylide in the presence of ethenylidene-1,1-bisphosphonate to form 1,3-cycloaddition adduct. The reaction is mild and proceeds in good yields and with high regioselectivity. For example, when compound **78a** was treated with an aldehyde and diethyl aminomalonate in tetrahydrofuran at room temperature, compounds **154-156** were produced in 50, 68 and 40% yield, respectively.

CONCLUDING REMARKS

During the past two decades, there has been a remarkable development in the chemistry of α-heteroatom substituted (α-OH, α-NR$_2$, α-SH) *gem*-bisphosphonates. In contrast, α-organyl-substituted methylenebisphosphonates not having the heteroatom substituents at the geminal carbon have not been as extensively studied as their α-hydroxy or α-amino substituted analogues. Nevertheless, as already mentioned in the introduction, some non-α-heteroatom containing bisphosphonates showed an unprecedented biological activity, and their use as enzyme inhibitors is of considerable interest.

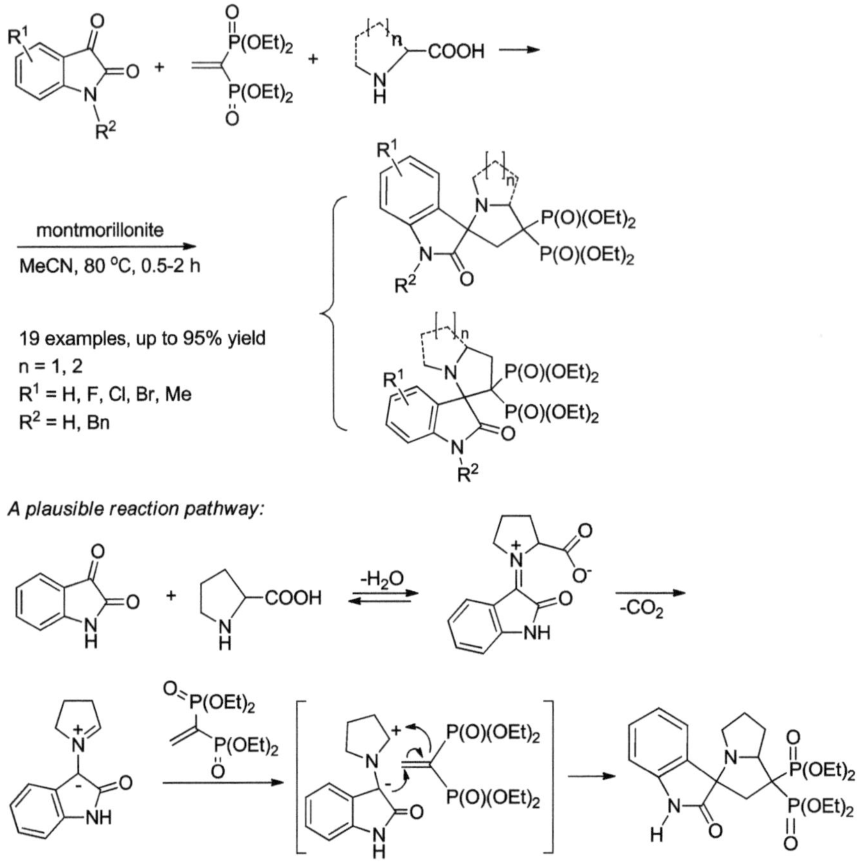

Scheme 56. Montmorillonite-promoted synthesis of spiro[indole-pyrrolizine], spiro[indole-indolizine], and spiro[indole-pyrrolidine] bisphosphonates. A plausible reaction pathway is shown for the multicomponent reaction of isatin, proline, and tetraethyl ethenylidenebisphosphonate.

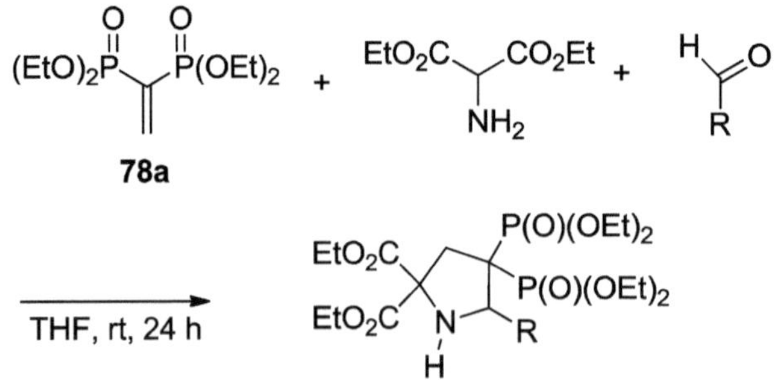

154, R = *n*-pentyl
155, R = Ph
156, R = 4-MeOC$_6$H$_4$

Scheme 57. Grigg azomethine ylide 1,3-dipolar cascade cycloaddition.

As presented in this report, a variety of methods is available for the synthesis of α-organyl *gem*-bisphosphonates, but three groups of methods are most frequently used. The first group includes synthesis *via* phosphites and monophosphonates; the second is deprotonation-alkylation of methylenebisphosphonate anions; and the third is the synthesis *via* alkenylidene-1,1-bisphosphonates. Landmarks in the development of contemporary chemistry of α-organyl *gem*-bisphosphonates include synthesis of 2-alkylaminoethyl-1,1-bisphosphonic acids as potent inhibitors of the enzymatic activity of *Trypanosoma cruzi* squalene synthase (Rodriguez, 2008), synthesis of isoprenoid bisphosphonates bearing one or more isoprenoid substituents as potent geranylgeranyl diphosphate synthase inhibitors (Wiemer, 2008), synthesis and biological evaluation of sulfur-containing bisphosphonic acids as potent antiparasitic agents (Rodriguez, 2013), the catalytic asymmetric synthesis of nitrogen-containing bisphosphonates using dinuclear Ni_2-Shiff base complex (Matsunaga and Shibasaki, 2009), development of a "click" methodology for the synthesis of functionalized bisphosphonates (Roschenthaler, 2007), development of a practical radical based approach to xanthate derivatives of functionalized bisphosphonates (Zard, 2003), the asymmetric synthesis of γ-keto bisphosphonates by Michael addition (Barros and Phillips, 2008), synthesis of 2-*C*-substituted benzotriazoles containing bisphosphonates *via* a copper-promoted three-component reaction (Yang, 2014), synthesis of a high-affinity, bone-targeted Src SH2 bisphosphonate inhibitor (Bohacek, 2001). However, several challenges remain, including the development of efficient methods for preparation of fluorine-containing α-organyl *gem*-bisphosphonates because these compounds may possess biological properties superior to those of non-fluorinated analogues. Other approaches to the *gem*-bisphosphonates with complex side chains bearing multiple functional groups are worth developing. Among them, the diazobisphosphonate pathway seems particularly attractive. The synthesis of bisphosphonates on a macrocyclic platform (dendrimers, polyazamacrocycles) also deserves undoubted interest. Finally, the direct introduction of a bisphosphonate moiety into organic molecules through the formation of a carbon-carbon bond by reactions of α-halogeno substituted bisphosphonates catalyzed by transition-metal complexes remains an unresolved issue.

CONSENT FOR PUBLICATION

Not applicable.

CONFLICT OF INTEREST

The author confirms that this chapter content has no conflict of interest.

ACKNOWLEDGEMENTS

This work was financially supported by the National Academy of Science of Ukraine (grant 04-03-18).

REFERENCES

[1] Fleisch, H. *Bisphosphonates in Bone Disease: From the Laboratory to the Patient,* 4th ed; Academic Press: San Diego, **2000**.

[2] Bartl, R.; Fritsch, B.; von Tresckow, E.; Bartl, C. *Bisphosphonates in Medical Practice*; Springer: Berlin, **2007**.
[http://dx.doi.org/10.1007/978-3-540-69870-8]

[3] Zhang, S.; Gangal, G.; Uludağ, H. "Magic bullets" for bone diseases: Progress in rational design of bone-seeking medicinal agents. *Chem. Soc. Rev.,* **2007**, *36*, 507-531.
[http://dx.doi.org/10.1039/B512310K]

[4] Abdou, W.M.; Shaddy, A.A. The development of bisphosphonates for therapeutic uses, and bisphosphonate structure-activity consideration. *ARKIVOC,* **2009**, *ix*, 143-182.

[5] Rosso, V.S.; Szajnman, S.H.; Malayil, L.; Galizzi, M.; Moreno, S.N.J.; Docampo, R.; Rodriguez, J.B. Synthesis and biological evaluation of new 2-alkylaminoethyl-1,1-bisphosphonic acids against *Trypanosoma Cruzi* and *Toxoplasma Gondii* targeting farnesyl diphosphate synthase. *Bioorg. Med. Chem.,* **2011**, *19*, 2211-2217.
[http://dx.doi.org/10.1016/j.bmc.2011.02.037]

[6] Szajnman, S.H.; García Liñares, G.E.; Li, Z-H.; Jiang, C.; Galizzi, M.; Bontempi, E.J.; Ferella, M.; Moreno, S.N.J.; Docampo, R.; Rodriguez, J.B. Synthesis and biological evaluation of 2-alkylaminoethyl-1,1-bisphosphonic acids against *Trypanosoma Cruzi* and *Toxoplasma Gondii* targeting farnesyl diphosphate synthase. *Bioorg. Med. Chem.,* **2008**, *16*, 3283-3290.
[http://dx.doi.org/10.1016/j.bmc.2007.12.010]

[7] Rodrígues-Poveda, C.A.; González-Pacanowska, D.; Szajnman, S.H.; Rodríguez, J.B. 2-Alkylaminoethyl-1,1-bisphosphonic acids are potent inhibitors of the enzymatic activity of *Trypanosoma Cruz*i Squalene Synthase. *Antimicrob. Agents Chemother.,* **2012**, *56*, 4483-4486.
[http://dx.doi.org/10.1128/AAC.00796-12]

[8] Wiemer, A.J.; Yu, J.S.; Lamb, K.M.; Hohl, R.J.; Wiemer, D.F. Mono- and dialkyl isoprenoid bisphosphonates as geranylgeranyl diphosphate synthase inhibitors. *Bioorg. Med. Chem.,* **2008**, *16*, 390-399.
[http://dx.doi.org/10.1016/j.bmc.2007.09.029]

[9] Kafarski, P.; Lejczak, B. Aminophosphonic acids of potential medical importance. *Curr. Med. Chem. Anticancer Agents,* **2001**, *1*, 301-312.
[http://dx.doi.org/10.2174/1568011013354543]

[10] Lecouvey, M.; Leroux, Y. Synthesis of 1-hydroxy-1,1-bisphosphonates. *Heteroatom Chem.,* **2000**, *11*, 556-561.
[http://dx.doi.org/10.1002/1098-1071(2000)11:7<556::AID-HC15>3.0.CO;2-N]

[11] Romanenko, V.; Kukhar, V. Progress in the development of pyrophosphate bioisosteres: Synthesis and biomedical potential of 1-fluoro- and 1,1-difluoromethylene-1,1-bisphosphonates. *Curr. Org. Chem.,* **2014**, *18*, 1491-1512.
[http://dx.doi.org/10.2174/1385272818111408151 24708]

[12] Romanenko, V.D.; Kukhar, V.P. 1-Amino-1, 1-bisphosphonates. Fundamental syntheses and new developments. *ARKIVOC,* **2012**, (iv), 127-166.

[13] Keglevich, G.; Grun, A.; Kovacs, R.; Garadnay, S.; Greiner, I. Rational synthesis of ibandronate and alendronate. *Curr. Org. Synth.,* **2013**, *10*, 640-645.

[http://dx.doi.org/10.2174/1570179411310040007]

[14] Keglevich, G.; Grün, A.; Garadnay, S.; Greiner, I. Rational synthesis of dronic acid derivatives. *Phosphorus Sulfur Silicon Relat. Elem.*, **2015**, *190*, 2116-2124.
[http://dx.doi.org/10.1080/10426507.2015.1072194]

[15] Demmer, C.S.; Krogsgaard-Larsen, N.; Bunch, L. Review on modern advances of chemical methods for the introduction of a phosphonic acid group. *Chem. Rev.*, **2011**, *111*, 7981-8006.
[http://dx.doi.org/10.1021/cr2002646]

[16] Prishchenko, A.A.; Livantsov, M.V.; Novikova, O.P.; Livantsova, L.I.; Milyaeva, E.R. Synthesis of 2-pyridylethylphosphinates containing 2,6-di-tert-butyl-4-methylphenol fragments. *Russ. J. Gen. Chem.*, **2008**, *78*, 2150-2153.
[http://dx.doi.org/10.1134/S1070363208110297]

[17] Prishchenko, A.A.; Livantsov, M.V.; Novikova, O.P.; Livantsova, L.I.; Petrosyan, V.S.; Milaeva, E.R. Synthesis of new functionalized mono- and bisphosphinates with 2, 6-di-tert-butyl-4- methylphenol fragments. *Heteroatom Chem.*, **2008**, *19*, 562-568.
[http://dx.doi.org/10.1002/hc.20475]

[18] Prishchenko, A.A.; Livantsov, M.V.; Novikova, O.P.; Livantsova, L.I.; Shpakovskii, D.B.; Milaeva, E.R. Synthesis of phosphorus derivatives of 2,6-di-tert-butyl-4-methylphenol. *Russ. J. Gen. Chem.*, **2006**, *76*, 1753-1756.
[http://dx.doi.org/10.1134/S1070363206110132]

[19] Pieper, T.; Keppler, B.K. Preparation of tetraethyl-4-hydroxyphenylmethylene-1,1-bisphosphonate by hydroxy-de-diazoniation of the corresponding diazonium salt of tetraethyl-4-aminophenylmethylene-1,1-bisphosphonate. *Phosphorus Sulfur Silicon Relat. Elem.*, **2000**, *165*, 77-82.
[http://dx.doi.org/10.1080/10426500008076326]

[20] Abdou, W.M.; Ganoub, N.A.; Fahmy, A.F.; Shaddy, A.A. Symmetrical and asymmetrical bisphosphonate esters. Synthesis, selective hydrolysis and isomerization. *Monatsh. Chem.*, **2006**, *137*, 105-116.
[http://dx.doi.org/10.1007/s00706-005-0403-y]

[21] Chaleix, V.; Lecouvey, M. Synthesis of novel phosphonated tripodal ligands for actinides chelation therapy. *Tetrahedron Lett.*, **2007**, *48*, 703-706.
[http://dx.doi.org/10.1016/j.tetlet.2006.11.094]

[22] Bohacek, R.S.; Dalgarno, D.C.; Hatada, M.; Jacobsen, V.A.; Lynch, B.A.; Macek, K.J.; Merry, T.; Metcalf, C.A.; Narula, S.S.; Sawyer, T.K.; Shakespeare, W.C.; Violette, S.M.; Weigele, M. X-Ray structure of citrate bound to Src SH2 leads to a high-affinity, bone-targeted Src SH2 inhibitor. *J. Med. Chem.*, **2001**, *44*, 660-663.
[http://dx.doi.org/10.1021/jm0002681]

[23] Arstad, E.; Hoff, P.; Skattebøl, L.; Skretting, A.; Breistøl, K. Studies on the synthesis and biological properties of non-carrier-added [(125)I and (131)I]-labeled arylalkylidenebisphosphonates: Potent bone-seekers for diagnosis and therapy of malignant osseous lesions. *J. Med. Chem.*, **2003**, *46*, 3021-3032.
[http://dx.doi.org/10.1021/jm021107v]

[24] Grison, C.; Joliez, S.; De Clercq, E.; Coutrot, P. Monoglycosyl, diglycosyl, and dinucleoside methylenediphosphonates: Direct synthesis and antiviral activity. *Carbohydr. Res.*, **2006**, *341*, 1117-1129.
[http://dx.doi.org/10.1016/j.carres.2006.03.023]

[25] Du, Y.; Jung, K-Y.; Wiemer, D.F. A one-flask synthesis of α,α-bisphosphonates *via* enolate chemistry. *Tetrahedron Lett.*, **2002**, *43*, 8665-8668.
[http://dx.doi.org/10.1016/S0040-4039(02)02148-2]

[26] Greiner, I.; Grün, A.; Ludányi, K.; Keglevich, G. Solid-liquid two-phase alkylation of tetraethyl methylenebisphosphonate under microwave irradiation. *Heteroatom Chem.*, **2011**, *22*, 11-14.

[http://dx.doi.org/10.1002/hc.20648]

[27] Keglevich, G.; Grün, A.; Kovács, R.; Garadnay, S.; Greiner, I. Green chemical synthesis of bisphosphonic/dronic acid derivatives. *Phosphorus Sulfur Silicon Relat. Elem.*, **2015**, *190*, 664-667.
[http://dx.doi.org/10.1080/10426507.2014.984024]

[28] Moreau, P.; Maffei, M. A Stereoselective alladium-catalyzed synthesis of amino alkenyl geminal bisphosphonates. *Tetrahedron Lett.*, **2004**, *45*, 743-746.
[http://dx.doi.org/10.1016/j.tetlet.2003.11.031]

[29] Shull, L.W.; Wiemer, D.F. Copper-mediated displacements of allylic THP ethers on a bisphosphonate template. *J. Organomet. Chem.*, **2005**, *690*, 2521-2530.
[http://dx.doi.org/10.1016/j.jorganchem.2004.10.013]

[30] Shull, L.W.; Wiemer, A.J.; Hohl, R.J.; Wiemer, D.F. Synthesis and biological activity of isoprenoid bisphosphonates. *Bioorg. Med. Chem.*, **2006**, *14*, 4130-4136.
[http://dx.doi.org/10.1016/j.bmc.2006.02.010]

[31] Holstein, S.A.; Wohlford-Lenane, C.L.; Wiemer, D.F.; Hohl, R.J. Isoprenoid pyrophosphate analogues regulate expression of ras-related proteins. *Biochemistry*, **2003**, *42*, 4384-4391.
[http://dx.doi.org/10.1021/bi027227m]

[32] Maalouf, M.A.; Wiemer, A.J.; Kuder, C.H.; Hohl, R.J.; Wiemer, D.F. Synthesis of fluorescently tagged isoprenoid bisphosphonates that inhibit protein geranylgeranylation. *Bioorg. Med. Chem.*, **2007**, *15*, 1959-1966.
[http://dx.doi.org/10.1016/j.bmc.2007.01.002]

[33] Li, C.; Yuan, C. Studies on organophosphorus compounds. A novel synthetic approach to substituted cyclopentane-1,1-diylbisphosphonates *via* Pd(0) catalyzed enyne cyclization. *Heteroatom Chem.*, **1993**, *4*, 517-520.
[http://dx.doi.org/10.1002/hc.520040518]

[34] Skarpos, H.; Osipov, S.N.; Vorob'eva, D.V.; Odinets, I.L.; Lork, E.; Röschenthaler, G-V. Synthesis of functionalized bisphosphonates *via* click chemistry. *Org. Biomol. Chem.*, **2007**, *5*, 2361-2367.
[http://dx.doi.org/10.1039/B705510B]

[35] Massarenti, C.; Bortolini, O.; Fantin, G.; Cristofaro, D.; Ragno, D.; Perrone, D.; Marchesi, E.; Toniolo, G.; Massi, A. Fluorous-tag assisted synthesis of bile acid – bisphosphonate conjugates *via* orthogonal click reactions: An access to potential anti-resorption bone drugs. *Org. Biomol. Chem.*, **2017**, *15*, 4907-4920.
[http://dx.doi.org/10.1039/C7OB00774D]

[36] Artamkina, G.A.; Tarasenko, E.A.; Lukashev, N.V.; Beletskaya, I.P. Synthesis of perhaloaromatic diethyl methylphosphonates containing electron-withdrawing groups. *Tetrahedron Lett.*, **1998**, *39*, 901-904.
[http://dx.doi.org/10.1016/S0040-4039(97)10651-7]

[37] Makarov, M.V.; Rybalkina, E.Y.; Klementova, Z.S.; Roschenthaler, G-V. 3,-5 Bis(arylidene)--piperidinones modified with bisphosphonate groups using a 1,2,3-triazole ring: Synthesis and antitumor properties. *Russ. Chem. Bull.*, **2014**, *63*, 2388-2394.
[http://dx.doi.org/10.1007/s11172-014-0752-y]

[38] Makarov, M.V.; Rybalkina, E.Y.; Brel, V.K. 3,5-Bis(arylidene)-4-piperidones modified by bisphosphonate groups as novel anticancer agents. *Phosphorus Sulfur Silicon Relat. Elem.*, **2015**, *190*, 741-746.
[http://dx.doi.org/10.1080/10426507.2014.976338]

[39] Zhou, X.; Hartman, S.V.; Born, E.J.; Smits, J.P.; Holstein, S.A.; Wiemer, D.F. Triazole-based inhibitors of geranylgeranyltransferase II. *Bioorg. Med. Chem. Lett.*, **2013**, *23*, 764-766.
[http://dx.doi.org/10.1016/j.bmcl.2012.11.089]

[40] Zhou, X.; Ferree, S.D.; Wills, V.S.; Born, E.J.; Tong, H.; Wiemer, D.F.; Holstein, S.A. Geranyl and

neryl triazole bisphosphonates as inhibitors of geranylgeranyl diphosphate synthase. *Bioorg. Med. Chem.,* **2014**, *22*, 2791-2798.
[http://dx.doi.org/10.1016/j.bmc.2014.03.014]

[41] Wills, V.S.; Allen, C.; Holstein, S.A.; Wiemer, D.F. Potent triazole bisphosphonate inhibitor of geranylgeranyl diphosphate synthase. *ACS Med. Chem. Lett.,* **2015**, *6*(12), 1195-1198.
[http://dx.doi.org/10.1021/acsmedchemlett.5b00334]

[42] Gagosz, F.; Zard, S.Z. A Practical radical based access to functionalised geminal bisphosphonates. *Synlett,* **2003**, 387-389.

[43] Vovk, A.I.; Kalchenko, V.I.; Cherenok, S.A.; Kukhar, V.P.; Muzychka, O.V.; Lozynsky, M.O. Calix[4]arene methylenebisphosphonic acids as calf intestine alkaline phosphatase inhibitors. *Org. Biomol. Chem.,* **2004**, *2*, 3162-3166.
[http://dx.doi.org/10.1039/b409526j]

[44] Cherenok, S.; Vovk, A.; Muravyova, I.; Marcinowicz, A.; Poznanski, J.; Muzychka, O.; Kukhar, V.; Zielenkiewicz, W.; Kalchenko, V. Bio-relevant derivatives of calixarene phosphonic acids. *Phosphorus Sulfur Silicon Relat. Elem.,* **2008**, *183*, 638-639.
[http://dx.doi.org/10.1080/10426500701795126]

[45] Vovk, A.; Kalchenko, V.; Muzychka, O.; Tanchuk, V.; Muravyova, I.; Shivanyuk, A.; Cherenok, S.; Kukhar, V. Calixarene methylenebisphosphonic acids: Alkaline phosphatase inhibition and docking studies. *Phosphorus Sulfur Silicon Relat. Elem.,* **2008**, *183*, 625-626.
[http://dx.doi.org/10.1080/10426500701793311]

[46] Berchel, M.; Salaün, J-Y.; Couthon-Gourvès, H.; Haelters, J-P.; Jaffrès, P-A. An unexpected base-induced [1,4]-phospho-Fries rearrangement. *J. Chem. Soc., Dalton Trans.,* **2010**, *39*, 11314-11316.
[http://dx.doi.org/10.1039/c0dt00880j]

[47] Abdou, W.M.; Ganoub, N.A.F.; El-Khoshnieh, Y.O. *Synthesis of a new type of 1,1-bisphosphonates bearing S-, and N-heterocycles, based on the reactions of methylenebisphosphonate with alkenes*; Syntlett, **2003**, pp. 785-790.

[48] Abdou, W.M.; Ganoub, N.A.; Geronikaki, A.; Sabry, E. Synthesis, properties, and perspectives of gem-diphosphono substituted-thiazoles. *Eur. J. Med. Chem.,* **2008**, *43*, 1015-1024.
[http://dx.doi.org/10.1016/j.ejmech.2007.07.005]

[49] Abdou, W.M.; Khidre, M.D.; Sediek, A.A. A practical synthesis of thio-bisphosphonic acids for the treatment of arthritis, based on the chemistry of tetraethyl methylene-1,1-bisphosphonate. *Lett. Org. Chem.,* **2006**, *3*, 634-639.
[http://dx.doi.org/10.2174/157017806778559536]

[50] Xiang, H.; Qi, J.; He, Q.; Jiang, M.; Yang, C.; Deng, L. Synthesis of 2-C-substituted benzothiazoles *via* a copper-promoted domino condensation/S-arylation/heterocyclization process. *Org. Biomol. Chem.,* **2014**, *12*, 4633-4636.
[http://dx.doi.org/10.1039/c4ob00564c]

[51] Rodriguez, J. Tetraethyl vinylidenebisphosphonate: A versatile synthon for the preparation of bisphosphonates. *Synthesis,* **2014**, *46*, 1129-1142.
[http://dx.doi.org/10.1055/s-0033-1340952]

[52] Janecki, T.; Kędzia, J.; Wąsek, T. Michael additions to activated vinylphosphonates. *Synthesis,* **2009**, 1227-1254.
[http://dx.doi.org/10.1055/s-0028-1088031]

[53] Szajnman, S. Bisphosphonates derived from fatty acids are potent inhibitors of *Trypanosoma Cruzi* farnesyl pyrophosphate synthase. *Bioorg. Med. Chem. Lett.,* **2003**, *13*, 3231-3235.
[http://dx.doi.org/10.1016/S0960-894X(03)00663-2]

[54] Lolli, M.L.; Lazzarato, L.; Di Stilo, A.; Fruttero, R.; Casco, A. Michael addition of Grignard reagents to tetraethyl ethenylidenebisphosphonate. *J. Organomet. Chem.,* **2002**, *650*, 77-83.

[http://dx.doi.org/10.1016/S0022-328X(02)01179-8]

[55] Gao, J.; Liu, J.; Qiu, Y.; Chu, X.; Qiao, Y.; Li, D. Multi-target-directed design, syntheses, and characterization of fluorescent bisphosphonate derivatives as multifunctional enzyme inhibitors in mevalonate pathway. *Biochim. Biophys. Acta*, **2013**, *1830*, 3635-3642.
[http://dx.doi.org/10.1016/j.bbagen.2013.02.011]

[56] Makarov, M.V.; Rybalkina, E.Y.; Roshenthaler, G-V. New 3, 5-bis(arylidene)-4-piperidones with bisphosphonate moiety : Synthesis and antitumor activity. *Russ. Chem. Bull.,* **2014**, *63*, 1181-1186.
[http://dx.doi.org/10.1007/s11172-014-0569-8]

[57] Lazzarato, L.; Rolando, B.; Lolli, M.L.; Tron, G.C.; Fruttero, R.; Gasco, A.; Deleide, G.; Guenther, H.L. Synthesis of NO-donor bisphosphonates and their in-vitro action on bone 5resorption. *J. Med. Chem.,* **2005**, *48*, 1322-1329.
[http://dx.doi.org/10.1021/jm040830d]

[58] Inoue, S.; Okauchi, T.; Minami, T. New synthesis of gem-bis(phosphono)ethylenes and their applications. *Synthesis,* **2003**, 1971-1976.

[59] Bianchini, G.; Scarso, A.; Chiminazzo, A.; Sperni, L.; Strukul, G. Water enhanced synthesis of gem-bisphosphonates *via* Rh(I) mediated 1,4-conjugate addition of aryl boronic acids to vinylidenebisphosphonate esters. *Green Chem.,* **2013**, *15*, 656-662.
[http://dx.doi.org/10.1039/c2gc36800e]

[60] Sturtz, G.; Guervenou, J. Synthesis of novel functionalized gem-bisphosphonates. *Synthesis*, **1991**, 661-662.
[http://dx.doi.org/10.1055/s-1991-26539]

[61] Page, P.C.B.; McKenzie, M.J.; Gallagher, J.A. novel synthesis of bis(phosphonic acid)-steroid conjugates. *J. Org. Chem.,* **2001**, *66*, 3704-3708.
[http://dx.doi.org/10.1021/jo001489h]

[62] Page, P.C.B.; Moore, J.P.G.; Mansfield, I.; McKenna, M.J.; Bowler, W.B.; Gallagher, J.A. Synthesis of bone-targeted oestrogenic compounds for the inhibition of bone resorption. *Tetrahedron,* **2001**, *57*, 1837-1847.
[http://dx.doi.org/10.1016/S0040-4020(00)01164-9]

[63] Dzięgielewski, M.; Hejmanowska, J.; Albrecht, Ł. A Convenient approach to a novel group of quaternary amino acids containing a geminal bisphosphonate moiety. *Synthesis,* **2014**, *46*, 3233-3238.
[http://dx.doi.org/10.1055/s-0034-1378997]

[64] Palmerini, C.A.; Tartacca, F.; Mazzoni, M.; Granieri, L.; Goracci, L.; Scrascia, A.; Lepri, S. Synthesis of new indole-based bisphosphonates and evaluation of their chelating ability in PE/CA-PJ15 cells. *Eur. J. Med. Chem.,* **2015**, *102*, 403-412.
[http://dx.doi.org/10.1016/j.ejmech.2015.08.019]

[65] Sulzer-Mossé, S.; Tissot, M.; Alexakis, A. First enantioselective organocatalytic conjugate addition of aldehydes to vinyl phosphonates. *Org. Lett.,* **2007**, *9*, 3749-3752.
[http://dx.doi.org/10.1021/ol7015498]

[66] Capuzzi, M.; Perdicchia, D.; Jørgensen, K.A. Highly enantioselective approach to geminal bisphosphonates by organocatalyzed Michael-type addition of beta-ketoesters. *Chemistry,* **2008**, *14*, 128-135.
[http://dx.doi.org/10.1002/chem.200701317]

[67] Barros, M.T.; Faísca Phillips, A.M. Enamine catalysis in the synthesis of chiral structural analogues of gem-bisphosphonates known to be biologically active. *Eur. J. Org. Chem.,* **2008**, 2525-2529.
[http://dx.doi.org/10.1002/ejoc.200800170]

[68] Kato, Y.; Chen, Z.; Matsunaga, S.; Shibasaki, M. Catalytic asymmetric synthesis of nitrogen-containing gem-bisphosphonates using a dinuclear Ni_2-Schiff base complex. *Synlett,* **2009**, 1635-1638.

[69] Xue, Z-Y.; Li, Q-H.; Tao, H-Y.; Wang, C-J. A facile Cu(I)/TF-BiphamPhos-catalyzed asymmetric

approach to unnatural α-amino acid derivatives containing gem-bisphosphonates. *J. Am. Chem. Soc.,* **2011**, *133*, 11757-11765.
[http://dx.doi.org/10.1021/ja2043563]

[70] Kimura, M.; Tada, A.; Tokoro, Y.; Fukuzawa, S. Silver-catalyzed asymmetric Michael addition of azomethine ylide to arylidene diphosphonates using ThioClickFerrophos ligand. *Tetrahedron Lett.,* **2015**, *56*, 2251-2253.
[http://dx.doi.org/10.1016/j.tetlet.2015.03.024]

[71] Szajnman, S.H.; García Liñares, G.; Moro, P.; Rodriguez, J.B. New insights into the chemistry of gem-bis(phosphonates): unexpected rearrangement of Michael-type acceptors. *Eur. J. Org. Chem.,* **2005**, 3687-3696.
[http://dx.doi.org/10.1002/ejoc.200500097]

[72] Kotsikorou, E.; Song, Y.; Chan, J.M.W.; Faelens, S.; Tovian, Z.; Broderick, E.; Bakalara, N.; Docampo, R.; Oldfield, E. Bisphosphonate inhibition of the exopolyphosphatase activity of the *Trypanosoma brucei* soluble vacuolar pyrophosphatase. *J. Med. Chem.,* **2005**, *48*, 6128-6139.
[http://dx.doi.org/10.1021/jm058220g]

[73] Ghosh, S.; Chan, J.M.W.; Lea, C.R.; Meints, G.A.; Lewis, J.C.; Tovian, Z.S.; Flessner, R.M.; Loftus, T.C.; Bruchhaus, I.; Kendrick, H.; Craft, S.L.; Kemp, R.G.; Kobayashi, S.; Nozaki, T.; Oldfield, E. Effects of bisphosphonates on the growth of *Entamoeba histolytica* and *Plasmodium* species *in vitro* and *in vivo*. *J. Med. Chem.,* **2004**, *47*, 175-187.
[http://dx.doi.org/10.1021/jm030084x]

[74] Couthon-Gourvès, H.; Simon, G.; Haelters, J-P.; Corbel, B. Synthesis of novel diethyl indolylphosphonates and tetraethyl indolyl-1,1-bisphosphonates by Michael addition. *Synthesis,* **2006**, *2006*, 81-88.
[http://dx.doi.org/10.1055/s-2005-921757]

[75] Gritzalis, D.; Park, J.; Chiu, W.; Cho, H.; Lin, Y-S.; De Schutter, J.W.; Lacbay, C.M.; Zielinski, M.; Berghuis, A.M.; Tsantrizos, Y.S. Probing the molecular and structural elements of ligands binding to the active site *versus* an allosteric pocket of the human farnesyl pyrophosphate synthase. *Bioorg. Med. Chem. Lett.,* **2015**, *25*, 1117-1123.
[http://dx.doi.org/10.1016/j.bmcl.2014.12.089]

[76] Teixeira, F.C.; Rangel, C.M.; Teixeira, A.P.S. Synthesis of new azole phosphonate precursors for fuel cells proton exchange membranes. *Heteroatom Chem.,* **2015**, *26*, 236-248.
[http://dx.doi.org/10.1002/hc.21254]

[77] Grigor'ev, I.A.; Morozov, D.A.; Svyatchenko, V.A.; Kiselev, N.N.; Loktev, V.B.; Luk'yanets, E.A.; Vorozhtsov, G.N. Synthesis and study of antitumor activity of tetraethyl 2-(2',2',6',6-tetramethylpiperidin-4'-ylamino)ethylidene-1,1-bisphosphonate. *Dokl. Chem.,* **2014**, *457*, 137-140.
[http://dx.doi.org/10.1134/S0012500814070064]

[78] Lv, M.; Wang, M.; Lu, K.; Peng, L.; Zhao, Y. An efficient synthesis of 2-aminoethylidene-1-1-bisphosphonates derivatives *via* Michael addition reaction. *Phosphorus Sulfur Silicon Relat. Elem.,* **2018**, *193*, 149-154.
[http://dx.doi.org/10.1080/10426507.2017.1393421]

[79] Shevchuk, M.; Sotiropoulos, J-M.; Miqueu, K.; Romanenko, V.; Kukhar, V. Tetrakis(trimethylsilyl) ethenylidene-1,1-bisphosphonate: A mild and convenient Michael acceptor for the synthesis of 2-aminoethylidene-1,1-bisphosphonic acids and their potassium salts. *Synlett,* **2011**, *2011*, 1370-1374.
[http://dx.doi.org/10.1055/s-0030-1260566]

[80] Sawicki, M.; Lecerclé, D.; Grillon, G.; Le Gall, B.; Sérandour, A-L.; Poncy, J-L.; Bailly, T.; Burgada, R.; Lecouvey, M.; Challeix, V.; Leidier, A.; Pellet-Rostaing, S.; Ansoborlo, E.; Taran, F. Bisphosphonate sequestering agents. Synthesis and preliminary evaluation for *in vitro* and *in vivo* uranium(VI) chelation. *Eur. J. Med. Chem.,* **2008**, *43*, 2768-2777.
[http://dx.doi.org/10.1016/j.ejmech.2008.01.018]

[81] Herczegh, P.E.; Buxton, T.B.; McPherson, J.C., III; Kovacs-Kulyassa, A.; Sztaricskai, F.; Stroebel, G.G.; Plowman, K.M.; Farcasiu, D.; Hartnamm, J.F. Osteoadsorptive bisphosphonate derivatives of fluoroquinolone antibacterials. *J. Med. Chem.*, **2002**, *45*, 2338-2341.
[http://dx.doi.org/10.1021/jm0105326]

[82] Bailly, T.; Burgada, R. Etude par RMN de l'addition des fonctions -NH, -OH et PH sur l'éthénylidène bis phosphonate de diéthyle. Synthèse de gem bis phosphonates fonctionnalisés. *Phosphorus Sulfur Silicon Relat. Elem.*, **1994**, *86*, 217-228.
[http://dx.doi.org/10.1080/10426509408018407]

[83] Alferiev, I.S.; Connolly, J.M.; Levy, R.J. A Novel mercapto-bisphosphonate as an efficient anticalcification agent for bioprosthetic tissues. *J. Organomet. Chem.*, **2005**, *690*, 2543-2547.
[http://dx.doi.org/10.1016/j.jorganchem.2004.10.011]

[84] Granchi, D.; Scarso, A.; Bianchini, G.; Chiminazzo, A.; Minto, A.; Sgarbossa, P.; Michelin, R.A.; Di Pompo, G.; Avnet, S.; Strukul, G. Low toxicity and unprecedented anti-osteoclast activity of a simple sulfur-containing gem-bisphosphonate: A comparative study. *Eur. J. Med. Chem.*, **2013**, *65*, 448-455.
[http://dx.doi.org/10.1016/j.ejmech.2013.04.032]

[85] Recher, M.; Barboza, A.P.; Li, Z-H.; Galizzi, M.; Ferrer-Casal, M.; Szajnman, S.H.; Docampo, R.; Moreno, S.N.J.; Rodriguez, J.B. Design, synthesis and biological evaluation of sulfur-containing 1,1-bisphosphonic acids as antiparasitic agents. *Eur. J. Med. Chem.*, **2013**, *60*, 431-440.
[http://dx.doi.org/10.1016/j.ejmech.2012.12.015]

[86] Kultyshev, R.G.; Liu, J.; Liu, S.; Tjarks, W.; Soloway, A.H.; Shore, S.G. S-Alkylation and S-amination of methyl thioethers - derivatives of *closo*-$[B_{12}H_{12}]^{2-}$. Synthesis of a boronated phosphonate, gem-bisphosphonates, and dodecaborane-*ortho*-carborane oligomers. *J. Am. Chem. Soc.*, **2002**, *124*, 2614-2624.
[http://dx.doi.org/10.1021/ja0123857]

[87] Guervenou, J.; Gourvès, J-P.; Couthon, H.; Corbel, B.; Sturtz, G.; Kervarec, N. Phtalimide derivatives towards phosphorylated reactants: Unexpected, but interesting, reactions. *Phosphorus Sulfur Silicon Relat. Elem.*, **2000**, *156*, 107-124.
[http://dx.doi.org/10.1080/10426500008044996]

[88] Guervenou, J.; Sturtz, G. Synthesis of some diethylphosphono substituted 2,5-dihydrofuran, 2H--benzopyrans and 3H-naphto[2,1-b]pyran. *Phosphorus Sulfur Silicon Relat. Elem.*, **1992**, *70*, 255-261.
[http://dx.doi.org/10.1080/10426509208049174]

[89] Delain-Bioton, L.; Turner, A.; Lejeune, N.; Villemin, D.; Hix, G.B.; Jaffrès, P-A. Michael addition of phosphorus derivatives on tetraethyl ethylidenediphosphonate. *Tetrahedron*, **2005**, *61*, 6602-6609.
[http://dx.doi.org/10.1016/j.tet.2005.04.045]

[90] Ruzziconi, R.; Ricci, G.; Gioiello, A.; Couthon-Gourvès, H.; Gourvès, J-P. First general approach to cyclohex-3-ene-1,1-bis(phosphonates) by Diels−Alder cycloaddition of tetraethyl vinylidenebis(phosphonate) to 1,3-dienes. *J. Org. Chem.*, **2003**, *68*, 736-742.
[http://dx.doi.org/10.1021/jo0205154]

[91] Nugent, R.A.; Murphy, M.; Schlachter, S.T.; Dunnj, C.J.; Smith, R.J.; Staite, N.D.; Galinetf, L.A.; Shields, S.K.; Aspar, D.G.; Richard, K.A.; Rohloff, N.A. Pyrazoline bisphosphonate esters. *J. Med. Chem.*, **1993**, *36*, 134-139.
[http://dx.doi.org/10.1021/jm00053a017]

[92] Bortolini, O.; Mulani, I.; De Nino, A.; Maiuolo, L.; Nardi, M.; Russo, B.; Avnet, S. Efficient synthesis of isoxazolidine-substituted bisphosphonates by 1,3-dipolar cycloaddition reactions. *Tetrahedron*, **2011**, *67*, 5635-5641.
[http://dx.doi.org/10.1016/j.tet.2011.05.098]

[93] Ye, Y.; Xu, G-Y.; Zheng, Y.; Liu, L-Z. Cycloaddition of nitrile oxides to substituted vinylphosphonates. *Heteroatom Chem.*, **2003**, *14*, 309-311.
[http://dx.doi.org/10.1002/hc.10149]

[94] Ferrer-Casal, M.; Barboza, A.; Szajnman, S.; Rodriguez, J. 1,3-Dipolar cycloadditions of the versatile intermediate tetraethyl vinylidenebisphosphonate. *Synthesis*, **2013**, *45*, 2397-2404.
[http://dx.doi.org/10.1055/s-0033-1338498]

[95] Chen, X.; Li, X.; Yuan, J.; Qu, L.; Wang, S.; Shi, H.; Tang, Y.; Duan, L. Simple, efficient one-pot method for synthesis of novel *N*-attached 1,2,3-triazole containing bisphosphonates. *Tetrahedron*, **2013**, *69*, 4047-4052.
[http://dx.doi.org/10.1016/j.tet.2013.03.078]

[96] Jiang, Q.; Yang, L.; Hai, L.; Wu, Y. Synthesis of melphalan-gem-bisphosphonate conjugation to bone tumors and study of affinity to hydroxyapatite *in vitro*. *Lett. Org. Synth.*, **2008**, *86*, 229-233.

[97] Li, G.; Wu, M.; Liu, F.; Jiang, J. One-pot, highly regioselective 1,3-dipole cycloaddition promoted by montmorillonite for the synthesis of spiro[indole-pyrrolizine], spiro[indole-indolizine], and spiro[indole-pyrrolidine] gem-bisphosphonates. *Synthesis*, **2015**, *47*, 3783-3796.
[http://dx.doi.org/10.1055/s-0035-1560463]

SUBJECT INDEX

A

Acetate 12, 13, 31, 40
   α-tocopheryl 31
   geranylgeranyl 12, 13, 40
Acylating agents 151, 157, 159, 167, 173, 174, 175, 177, 179, 184
Acylation 151, 156, 157, 160, 161, 162, 164, 165, 167, 169, 170, 171, 172, 173, 174, 176, 178, 180, 181, 182, 184, 185, 186, 187, 188
   of amines 161, 165, 167, 174
   of racemic 170, 181
   of racemic 3-benzylmorpholine 171, 172
   of racemic amines 165, 176, 178, 180, 181, 188
   product 157, 162, 188, 190
   reaction 151, 152, 175, 176
   stereoselectivity of 157, 178, 181
   selectivity 162, 167, 169
Acyl chlorides 151, 161, 162, 163, 164, 165, 167, 175, 176, 177, 178, 179, 180, 181, 183, 184, 185, 186, 188, 189
   amino 178, 179, 180, 185
Addition reactions of tetraethyl 223, 232
   ethenylidenebisphosphonate 223, 232
Alcohols 101, 131, 132, 139
   α-methylbenzyl 131, 132
   secondary 10, 131, 132, 139
Alkenylidene-1,1-Bisphosphonates 200, 221, 244
Alkylation 5, 14, 15, 19, 21, 25, 33, 34, 37, 43, 50, 51, 55, 134, 200, 209, 210, 211
All-*trans* heptaprenol 19, 20
Allyl barium reagents 22, 46
Allyl halides 15, 46, 47
Allylic 1, 8, 9, 10, 15, 18, 20, 21, 22, 23, 24, 25, 26, 38, 39, 42, 46, 49, 50, 52, 56, 131, 139, 209, 210
   alcohols 10, 38, 42, 56, 131
   bromides 22, 209, 210

   carbonate 15, 25, 26
   chloride 8, 15, 18, 23, 39, 49, 50, 52
   halides 9, 15, 20, 21, 46, 49
   phosphate 23, 24
   secondary 22, 23
   reductive coupling 1
   sulfides 139
Amides 157, 165, 175, 183, 186, 187, 190
   acidic hydrolysis of 183, 186, 187
   diastereoisomeric 157, 165, 175
   enriched 157, 190
Amines 164, 165, 177, 187
   -enantiomer of 162, 163, 164, 165, 177, 180, 181, 183, 187
   -enantiomers of 162, 180, 181, 183
   preparation of 163, 164
Amino acids 151, 157, 177, 183, 186, 241
Anhydrides, mixed 157, 158
Aniline 130, 167
Anionic 117, 134, 136, 140, 141
   cascade 140, 141
   rearrangements 117, 134, 136
Anisotropic 81, 87, 88, 96, 99
   molecular structure 81, 87
   shape 88, 96, 99
Aromatic 25, 48, 84, 81, 83, 84, 87, 88, 90, 94, 99, 101, 102, 103, 106, 108, 111, 112, 121, 123, 127, 128, 138, 143, 179, 235
   cycles 90
   rings 81, 83, 84, 87, 99, 101, 102, 103, 108, 111, 112
Arylation 47, 49, 51
Arylboronic acids 222, 224
Aryl diazonium salt 130
Arylidene 213, 214
Asymmetric 152, 200, 229, 230, 231, 244
   Michael addition 230, 231
   synthesis 152, 200, 229, 244
Azlactones, α-substituted 225, 226
Azo-aromatic compounds 81

B

Benzoxazine 163, 164, 165, 178, 185
Biocatalytic syntheses 1
Biomedical Applications 200
Biosynthesis 5, 29, 41
    bacterial peptidoglycan 41
    blocks 29
    enhancing 5
Bisphosphonates 200, 201, 204, 205, 206, 207, 209, 211, 212, 213, 214, 215, 216, 217, 218, 220, 222, 223, 224, 227, 229, 232, 235, 237, 240, 241, 243, 244
    α-propargyl-substituted 211, 212
    containing 201, 240, 241, 242, 244
    corresponding 212, 220, 222
    functionalized 214, 244
    substituted 211, 213, 218, 235, 244
    synthesis of 206, 207, 211, 213, 218, 223, 244
    synthesized 232
    thiol-containing 235
    xanthate 216, 217
Bisphosphonic acids 220
    synthesized 220
Bromide 20, 22, 23, 50, 51
    prenylmagnesium 20, 22, 23
    solanesyl 50, 51
Building blocks 6, 7, 8, 9
    bifunctional 6, 7, 8, 9
    monoisoprenyl bifunctional 7, 8, 9

C

Carbonyl compounds 207
Catalytic amount 10, 20, 25, 37, 46, 51, 53, 126, 238
Chain 56, 58, 59, 96, 97
    polyprenyl side 56, 58, 59
    terminal flexible 96, 97
Cholesteric mesophases 81, 103
Cholesterol biosynthesis 41
Cholesteryl 81, 83, 100, 101, 102, 103, 104, 105, 112
    esters 81
    units 83, 100, 101, 102, 103, 104, 105, 112
Coenzyme Q10 1, 3, 18, 16, 24, 25, 26, 48, 49, 50
    synthesis of 16, 24
Copper 126, 127, 128, 130, 133
    -catalysis 126, 127, 130, 133
    iodide 126, 127, 128
Coupling 1, 24, 32, 35, 37, 43, 46, 49, 52, 62, 123, 124, 125, 127, 129, 204, 225
    cross-dehydrogenative 125
    nickel-catalyzed 37, 62
Cross-metathesis 31, 62
Crystalline 83, 87, 89, 109
    structures 83
    transitions 89
Cyanocuprate 20, 23, 25
Cycloaddition reactions 215, 238, 239
Cyclopropane 222, 224
Cystamine 235

D

Dendrolasin 39
Deprotonation 15, 17, 33, 134, 135, 167, 204, 211
Derivatives, tetrahydroquinoline 181
Dialkyl phosphites 127, 130
Diaryliodonium 130
Diastereoselective acylation 157, 158, 160, 164, 165, 167, 171, 173, 174, 183, 185, 186
    of racemic 164, 186
Diaziridines 160, 161
Dichloromethane 157, 162, 163, 164, 165, 175, 178, 181, 182, 188
Diethyl chlorophosphate 204, 205
Dihydroxylation 13, 42
Diphosphate 2, 8, 27, 28, 29
    dimethylallyl 2, 27, 29

Subject Index

isopentenyl 2, 27, 28
Direct alkylation reactions 54
Display, cyclic compounds 4
Di-tert-butyl peroxide (DTBP) 124
DSC thermograms 89, 97, 98
Dynamic kinetic resolution (DKR) 155, 156, 175

E

Electrophiles 83, 119, 123, 125
Electrophilic substitution reactions 48, 49
Enantiomeric excess 153, 154
Enolate chemistry 207
Enol 33, 34
   phosphates 33, 34
   triflates 33, 34
Epoxidation of geraniol 10
Escherichia coli 28, 29, 232
Ethenylidene-1,1-bisphosphonate 211, 240, 241, 242
Ethenylidenebisphosphonates 222, 224, 227, 230, 232, 233, 234, 235, 236, 237, 238, 241
   esters 222, 224, 227, 238
   reaction of 234, 238

F

Farnesyl 11, 13, 19, 22, 27, 43, 46, 47, 51, 54
   acetate 11, 13, 27
   bromide 19, 43, 46, 47, 51, 54
   chloride 22, 46, 47
Ferrocene 85, 86, 92, 93, 96, 100, 101, 103, 106, 112
   compounds 96
   derivatives 85, 86, 92, 93, 100, 101, 103, 106, 112
Ferrocenomesogens 81, 83, 84, 85, 86, 92, 94, 99, 102, 106, 110, 112, 113
   monosubstituted 81, 83, 84, 85, 86, 92, 110
   synthesized 81, 86

Flexible aliphatic chains 81, 83
Fluoroquinolone antibacterials 234, 235

G

Gembisphosphonates 242, 244
Geranyl 10, 11, 12, 13, 20, 22, 25, 27, 33, 57, 215
   acetate 10, 12, 13, 20, 22, 57
   bromide 25, 27, 33, 215
   derivatives 11, 12
Geranylgeraniol 1, 3, 4, 15, 27, 29, 215
   diphosphate synthase 215
   synthesis of 15, 27
Grignard reagents 20, 22, 24, 37, 51, 52, 62, 133, 143

H

Heating allylic alcohol 38
Heterocyclic amines 183, 232
H-phosphine oxides 119, 122, 124
Hydrolysis 58, 156, 157, 186, 204, 232, 235

I

Ibuprofen chloride 167
Imino ester, glycine 230, 231
Industrial chemical synthesis 29
Isomeric ballast 154, 164, 165
Isoprene monoepoxide 7, 25
Isoprenoid 1, 39, 40, 47, 48, 49, 201, 209, 210, 212, 244
   bisphosphonates 201, 209, 210, 212, 244
   chains 39, 40, 209
   quinones 1, 47, 48, 49

K

Ketal Claisen rearrangement 38
KR of racemic 152, 153, 158, 160, 161, 162, 163, 176, 177, 178, 181, 189
   1-phenylethylamine 158, 177

amines 152, 161, 162, 163, 176, 178, 181, 189
compounds 153
diaziridine 160, 161

L

Lateral interactions 99, 101, 106
Liquid crystalline 83, 84, 85, 86, 87, 88, 90, 92, 96, 98, 100, 101, 104, 105, 106, 109, 110, 112
   behavior 85, 88, 101, 105, 106
   properties 83, 84, 86, 87, 90, 92, 96, 98, 100, 101, 104, 105, 106, 109, 110, 112

M

Maleic anhydride 7
Membrane constituents 1
Menaquinone 1, 4, 35, 47, 48, 49, 50, 51, 52
Mercaptophosphonate 117, 134, 136, 138
Mesomorphic properties relationship 81, 82
Mesophases 81, 82, 84, 86, 87, 89, 90, 92, 93, 96, 97, 105, 106, 108, 109, 112
   nematic 90, 92, 93
Metallomesogens 81, 82
Metathesis, ring closure 31, 32
Methodologies, deprotonation-alkylation 210, 212
Methyl acetoacetate 33, 34
Michael addition 220, 225, 226, 227, 229, 233, 239, 244
Michaelis-Arbuzov 120, 200, 202
   Reaction 200, 202
   type reaction 120
Michael-type additions 222, 237
Monophosphonates 200, 202, 204, 206, 207, 220, 244
Monotropic phase 88
Multifunctional materials 81, 82

N

Nanoparticle carrier material 1
Naproxen chloride 161, 162, 163, 164, 165, 167, 178, 180, 185, 186
Nematic 81, 84, 89, 90, 91, 93, 98
   domain 89, 90
   droplets 90, 91, 93
Nitroxide-mediated polymerization (NMP) 61
NMR spectroscopy 157, 160, 164
Nucleophiles 119, 123, 125, 137, 222
Nucleophilic aromatic substitutions 212, 213

O

O-dialkyl phosphorothioates 135, 138
Optical 81, 86, 91, 94, 95, 97, 98, 110, 151, 156, 157, 161, 187, 188
   photomicrographs 91, 94, 95, 97, 98, 110
   polarized microscopy (OPM) 81, 86
   purity 151, 156, 157, 161, 187, 188
Organocopper compounds 23
Organometallic reagents 33, 34, 222, 223
Ortho-lithiation 136, 137

P

Pd-catalyzed thiophosphorylation 142
Pentafluorophenyl esters, active 171
Peptide synthesis 159
Phenylalanine 159, 168, 180
Phosphate mimics 200
Phosphonate carbanions 200
Phosphoramide series 137
Phosphorodithioates 117, 118, 141
Phosphorothioate salts 132, 133
Phosphorothioic 117, 119, 132, 132, 133, 139
   acids 131, 132, 139
   esters 117, 119, 132, 133
Phosphorothionates 117, 118
Polyisoprenoid compounds 1, 4, 5, 9, 19, 27, 29, 32, 55
   acyclic 1, 4
   long chain 19

Polyisoprenoids 1, 2, 3, 6, 15, 19, 28, 30, 31, 32, 38
    system of 38
Polyisoprenyl chains 4, 6, 19, 22, 31, 33, 56
Polymorphic behaviors 94, 96, 103
Polyprenoid compounds 2
Polyprenol 6, 12, 13, 27, 41, 47, 49, 51
    derivatives 12, 41
    synthesis 27
    chains 6, 13, 47, 49, 51
Prolyl chloride 160, 161, 181, 182, 183, 184, 186, 188
Propargyl bromide 211
Properties, mesomorphic 82, 85, 96, 103, 106
Protein 1, 209, 210
    geranylgeranylation 209, 210
    modulators 1
Pyrophosphate Bioisosteres 200

Q

Quinone 4, 48, 51, 53
    chloromethylated 53

R

Racemates 152, 153, 162, 164
    resolution of 152
    starting 153, 162, 164
Racemic 131, 156, 158, 159, 165, 164, 166, 167, 169, 170, 171, 172, 173, 174, 176, 177, 178, 181, 185, 186, 187
    1-phenylethylamine 158, 177, 178
    3-benzylmorpholine 171, 172, 173, 174
    acyl chlorides 165, 166, 176
    amine acylating 177
    stereoselectivity of acylation of 167, 185
Racemic amines 151, 152, 157, 158, 161, 162, 163, 164, 165, 166, 175, 176, 177, 178, 179, 180, 181, 182, 184, 188, 189
    acylative KR of 157, 164

diastereoselective acylation of 158, 162, 163, 164, 165, 166, 167, 168, 169, 170, 171, 175, 177, 178, 179, 180, 181, 182, 184, 185
diaziridines 160, 161
heterocyclic amines 165, 178, 185
hydrochloride of 164
oxazolidin-2-ones 167, 168, 169, 170
oxazolidinones 170, 171
stereoselectivity of acylation of 180, 181
Reagents, organozinc 23, 35, 36
Rearrangement 138, 139, 218, 219
    phosphoroamidate 138, 139
Rieke barium 22, 47
Ring closure metathesis (RCM) 31, 32

S

Saccharomyces cerevisiae 29
Schlieren textures 90, 91, 94, 95
    characteristic nematic 91, 94
Sigmatropic rearrangements 1, 11, 37, 38, 39, 40, 56, 58, 117
Sodium 232, 237, 238
    alkoxides 232, 238
    phthalimide 237, 238
Sodium sulfinates 9, 123
Squalene 43, 44, 45, 46
    all-trans 43, 44, 46
    central segment of 45
    synthesis, early 43
Stereoselective synthesis 6, 16, 25, 43, 215
Steric repulsions 83, 90, 106
Sulfinic acids 121, 122
Sulfone 9, 15, 16, 17, 19, 20, 23, 41, 44, 49, 50
    alkylation 15, 16
    chemistry 9, 19, 41, 49, 50
Sulfur 17, 117, 119, 129, 130, 133, 136, 138, 141
    elemental 119, 129, 130
    atom 17, 117, 133, 136, 138, 141

heterocycles 117
Syntheses (synthesis) 16, 19, 22, 23, 27, 30, 31, 32, 36, 49, 53, 61, 140, 141, 207, 210, 225, 226, 233, 234
   of 2-aminoethylenebisphosphonic acids 233
   of CoQ10 and vitamin K2 analogues 49
   of chiral thiolanes 140
   of farnesyl acetate 27
   of functionalized methylenebisphosphonates 210
   of functionalized polyisoprene derivatives 61
   of functionalized polyisoprene derivatives by ROP 61
   of geranyl acetate 22
   of jaspaquinol by Stille cross-coupling reaction 53
   of lactam bisphosphonates 207
   of library bisphosphonate ligands 234
   of long chain polyisoprenoids 36
   of oligoprenoids by sp3-sp3 carbon bond formation 14
   of phosphorothioic esters 119
   of polyisoprenoid ansa-chains 32
   of Polyisoprenoids by sp^2-sp^3 Carbon Bond Formation 32
   of polyisoprenoids by wittig reaction 30
   of regiosubstituted indole-based bisphosphonates 226
   of solanesol 16, 19
   of steroid-bisphosphonate conjugates 225
   thiopyrans 141
Synthetic 43, 44, 144, 151
   approaches to squalene 43, 44
   transformations 144, 151

T

Temperature 90, 99, 108, 109, 112
   isotropization 90, 99, 108, 109

melting 112
Terminal 11, 12, 30, 86, 89, 92, 96, 100, 142
   alkynes 30, 142
   chain, flexible 86, 89, 92, 96, 100
   double bond of geranyl derivatives 11, 12
   double bond of polyprenol derivatives 12
Tetraethyl 209, 211, 220, 223, 224, 225, 226, 232, 238, 239, 240, 243
   ethenylidene-1,1-bisphosphonate 239, 240
   ethenylidenebisphosphonate 223, 224, 225, 226, 232, 238, 243
   methylenebisphosphonate 209, 211, 220
Textures 95, 96, 97
   mosaic 95, 96, 97
   nematic marble 95
Thermal 81, 84, 86, 100, 102, 106, 112
   degradation 81, 86, 102, 106
   stability 84, 100, 102, 106, 112
Thermostability 86, 102, 112, 113
Thiocyanates 119, 120
Thioiminium salt 132
Traditional KR 156
Transition temperatures 101, 108, 112
Trialkyl phosphites 202, 203
Triethylamine 11, 127, 129, 232, 234
Triethyl orthoacetate 38, 44, 45
Tsuji-Trost reactions 26, 27, 62
Two-directional synthesis, efficient 44

V

Vitro biosynthesis 28

X

X-ray diffraction (XRD) 86, 108, 160, 161, 164, 165
   analysis 160, 161, 164, 165

www.ingramcontent.com/pod-product-compliance
Lightning Source LLC
Chambersburg PA
CBHW051144220526
45473CB00003B/649